Optimal Control Theory

Zhongjing Ma · Suli Zou

Optimal Control Theory

The Variational Method

 Springer

Zhongjing Ma
School of Automation
Beijing Institute of Technology
Beijing, China

Suli Zou
School of Automation
Beijing Institute of Technology
Beijing, China

ISBN 978-981-33-6294-9   ISBN 978-981-33-6292-5 (eBook)
https://doi.org/10.1007/978-981-33-6292-5

This Springer imprint is published by the registered company Springer Nature Singapore Pte Ltd.
The registered company address is: 152 Beach Road, #21-01/04 Gateway East, Singapore 189721,
Singapore

Preface

Many systems, like electrical, mechanical, chemical, aerospace, economic, and so on, can be mathematically modeled as linear/nonlinear deterministic/stochastic differential/difference state equations. The state systems evolve with time and possibly with other variables under certain specified dynamical relations with each other.

The underlying systems might be driven from a specific state to another one by applying some external controls. In case there exist many different ways to implement the same given task, one of them may be best in some sense. For instance, there may be a typical way to drive a vehicle from an initial place to the destination in a minimum time or with a minimum consumed fuel. The applied control corresponding to the best solution is called an optimal control. The measure of the performance is called the cost function.

It has been briefly introduced an optimal control problem by putting the above together. This book mainly focuses on how to implement the optimal control problems via the variational method. More specifically,

- It is studied how to implement the extrema of functional by applying the variational method. The extrema of functional with different boundary conditions, involving multiple functions and with certain constraints, etc., are covered.
- It is given the necessary and sufficient condition for the (continuous-time) optimal control solution via the variational method, the optimal control problems with different boundary conditions is solved, and the linear quadratic regulator and tracking problems are analyzed, respectively, in detail.
- It is given the solution of optimal control problems with state constraints by applying the Pontryagin's minimum principle, which is developed based on the calculus of variations. And the developed results are applied to implement several classes of popular optimal control problems, say minimum-time, minimum-fuel, minimum-energy problems, and so on.

This book is aimed at senior undergraduate students or graduate students in electrical, mechanical, chemical, and aerospace engineering, operation research and applied mathematics, etc. This book contains the stuff which can be covered in a one-semester course and it requires the students to have a background on control systems or linear systems theory. This book can also be used by professional researchers and engineers working in a variety of fields.

Beijing, China   Zhongjing Ma
October 2020   Suli Zou

Acknowledgements

Parts of this book are mainly based on the lecture materials that the first author has organized during the past 10 years for the graduate course, Optimal and Robust Control, at the Beijing Institute of Technology (BIT).

They would like to express their sincere gratitude to many of their colleagues and students for their support and encouragement.

First of all, the authors thank Prof. Junzheng Wang, Prof. Yuanqing Xia, and Prof. Zhihong Deng for their advice to the first author of this book to take the graduate course of Optimal and Robust Control in English at the school of Automation, BIT, from the winter term of 2010–2011 when he just joined the BIT. Without taking this course, it may not be possible for the authors to organize this book.

The first author of the book also expresses the deepest gratitude to his Ph.D. advisors, Prof. Peter Caines and Prof. Roland Malhame, and his postdoctoral advisors, Prof. Duncan Callaway and Prof. Ian Hiskens. Their enthusiasm, patience, and inspiring guidance have driven my research career.

Moreover, the authors also thank their graduate students, Peng Wang, Xu Zhou, Dongyi Song, Fei Yang, Tao Yang, Yajing Wang, Yuanming Sun, Jing Fan, and other students, for their efforts on the simulations of parts of numerical examples given in this book. Besides, they would like to thank the colleagues, Prof. Zhigang Gao, Dr. Liang Wang, and Dr. Hongwei Ma who have provided with their valuable suggestions for this book.

In addition, they would like to thank the editors, the reviewers, and the staffs at the Springer Nature for their assistance.

The authors also thank the financial support from the National Natural and Science Foundation, China (NNSFC), and the Xuteli Grant, BIT.

Last not least, they would like to express their deepest thanks to the family members who have always provided them with endless encouragement and supports behind us.

Contents

List of Figures

List of Tables

Chapter 1
Introduction

1.1 Backgrounds and Motivation

This book originates from parts of the lecture notes of the graduate course "Optimal and Robust Control" given at Beijing Institute of Technology since 2011. The purpose of the course is to provide an extensive treatment to the optimal and robust control in the modern control theory for the complex, multiple inputs and multiple outputs systems, to meet radically different criteria of the performance from the classical control theory.

This book contains some classical materials for the optimal control theory, including variational method, optimal control based upon the variational method and Pontryagin's minimum principle, with lots of numerical examples. The authors appreciate being informed of errors or receiving other comments about this book.

In this first chapter, the motivations to do researches from the classical control theory to the optimal control theory are introduced, and an explicit formulation for optimal control problems is provided.

In classical control theory, the analysis and design of control systems mainly depend on the concept of transfer function or the theory of Laplace transforms. Due to the convolution property of Laplace transforms, a convenient representation of a control system is the block diagram configuration that is illustrated in Fig. 1.1.

In such a block diagram representation, each block contains the Laplace transform of the differential equation and the component of the control system that relates the input to the block to its output is represented. The overall transfer function giving the ratio of the output and input will be yielded through simple algebra. That is, the classical control theory takes the input and output characteristics as the mathematical model of the system.

The classical control theory has been covered in many textbooks for senior undergraduate or graduate students, e.g., [1–12]. The commonly used analysis methods include frequency response analysis, root locus, description function, phase plane

© The Author(s), under exclusive license to Springer Nature Singapore Pte Ltd. 2021
Z. Ma and S. Zou, *Optimal Control Theory*,
https://doi.org/10.1007/978-981-33-6292-5_1

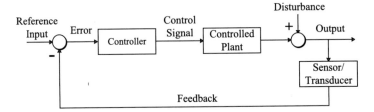

Fig. 1.1 The block diagram form of classical control

and Popov method, etc., and the control is limited to feedback control, PID control, and so on. By applying these techniques, it mainly studies

- The characteristics of the system in the time domain and frequency domain, such as rise time, peak overshoot, gain and phase margin, and bandwidth;
- The stability of the system;
- The design and the correction methods of control systems.

The control plants concerned in the classical control are usually single-input and single-output (SISO) systems, especially linear time-invariant (LTI) systems. These methods of analysis are difficult to apply to multi-input and multi-output (MIMO) systems.

In contrast to the classical control materials, generally speaking, the modern control theory is a time-domain control approach that is amenable to MIMO systems, and is based upon the state-space method to characterize the control plant in terms of a set of first-order differential equations [13–19].

For example, an LTI system could be expressed by

$$\dot{x}(t) = Ax(t) + Bu(t),$$
$$y(t) = Cx(t) + Du(t),$$

where $x(t)$ is the system state, $u(t)$ and $y(t)$ are the control input and output vectors, respectively, and the matrices A, B, C, and D are the corresponding state, input, output, and transfer matrices, respectively. A nonlinear system is characterized by

$$\dot{x}(t) = f(x(t), u(t), t),$$
$$y(t) = \widehat{f}(x(t), u(t), t).$$

The state variable representation could uniquely specify the transfer function while this does not hold vice vera. Therefore, modern control theory can deal with much more extensive control problems than classical control theory, including linear and nonlinear system, time-invariant and time-variant system, single-variable and multivariable system. Moreover, it provides the possibility to design and construct the optimal control system with a specified performance cost function.

1.2 Optimal Control Theory

When facing with a task, the objective of control is to find the control scheme among all possible ones that will cause a process/plant to satisfy some physical constraints and at the same time maximize or minimize a chosen performance criterion.

As one of the major branches of the modern control theory, the optimal control theory covered in this book targets on finding optimal ways to control a dynamic system that evolves over time in the way of continuous-time systems.

It also studies the control and its synthesis method when the controlled system achieves the optimal performance cost function. This could be the consumed time, cost, or the error between the actual and the expected. The problems studied by the optimal control theory could be summarized as follows: For a controlled dynamic system or motion process, an optimal control scheme is found out from a class of allowed control schemes, so that the performance value of the system is optimal when the motion of the system is transferred from an initial state to a specified target state.

This kind of problems widely exists in the field of technology or social problems. For example, to determine an optimal attitude control to minimize the fuel expenditure in the process of a spacecraft changing from one orbit to another, to select a regulation law of temperature and the corresponding raw material ratio to maximize the output of the chemical reaction process, and to formulate a most reasonable population policy to optimize the aging, dependency, and labor performances in the process of population development. All these are typical optimal control problems.

Consider an explicit example of designing a control for an unmanned vehicle. The control objective is to reach the expected target position along some certain trajectory. To complete this control task, the first thing is to get the current state of the vehicle and how it changes under the input control signal. In this problem, the states that are concerned most are the current position and speed of the vehicle.

The control behaviors could be controlling the throttle to accelerate the vehicle, or controlling the brake to decelerate. The running speed of the vehicle will further affect the position in the next period of time. At the same time, it should ensure that the vehicle speed would not be too fast to violate the traffic rules. When the fuel on the vehicle is limited, it is also necessary to keep the fuel consumption to finish the journey not greater than the fuel amount.

The above example is a typical control task. An experienced driver may have a variety of control methods to get there. However, if the objective is to reach the goal in the shortest period of time, the control is not intuitive: increasing the throttle could reduce the time to reach the destination, while it may cause over-speed problems; and even fail to reach the destination because the fuel is consumed in advance. Besides, unexpected disturbances may occur during the journey.

From the above example, the fundamental elements related to optimal control problems can be given as follows:

- State-space equations describing the dynamic system. That is, the control input $u(\cdot)$ depending on the time influences the state variable $x(\cdot)$, which is usually represented by differential equations in case of continuous-time or difference equations in case of discrete time.

- Admissible control set describing the constraints satisfied by the input and state variables.
- Specific conditions of the final state at a final time which may be or may not be fixed.
- Performance cost which is used to measure the performance of the control tasks when the objective is achieved.

The detailed formulations of the optimal control problems in cases of continuous time and discrete time are given in Sects. 1.4 and 1.5, respectively. Therefore, from the mathematical point of view, the determination of the optimal control problem could be expressed as follows: Under the constraints of the dynamic equation and the admissible control range, the extreme value of the performance cost function related to the control and state variables can be calculated.

The realization of optimal control is inseparable from optimization technology, which is a subject of studying and solving optimization problems. It studies how to find the optimal solution that optimizes the objective from all the possible ones. That is to say, the optimization technology is to study how to express the optimization problem as a mathematical model and how to effectively figure out the optimal solution via the mathematical model. Generally speaking, solving practical engineering problems with the optimization method can be divided into the following procedure:

- Establish the mathematical model of the optimization problem, determine the variables, and list the constraints and objective function for the proposed optimization problem;
- Analyze the mathematical model in detail, and select the appropriate optimization method;
- Implement the optimal solution by proceeding the algorithm of the optimization method, and evaluate the convergence, optimality, generality, simplicity and computational complexity of the proposed algorithm.

After the mathematical model of the optimization problem is established, the main problem is how to solve the optimization problem by different methods. In the following, it briefly introduces those methods for solving the optimal control problems in the literature.

For optimal control problems with simple and explicit mathematical expression of objective functions and constraints, analytical methods could be applied to solve them. Generally, the way to find the analytical solutions is to first find out the necessary conditions of the optimal control by the derivative or variational methods which has been covered in many classical textbooks published in the past decades, [20–28].

The creation of the calculus of variations occurred almost immediately after the invention or formalization of calculus by Newton and Leibniz. An important problem in calculus is to find an argument of a function at which the function takes on its extrema, say maxima or minima.

The extension of the problem posed in the calculus of variations is to find a function that maximizes or minimizes the value of an integral or functional of that function.

Fig. 1.2 An illustration of
the Brachistochrone problem

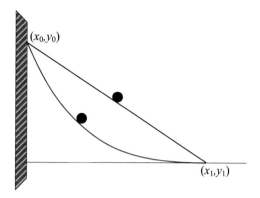

Due to the infinite dimension of a function to be implemented, it is well expected
that the extremum problem of the calculus of variations is much more challenging
than the extremum problem of the calculus.

It has been widely known that the calculus of variations had been considered as
a key mathematical branch after Leonhard Euler published the famous monograph,
Elementa Calculi Variationum, in 1733, and a method for finding curved lines enjoying properties of maximum or minimum, or solution of isoperimetric problems in
the broadest accepted sense in 1744. The variational method is a powerful mathematical tool to deal with the implementation of the extrema (maxima or minima)
of a function. In his book on the calculus of variations, Euler extended the known
method of the calculus of variations to form and solve differential equations for the
general problem of optimizing single-integral variational quantities.

Nevertheless, it is worth to state, before Euler studied the variational method in
a systematic way, quite a few specific optimization problems had been essentially
solved by using the variational principles. Queen Dido faced with the problem to find
the closed curve with a fixed perimeter that encloses the maximum area. Certainly,
the extremal solution is a circle which can be obtained by applying the variational
method.

Another problem was from Isaac Newton who designed the shape of a body
moving in the air with the least resistance. However, a first problem solved by the
method of calculus of variations was the path of least time or the Brachistochrone
problem, proposed by Johann Bernoulli at the end of the seventeenth century, which
is shown as an illustration in Fig. 1.2, which was solved to be a cycloid by Jacob
Bernoulli, Isaac Newton, L'Hospital, and himself.

It involves finding a curve connecting the two points (x_0, y_0) and (x_1, y_1) in a
vertical plane with the proof that a bead sliding along a curve driven under the force
of gravity will move from (x_0, y_0) to (x_1, y_1) in a shortest period of time.

Besides Euler, Lagrange, Newton, Leibniz, and the Bernoulli brothers also gave
great contributions to the early development of the field. In the nineteenth and early
twentieth centuries, many mathematicians such as Hamilton, Jacobi, Bolza, Weier-
strass, Caratdory, and Bliss also contributed much to the theory of the solution of

variations problems, see [29, 30] for a historical view on the topics of the method of calculus of variations.

The initial stage of modern control theory was the publication of the well-known minimum principle in Russian in the later 1950s, e.g., [31–37], and in 1962 in English, of the book, the Mathematical Theory of Optimal processes, [38], by Russian mathematicians, Pontryagin and his collaborators, such as Boltyanskii, Gamkrelidze, and Mischenko. Besides, there are American researchers who made many contributions to these topics are Valentine, McShane, Hestenes, Berkovitz, and Neustadt et al.

The key contributions of the work by Pontryagin and his collaborators include not only a rigorous formulation of calculus of variations problem with constrained control variables, but also a mathematical proof of the minimum principle for optimal control problems.

In this book, it will be mainly studied how to apply the minimum principle in its various forms to implement the solutions of various optimal control problems in the fields of electrical engineering, chemical engineering, etc.

Basically, it determines the optimal solution according to a set of equations or inequalities. This kind of method fits the problems with obvious analytical expressions of performance cost function and constraints.

More especially, when the control vector is not constrained in a set, the Hamiltonian (function) is introduced to solve the optimal control problem, and the necessary conditions of optimal control, i.e., regular equation, control equation, boundary condition, and cross-sectional condition can be derived by using the variational method.

The variational method works on the premise that the control vector is not subject to any restrictions, that is, the admissible control set can be regarded as the open set of the whole control space and the control variable is arbitrary. At the same time, the Hamiltonian is assumed to be continuously differentiable on the control variable.

However, in practical problems, the control variable is often limited to a certain range, which motivates the minimum principle. In fact, this method is extended from the variational method [39], but it can be applied to the case that the control variable is limited by the boundary, and does not require the Hamiltonian output to be continuously differentiable on the control input. Those problems with constraints can be solved by using Pontryagin's minimum principle, [40, 41]. There have been many textbooks dedicated to introducing this kind of optimal control method, e.g., [39, 42–56][?] and references therein. Many of them focus on the applications of the Pontraygin's minimum principle in various practical fields such as aerospace engineering [57–61], mechanical engineering [62–66], electrical engineering [67–72], chemical engineering [73–76], management science, economics [77–83], social science [84, 85], etc.

This book mainly introduces the materials introduced above, say the extrema of the functionals via the variational method, the optimal control by the variational method, optimal control with constraints via Pontryagin's minimum principle.

Besides, the dynamic programming method, initialized by Richard Bellman and his collaborators, [86–88], is another branch of key analytical optimal control methods to solve the optimal control problems, e.g., [89–94]. Like the minimum principle, it is an effective way to deal with the optimal control problem in which the control vector is constrained to a certain closed set. It transforms the complex optimal control problem into a recursive function relation of the multi-stage decision-making process.

Regardless of the initial state and initial decision, the decision must be optimal for this stage and the following stages when given a stage and a state as the initial stage and initial state. Therefore, by using this principle of optimality, a multi-stage decision-making problem can be transformed into multiple optimal single-stage decision-making problems. The decision-making of this stage is independent of the previous ones, and only related to the initial state and initial decision of this stage. As dynamic programming is used to solve the optimal control problem of continuous systems, the continuous system could be discretized first, and the continuous equation can be approximately replaced by the finite difference equations.

Besides the above methods to implement the optimal solution in a centralized way, differential games have been employed since 1925 by Charles F. Roos [95]. Nevertheless, Rufus Isaacs was the first to study the formal theory of differential games in 1965 [96]. He formulated so-called two-person (zero-sum) pursuit-evasion games. And Richard Bellman made a similar work via the dynamic programming method in 1957, [86]. See [97] for a survey of pursuit-evasion differential games.

Such differential games are a group of problems related to the modeling and analysis of conflict in the context of a dynamic system, [98, 99]. More specifically, state variables evolve over time according to certain differential equations.

Distinct from the early analyses of differential games which reflected military interests by considering two players, say the pursuer and the evader, with opposed goals. Nowadays, more and more analyses mainly reflect engineering, economic, or social considerations, [100, 101].

Some research work considers adding randomness to differential games and the derivation of the stochastic feedback Nash equilibrium (SFNE), e.g., the stochastic differential game of capitalism by Leong and Huang, [102].

In 2016, Yuliy Sannikov received the Clark Medal from the American Economic Association for his contributions to the analysis of differential games by applying stochastic calculus methods, [97, 103]

Differential games are related to optimal control problems. As stated earlier, in an optimal control problem there are single control u and a single performance criterion to be optimized; while differential game generalizes this to several controls $u_1, u_2, \ldots, u_i$ and several performance criteria as well. Each individual player attempts to control the state of the system so as to achieve his own goal; the system responds to the inputs of all players.

More recently it has been developed a so-called mean-field game theory by Minyi Huang, Peter E. Caines, and Roland Malhame [104–109], who solved the optimal control problems in the field of engineering, say telecommunication problems, with a large population of individual agents that mutually interact with each other and each of which has negligible effects on the system which tends to vanish as the population size goes to infinity, and independently around the same time by Jean-Michel Lasry and Pierre-Louis Lions [110–112, 112], who studied strategic decision-making in a large population of individual interacting players. Mean-field game theory has been extended and applied to many fields.

The term "mean-field" is inspired by mean-field theory in physics, which considers the behavior of systems of large numbers of particles where each of the individual particles has a negligible impact on the whole system.

In continuous time, a mean-field game is typically composed of a Hamilton–Jacobi–Bellman equation that describes the optimal control problem of an individual and a Fokker–Planck equation that describes the dynamics of the aggregate distribution of agents.

In continuous time case, a mean-field game is typically composed of a Hamilton–Jacobi–Bellman equation that describes the optimal control problem of an individual and a Fokker–Planck equation that describes the dynamics of the aggregate distribution of agents. Under certain general assumptions, it can be shown that a class of mean-field games is the limit of a N-player Nash equilibrium as the population size N goes to infinity [107, 108, 113].

A related concept to that of mean-field games is "mean-field-type control". In this case, a social planner controls distribution of states and chooses a control strategy. The solution to a mean-field-type control problem can typically be expressed as a dual adjoint Hamilton–Jacobi–Bellman equation coupled with the Kolmogorov equation. Mean-field-type game theory [114–117] is the multi-agent generalization of the single-agent mean-field-type control [118, 119].

Basically, all the above methods described so far, rely upon the explicit formulation of the problems to be solved. Nevertheless, some methods, like the direct method, [120–123], can be applied to solve the optimization problems with complex objective function or without explicit mathematical expression. The basic idea of the direct method is to use the direct search method to generate the sequence of points through a series of iterations, such that it is gradually close to the best solution. Direct methods are often based on experiences or experiments. The numerical calculation method could be divided into the categories below:

- Interval elimination method, [124], which is also known as the one-dimensional search method, is mainly used for solving single-variable problems. For example, there are golden section method, polynomial interpolation method, and so on.
- Hill climbing method, [125], which is also known as a multidimensional search method, is mainly used for solving multivariable problems. For example, coordinate rotation method, step acceleration method, and so on.
- Gradient-type methods, [126] include unconstrained gradient methods, such as the gradient descent method and quasi-Newton method, and constrained gradient methods such as feasible direction method and gradient projection method.

In practice, in certain scenarios, it might be infeasible to apply the offline optimization methods, described briefly above, since they are usually relied upon the mathematical model of the problem. More especially, many factors like the changes in the environment, aging of the catalyst and equipment, etc., may introduce disturbances to the process even though the process is designed to operate continuously under certain normal working conditions. Consequently, the proposed optimal control solutions may be distinct with the actual optimal ones for the actual problems. There are quite a few online optimization methods in the literature to overcome the challenges.

The so-called local parameter optimization method [127, 128] is to adjust the adjustable parameters of the controller according to the difference between the reference model and the output of the controlled process, so as to minimize the integration of the square of the output error. In this way, the controlled process could track the reference model accurately as soon as possible.

The predictive control, which is also known as model-based control, is a type of optimal control algorithm rising in the late 1970s. Different from the usual discrete optimal control algorithm, it does not use a constant global optimization objective, but uses a receding finite-time-domain optimization strategy. This means that the optimization process is not carried out offline, but repeatedly online.

Due to the localization of the finite objectives, we can only obtain the solution with an acceptable performance in ideal situations, while the receding implementation could take into account the uncertainties caused by model mismatch, time variant, disturbances, and so on, by compensating them in real time. It always establishes the optimization based on the actual environments in order to keep the control input optimal in practice. This heuristic receding optimization strategy takes into account the influence of ideal optimization and actual uncertainty in the future.

In the complex industrial environment, it is more practical and effective than the optimal control based on ideal conditions. We can establish an optimization mode by applying predictive controls. In this way, we can deal with those problems with complex constraints, multi-objective and nonlinear components. It is promising to overcome the shortcomings of single model predictive control algorithm and attract attentions by introducing the ideas of hierarchical decision-making or artificial intelligence technique to the predictive control methods.

The decentralized control is commonly used in the control of large-scale systems. In this case, the computer online steady-state optimization often applies a hierarchical control structure. This structure has both a control layer and optimization layer, wherein the optimization layer is a two-level structure composed of local decision-maker and the coordinator. The optimization process is that each decision-maker responds to the subprocess optimization in parallel, and the coordinator coordinates the optimization processes. The optimal solution is then obtained through mutual iteration.

Due to the difficulty in having the accurate mathematical model of industrial processes, which tend to be nonlinear and time variant, the Polish scientist Findesien proposed that the solution obtained by using the model in the optimization algorithm is open loop, [129–131]. In the design stage of online steady-state control of large-scale industrial processes, the open-loop solution can be used to determine the optimal working point. However, in practice, this solution may not make the industrial process in the optimal condition, on the contrary, it even violates the constraints. Their new idea is to extract the steady-state information of the related variables from the actual process, and feed it back to the coordinator or local decision-makers.

The difficulty of steady-state hierarchical control is that the input and output characteristics of the actual process are unknown. The feedback correction mechanism proposed by Findesien could only get a suboptimal solution. But its main disadvantage is that it is difficult to accurately estimate the degree of suboptimal solution deviating from the optimal solution, and the suboptimal degree of suboptimal solution often depends on the selection of the initial point. A natural idea is to separate the optimization and parameter estimation and carry out them alternately until the iteration converges to a solution. In this way, the online optimization control of the computer includes two tasks: the optimization based on the rough model which is usually available, and the modification based on the set point. This method is called the integrated research method of system optimization and parameter estimation.

For more and more complex control plants, on one hand, the control performance required is no longer limited to one or two indices; on the other hand, all the above optimization methods are based on the accurate mathematical model of the optimization problem. But many practical engineering problems are very difficult or impossible to get its accurate mathematical model. This limits the practical application of the classical optimization method. With the development of fuzzy theory, neural network and other intelligent technology, and computer technology, the smart optimization method has been developed.

The research of artificial neural network originated from the work of Landahl, Mcculloch, and Pitts in 1943 [132]. In the aspect of optimization, in 1982, Hopfield first introduced the Lyapunov energy function to judge the stability of the network [133], and proposed Hopfield single-layer discrete model. This work has been extended by Hopfield and Tank in [134]. In 1986, Hopfield and Tank directly corresponded the electronic circuit with the Hopfield model and realized the hardware simulation [135]. Kennedy and Chua in [136] proposed the analog circuit model based on the nonlinear circuit theory and studied the stability of the electronic circuit using the Lyapunov function of the system differential equation.

All these works promote the research of neural network optimization. According to the theory of neural network, the minimum point of the energy function of the neural network corresponds to the stable equilibrium point of the system, so the problem is transformed into seeking the stable equilibrium point of the system. With the evolution of time, the orbit of the network always moves in the direction of decreasing the energy function, and finally reaches the equilibrium point of the system.

Therefore, if the stable attractor of the neural network system is considered as the minimum point of the appropriate energy function or augmented energy function, the optimal calculation will reach a minimum point along with the system flow from an initial point. If the concept of global optimization is applied to the control system, the objective function of the control system will eventually reach the desired minimum point. This is the basic principle of neural optimization [137]. Since the Hopfield model can be applied to both discrete and continuous problems, it is expected to effectively solve the nonlinear optimization problem of mixed discrete variables in control engineering.

Like the general mathematical programming, the neural network method also has the weakness of costing much. How to combine the approximate reanalysis and other structural optimization techniques to reduce the number of iterations is one of the directions for further research.

Genetic algorithm and genetic programming are new search and optimization techniques [138, 139]. It imitates the evolution and heredity of the organism, and according to the principle of "survival of the fittest", it makes the problem to be solved gradually approach the optimal solution from the initial solution. In many cases, the genetic algorithm is superior to the traditional optimization method. It allows the problem to be nonlinear and discontinuous, and can find the globally optimal solution and the suboptimal solutions from the whole feasible solution space, avoiding only getting the local optimal solution. In this way, we can provide more useful reference information for better system control. At the same time, the process of searching for the optimal solution is instructive, and may avoid the dimension disaster by applying a general optimization algorithm. With the development of computer technology, these advantages of the genetic algorithm will play an increasingly important role in the field of control. The results show that the genetic algorithm is a potential structural optimization method.

Optimal control problem is one of the most widely used fields of fuzzy theory. Since Bellman and Zadeh made pioneering work on this research in the early 1970s [140], their main research focuses on theoretical research in the general sense, fuzzy linear programming, multi-objective fuzzy programming, and the application of fuzzy programming theory in random programming and many practical problems.

The main research method is to use the membership function of the fuzzy set to transform the fuzzy programming problem into the classical one. The requirements of the fuzzy optimization method are the same as those of the ordinary optimization method. It is still to seek a control scheme (i.e., a set of design variables) to meet the given constraints, and optimize the objective function. The fuzzy optimization method can be summarized for solving a fuzzy mathematical programming problem including control variables, objective functions, and constraints. However, those control variables, objective functions, and constraints may be fuzzy, or some parts are fuzzy and the other parts are clear. For example, the fuzzy factors could be included in the constraints such as geometric constraints, performance constraints, and human constraints.

The basic idea of solving a fuzzy programming problem is to transform fuzzy optimization into an ordinary optimization problem. One way for solving fuzzy problems is to give a fuzzy solution; the other is to give a specific crisp solution. It must be pointed out that the above solutions are all for fuzzy linear programming.

Nevertheless, lots of practical engineering problems are described by nonlinear fuzzy programming. Therefore, some people put forward the level cut set method, the limit search method , and the maximum level method, and achieved some gratifying results. In the field of control, fuzzy control is integrated with a self-learning algorithm, fuzzy control, and genetic algorithm. By improving the learning algorithm and genetic algorithm, and according to the given optimization performance function, the controlled object is gradually optimized for learning, such that the structure and parameters of the fuzzy controller can be effectively determined.

There also exist many other smart optimization methods in the literature, e.g., ant colony optimization [141], particle swarm optimization [142], and simulated annealing algorithm [143].

In Sects. 1.4 and 1.5, it will then give the general formulation for optimal control problems. Before that in Sect. 1.3, some optimal control problems in different fields are introduced first.

1.3 Examples of Optimal Control Problems

Example 1.1 (*Minimum Time for An Unmanned Vehicle Driving*)

We first consider the example mentioned in the previous section for introducing the optimal control theory. Here consider a simple case, say the vehicle drives in a straight line from the parking point O to the destination point e, as illustrated in Fig. 1.3. A similar example is also specified in [144]. The objective is to make the vehicle reach the destination as quickly as possible.

Let $d(t)$ denote the distance of the vehicle from the starting point O at time t. As stated in the earlier part, the vehicle could be accelerated by using the throttle and decelerated by using the brake. Let $u(t)$ represent the throttle acceleration when it is positive valued and the brake deceleration when it is negative valued. Then the following equation holds:

$$\dot{d}(t) = u(t).$$

Selecting the position and velocity of the vehicle as the state variables, i.e., $x(t) = \begin{bmatrix} d(t) \\ \dot{d}(t) \end{bmatrix}$, and the throttle acceleration/brake deceleration as the control variables.

Hence we obtain the state dynamics differential equation as follows:

$$\dot{x}(t) = \begin{bmatrix} 0 & 1 \\ 0 & 0 \end{bmatrix} x(t) + \begin{bmatrix} 0 \\ 1 \end{bmatrix} u(t). \tag{1.1}$$

Let t_0 and t_f denote the departure time and arrival time of the vehicle, respectively. Since the vehicle is parking at O and it stops at e, it could get the boundary conditions of the state

$$x(t_0) = \begin{bmatrix} 0 \\ 0 \end{bmatrix},$$

$$x(t_f) = \begin{bmatrix} e \\ 0 \end{bmatrix}.$$

Fig. 1.3 An illustration of a simplified vehicle driving control

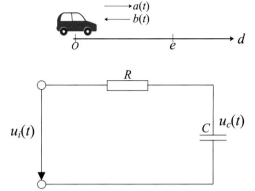
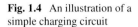

Fig. 1.4 An illustration of a simple charging circuit

Practically, the acceleration of a vehicle is bounded by some upper limit which depends on the capability of the engine, and the maximum deceleration is also limited by the braking system parameters. Denote the maximum acceleration and maximum deceleration by M_1 and M_2, respectively, with $M_1, M_2 > 0$, which gives the constraint for the control variable:

$$-M_2 \leq u(t) \leq M_1.$$

In addition, the vehicle has limited fuel, the amount of which is denoted by G, and there are no gas stations on the way, then another constraint is posed:

$$\int_{t_0}^{t_f} [k_1 a(t) + k_2 \dot{d}(t)] dt \leq G.$$

Now we can formulate the optimal control problem: for the system specified in (1.1), given t_0, $x(t_0)$, and $x(t_f)$, find the $u(t), t \in [t_0, t_f]$ under the underlying constraints to minimize the time used to reach the destination, i.e.,

$$J(u) \triangleq t_f - t_0.$$

$\square$

Example 1.2 (*Minimum Energy Consumption in An Electric Circuit*)

Consider the charging circuit shown in Fig. 1.4. Assume that a control voltage is applied to charge the capacitor to a given voltage within a given time period, and at the same time, minimize the electric energy consumed on the resistor.

Denote by $u_i(t)$ and $u_c(t)$ the control voltage and the voltage of the capacitor, respectively, $i(t)$ the charging current, R the resistance of the resistor and C the capacitance of the capacitor. Hence, the following equation holds for the control variable $u_i(t)$:

$$C\frac{du_c(t)}{dt} = \frac{1}{R}[u_i(t) - u_c(t)] = i(t).$$

That is, we get the state dynamics equation

$$\frac{du_c(t)}{dt} = -\frac{1}{RC}u_c(t) + \frac{1}{RC}u_i(t). \tag{1.2}$$

The power consumed on the resistor is

$$w_R(t) = \frac{1}{R}[u_i(t) - u_c(t)]^2.$$

Let t_0 and t_f denote the starting time and ending time of the charging process, respectively, and V_0 and V_f denote the starting voltage and ending voltage of the capacitor, respectively.

Similarly, the problem could be formulated as, for the system specified in (1.2), given t_0, t_f, $u_c(t_0) = V_0$, and V_f, find the $u_i(t), t \in [t_0, t_f]$ such that $u_c(t_f) = V_f$ and minimize the consumed power on the resistor

$$J(u) \triangleq \int_{t_0}^{t_f} \frac{1}{R}[u_i(t) - u_c(t)]^2 dt.$$

□

Next, it introduces a specific optimal problem below on how social insects, e.g., a population of bees, to determine the makeup of their society. This kind of problems is originally from Chap. 2 of the book "Caste and Ecology in Social Insects" by G. Oster and E.O. Wilson [145, 146].

Example 1.3 (*Reproductive Strategies in Social Insects*) [146]

Denote by t_f the length of the season starting from $t_0 = 0$ to t_f. Introduce $w(t)$ to represent the number of workers at time t, $q(t)$ the number of queens and $\alpha(t)$ the fraction of colony effort devoted to increasing workforce.

The control variable $\alpha(t)$ is constrained to

$$0 \le \alpha(t) \le 1.$$

We continue this model by introducing the state dynamics for the numbers of workers and the number of queens. The worker population evolves according to

$$\dot{w}(t) = -\mu w(t) + bs(t)\alpha(t)w(t),$$

where μ is a given constant death rate, b is another constant, and $s(t)$ is the known rate at which each worker contributes to the bee economy, and the initial state is $w(0) = w_0$.

Suppose also that the population of queens changes according to

$$\dot{q}(t) = -vq(t) + c[1 - \alpha(t)]s(t)w(t),$$

with constants v and c and initial $q(0) = q_0$.

The number of queens at the final time t_f is $q(t_f)$. Thus the problem is formulated as an optimization problem such that the bees attempt to maximize $q(t_f)$.

□

Example 1.4 (*Control of the Traffic Lights*)

Consider the road junction of two single-direction lanes which are called lane 1 and lane 2, respectively. As illustrated in Fig. 1.5, the lengths or the amounts of the waiting vehicles at the junction in the two lanes are denoted by $x_1(t)$ and $x_2(t)$, respectively, and the traffic flows to the junction are denoted by $v_1(t)$ and $v_2(t)$, respectively. Suppose that the maximum traffic flows for the two lanes are represented by a_1 and a_2, respectively.

Denote by $g_1(t)$ and $g_2(t)$ the lengths that the green lights are "on" in the two lanes, respectively. Suppose that the switching period of the traffic lights, denoted by t_f, is fixed. Suppose also fixed the time required for the vehicle to accelerate and the time of yellow lights, which aggregately is denoted by y.

It has the following:

$$g_1(t) + g_2(t) + y = t_f.$$

Let $u(t)$ represent the average traffic flow in lane 1 within one switching period of the traffic lights, i.e.,

$$u(t) = a_1 \frac{g_1(t)}{t_f},$$

Fig. 1.5 An illustration of a simplified traffic control at the junction

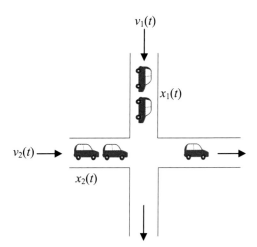

then the average traffic flow in lane 2 could be obtained

$$a_2 \frac{g_2(t)}{t_f} = -\frac{a_2}{a_1} u(t) + a_2 \left[1 - \frac{y}{t_f} \right].$$

Based on these analysis, the state dynamics equation is expressed as

$$\dot{x}_1(t) = v_1(t) - u(t),$$

$$\dot{x}_2(t) = v_2(t) + \frac{a_2}{a_1} u(t) - a_2 \left[1 - \frac{y}{t_f} \right].$$

There would be a constraint on $u(t)$ if the time length of green light in lane 1 has time limits:

$$u_{min} \le u(t) \le u_{max}.$$

Hence the optimal problem is formulated: given the initial states $x_1(t_0)$ and $x_2(t_0)$, determine the control $u(t), t \in [0, t_f]$ such that $x_1(t_f), x_2(t_f) = 0$, and at the same time minimize the waiting time of the vehicles

$$J(x, u) \triangleq \int_0^{t_f} [x_1(t) + x_2(t)] dt.$$

□

Example 1.5 (*Soft Landing of A Spacecraft*)
    It is considered that a spacecraft is required to softly land on the earth surface, that is the landing speed of the spacecraft should be as small as possible. In addition to the parachute deceleration during the return stage, the engine needs to be started in the final stage to reduce the landing speed into the allowable range. At the same time, in order to reduce the cost, the deceleration engine is required to consume as little fuel as possible during the landing process.
    In order to simplify the description of the problem, it may be considered the spacecraft as a particle, which is assumed to move along the vertical line of the surface at the last stage of the landing process. Then the state dynamics equation is given as

$$m(t) \frac{dv(t)}{dt} = p(t) + f(h, v) - m(t)g$$

$$\frac{dh(t)}{dt} = v(t)$$

$$\frac{dm(t)}{dt} = -\alpha p(t),$$

Fig. 1.6 An illustration of a
soft landing problem of a
spacecraft

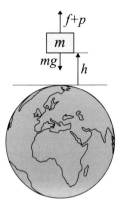

where $m(t)$ is the mass of the spacecraft, including the self-weight of the spacecraft
and the mass of the fuel carried, $h(t)$ is the distance from the spacecraft to the surface,
$v(t)$ is the velocity of spacecraft with the direction of vertical upward being positive,
$f(h, v)$ is the air resistance, g is the acceleration of the gravity, which is set as a
constant, α is the combustion coefficient of the engine, which is also a constant, and
$p(t)$ is the engine thrust, which is the control variable to be determined.

Suppose that the engine in the return capsule ignites at $t = 0$, and give the initials

$$v(0) = v_0,$$
$$h(0) = h_0,$$
$$m(0) = M_s + M_e,$$

where M_s and M_e represent the mass of the spacecraft itself and the total mass of
the fuel carried, respectively (Fig. 1.6).

If the soft landing time of the return capsule is t_f, then it has the requirement

$$v(t_f) = 0,$$
$$h(t_f) = 0.$$

The thrust $p(t)$ of the engine is always positive and the maximum value is set
to be p_M, that is, $0 \leq p(t) \leq p_M$. Hence, the problem of soft landing is formulated
as an optimization problem such that the engine thrust function $p(t)$ is designed to
transfer the spacecraft from the initial state $\begin{bmatrix} h(0) \\ v(0) \end{bmatrix}$ to the end state $\begin{bmatrix} h(t_f) \\ v(t_f) \end{bmatrix}$ under
the above constraints, and minimize the fuel consumption at the same time, that is,
maximize the mass at the final time

$$J \triangleq m(t_f).$$

In the following, a problem of a target interception in space is given.

Example 1.6 (*Interception of A Target in Space*)
Suppose that a space target moves at a certain speed, the thrust of the interceptor is fixed, and the space target and the interceptor move in the same plane. The direction of the thrust of the interceptor is the control variable. The goal is to control the thrust direction of the interceptor during the time interval $[t_0, t_f]$ in order to intercept the space target.

Then the simplified state dynamics equations of the relative motion of the space target and interceptor are given as

$$\ddot{x}(t) = \eta \cos(\alpha(t))$$
$$\ddot{y}(t) = \eta \sin(\alpha(t)),$$

where $x(t)$ and $y(t)$ are the coordinates of the relative positions of the space target and the interceptor, respectively, η is the thrust amplitude of the interceptor under the assumption that the mass of the interceptor is equal to 1, and $\alpha(t)$ is the thrust direction angle of the interceptor.

We hope to design the control for the thrust direction angle $\alpha(t)$ of the interceptor to achieve the goal of target interception, that is,

$$x(t_f) = 0,$$
$$y(t_f) = 0,$$

and meanwhile minimize the time cost to accomplish the interception

$$J \triangleq t_f - t_0.$$

□

Example 1.7 (*Optimal Consumption Strategy*)
Suppose a person holds a fund of the amount x_0 at initial time t_0. He plans to consume the fund within the time interval $[t_0, t_f]$. If considering the profit of the bank deposit, how can this person consume to obtain the maximum benefit?

Denote $x(t)$ as the fund and he has time t, including the interest he obtained from bank deposit, and $u(t)$ the consumption he makes. The interest rate of the bank deposit is fixed and denoted by α. During the process of consumption $u(t)$, the satisfaction or the utility of this person is

$$L(t) = \sqrt{u(t)} \exp(-\beta t),$$

where β is the discount rate, which represents the interest rate that the future consumption funds are discounted into the present value.

The optimal consumption problem can be described as the following. Consider the dynamic equation of the fund

$$\dot{x}(t) = \alpha x(t) - u(t),$$

and the initial and the end value of the fund

$$x(t_0) = x_0,$$
$$x(t_f) = 0.$$

To implement the optimal consumption strategy $u(t), t \in [t_0, t_f]$ such that the total utility is maximized

$$J(u) \triangleq \int_{t_0}^{t_f} \sqrt{u(t)} \exp(-\beta t) dt.$$

□

Example 1.8 (*Advertising Expense in A Market*)

Advertising expense is one of the important expenses of an enterprise, and its payment strategy is an important factor that determines the total income of an enterprise. When enterprises make the strategy of advertising expenses, on the one hand, they should increase the sales of their products through the role of advertising to avoid the market forgetting of their products; on the other hand, they should pay attention to the market saturation, that is, the demand for a product from customers is finite. When the actual sales go close to the maximum demand, the role of advertising will be reduced.

Based on the above considerations, the relationship between the product sales and advertising expenses can be described by the following dynamic model:

$$\dot{x}(t) = -\alpha x(t) + \beta u(t) \left[1 - \frac{x(t)}{x_M} \right],$$

which is the so-called Vidale–Wofle model [147], and where $x(t)$ and $u(t)$ are the sales volume of products and the payment of advertising expenses, respectively, α and β are constants, which, respectively, represent the role of market forgetfulness and advertising utility in increasing the sales volume, and x_M is the maximum demand of the product in the market.

Assuming the sale revenue of the unit amount of the product is q, the cumulative revenue within the time interval $[t_0, t_f]$ is

$$J(x, u) \triangleq \int_{t_0}^{t_f} \exp(-\beta t)[qx(t) - u(t)] dt,$$

where β represents a discount rate.

The optimal market advertising expense problem is that to find the advertising payment strategy $u(t), t \in [t_0, t_f]$, to make the sales volume increase from $x(t_0) = x_0$ to $x(t_f) = x_f$, and maximize the cumulative revenue J.

When considering the practical situation, it should be noted that the following constraints should be satisfied in the above optimal market advertising payment problem:

$$0 \leq x(t) \leq x_M,$$
$$0 \leq u(t) \leq u_M,$$

where u_M is a given upper limit of the advertising payment.

◻

Example 1.9 (*Mass and Spring*) [148]

Consider a mechanical system composed of two masses, m_1 and m_2, and two springs, such that one spring, with a spring constant k_1, connects the mass m_1 to a fixed place and the other, with a spring constant k_2, connects the two masses. A control input $u(t)$ is applied to the mass m_2. See an illustration displayed in Fig. 1.7.

Denote by $x_1(t)$ and $x_2(t)$) the displacements of the masses m_1 and m_2, respectively. And denote by $x_3(t)$ and $x_4(t)$ the velocities of these two masses, respectively. Thus the state dynamics of the underlying system is specified as

$$\dot{x}_1(t) = x_3(t),$$
$$\dot{x}_2(t) = x_4(t),$$
$$\dot{x}_3(t) = -\frac{k_1}{m_1}x_1(t) + \frac{k_2}{m_1}[x_2(t) - x_1(t)],$$
$$\dot{x}_4(t) = -\frac{k_2}{m_2}[x_2(t) - x_1(t)] + \frac{1}{m_2}u(t).$$

The performance for the underlying system could be to minimize the deviation from the desired displacement and velocity, and to minimize the control effort.

◻

Example 1.10 (*Vehicle Suspension Systems*)

Consider a simplified model of a vehicle suspension system. Denote by $x(t)$ the position, and by $u(t)$ the control input for the suspension system.

Fig. 1.7 An illustration of a mechanical system composed of two masses and two springs

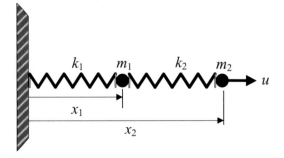

Fig. 1.8 An illustration of a simplified model of a vehicle suspension system

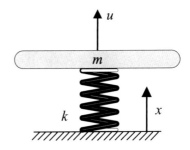

Thus the state equation is specified as

$$m\ddot{x}(t) = -kx(t) - mg + u(t),$$

where m and k represent the mass and the spring constant of the suspension system, respectively, and g is the acceleration of the gravity (Fig. 1.8).

Formulate the optimal control problem to consider a tradeoff between the minimization of the control energy and passengers' comfort.

□

Example 1.11 (*Chemical Processing*) [144, 149]

Consider a chemical processing system as displayed in Fig. 1.9. It supposes that the water liquid flows into tank I and tank II at rates of $w_1(t)$ and $w_2(t)$, respectively, and the chemical liquid flows into tank I with a rate of $z(t)$. The intersection areas of tank I and tank II are denoted by α_1 and α_2, respectively. Denote by $y_1(t)$ and $y_2(t)$ the liquid levels of tank I and tank II, respectively.

It also considers that there is a tunnel between the tanks such that the flow rate between the tanks, denoted by $v(t)$, is proportional to the difference between $y_1(t)$ and $y_2(t)$.

Assume that the mixtures in the tanks are homogeneous. And denote by $\theta_1(t)$ and $\theta_2(t)$ the volumes of chemical components in tank I and tank II, respectively.

Specify the state dynamics equations for the underlying system on variables of $y_1(t)$, $y_2(t)$, $\theta_1(t)$ and $\theta_2(t)$.

□

Next, a complex and famous optimal control problem is introduced—the inverted pendulum problem. The inverted pendulum system is a typical teaching experimental equipment in the field of automatic control. The research on inverted pendulum system can be summed up as the research on multivariable, strong coupling, absolutely unstable, and nonlinear system. Its control method has a wide range of uses in the fields of military industry, aerospace, robotics, and general industrial processes. For example, the balance control of the robot's walking process, the verticality control of rocket launching, and the attitude control of satellite flight are all related to inversion, which has become a research hotspot in the field of control. On one hand,

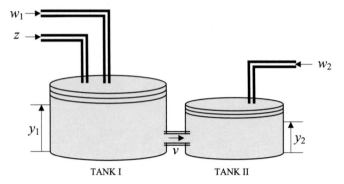

Fig. 1.9 An illustration of a simplified model of a chemical processing system

many control theories and methods have been put into practice here; on the other hand, in making efforts to study its stability control, people have been discovering new control methods and exploring new control theories.

Example 1.12 (*Inverted Pendulum Systems*)

Considering a single inverted pendulum system, it is required to keep the pendulum with a length l on a cart with mass M in a vertical position. A horizontal force $u(t)$ is applied to the cart. See an illustration displayed in Fig. 1.10.

For analytical simplicity, it supposes that there is a ball with mass m at the top of the pendulum, and the radius of the ball and the mass of the pendulum are all negligible

Denote by $x_1(t)$ and x_3 the horizontal displacement of the cart and the angular position of the pendulum from the vertical line. And denote by $x_2(t)$ and x_4 the velocity of the cart and the angular velocity of the pendulum from the vertical line.

Thus the linearized state dynamics of the underlying system can be specified as

$$\dot{x}_1(t) = x_2(t),$$
$$\dot{x}_2(t) = \frac{[M+m]g}{Ml} x_1(t) - \frac{1}{Ml} u(t),$$
$$\dot{x}_3(t) = x_4(t),$$
$$\dot{x}_4(t) = -\frac{mg}{M} x_1(t) + \frac{1}{M} u(t),$$

where g represents the gravitational constant.

The performance for the underlying system could be to keep the pendulum in the vertical position with as little control efforts or consumed energy as possible.

□

In the next, consider a differential game problem given in Example 1.13, which was proposed by Rufus Isaacs in [96].

Fig. 1.10 An illustration of
a single inverted pendulum
system

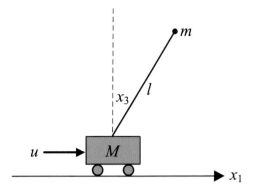

Example 1.13 (*Homicidal Chauffeur Problems*)

In the homicidal chauffeur problem, the pursuer is a driver of a vehicle which is fast but less maneuverable and the evader is a runner who can only move slowly, but is highly maneuverable, against the driver of a motor vehicle.

The pursuer wins in case the pursuer catches the evader; otherwise, the evader wins.

☐

Notice that the homicidal chauffeur problem introduced above is a classic differential game in continuous time and in a continuous state space.

In the following example, it specifies an optimal charging problem of electric vehicles (EVs) in power grid systems.

Example 1.14 (*Charging of Electric Vehicles in Power Systems*) [150, 151]

Consider the state of charge (SoC) of each of the electric vehicles EVs, with $n = 1, \ldots, N$, satisfying the following dynamics:

$$x_n(k+1) = x_n(k) + \beta_n u_n(k)$$

for all $k = k_0, \ldots, k_f - 1$, where $x_n(k)$ and $u_n(k)$ represent the SoC and the charging rate of EV n during time interval k, respectively, such that the performance cost function defined as

$$J(u) \triangleq \sum_{k=k_0}^{k_f-1} \left[\sum_{n=1}^{N} u_n(k) + D(k) \right]^2,$$

where $D(k)$ represents the total demand in the grid system at time k, is minimized.

It may also consider the following constraints for the system

$$x_n(k_f) = \Gamma_n,$$

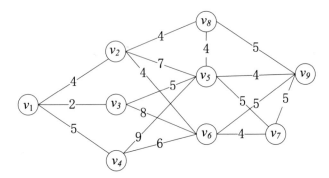

Fig. 1.11 Shortest path problems

for all n, with Γ_n representing the capacity of the battery of EV n, say, at the end of the charging interval t_f, all of the electric vehicles are fully charged.

In [152], it is considered an updated performance function such that

$$J(u) \triangleq \sum_{k=k_0}^{k_f-1} \left[\sum_{n=1}^{N} \gamma_n [u_n(t)]^2 + \left[\sum_{n=1}^{N} u_n(k) + D(k) \right]^2 \right],$$

where $\gamma_n [u_n(t)]^2$ represents the battery degradation cost subject to charging rate $u_n(t)$.

$\square$

In the next, consider a shortest path problem given in Example 1.15.

Example 1.15 (*Shortest Path Problems between Two Places*)

Figure 1.11, displays a collections of places, denoted by v_i, with $i = 1, \ldots, 9$, each of which is directly connected with some other places via certain edges, respectively.

It is also considered that each of the edges has a weighting factor. For instance, the edge between v_2 and v_5, denoted by (v_2, v_5), is valued as 7.

And a path between two places is composed of a collection of consecutive places which are connected by an appropriate edge and the cost of this path is the sum of the weighting factors of all the edges in this path. For instance, (v_1, v_3, v_5, v_9) is a path between v_1 and v_9 with the cost equal to $2 + 5 + 4 = 11$.

For the system specified above, the problem is to find the short path(s) for each pair of places with a minimum cost.

$\square$

Example 1.16 (*An Investment Problem*)

Give a certain amount of money to invest on items I, II, and III, and suppose that the investments must be made in integer amounts. It further considers that it has 8 units to invest in. Table 1.1, lists the profits with respect to allocations on each of the

Table 1.1 Profit with respect to investment amount on each item

Investment amount	Profit on I	Profit on II	Profit on III
0	0	0	0
1	3.0	2.5	1.8
2	4.6	4.1	2.5
3	6.7	5.5	4.0
4	8.0	6.5	5.0
5	9.2	7.5	6.5
6	10.2	8.0	7.3
7	11.3	8.5	8.2
8	13.2	8.8	9.0

items. For instance, for the allocation with 3 in item I, 1 in item II, 4 in item III; then
the total profit is given as $6.7 + 2.5 + 5.0$ which is equal to 14.2.

To find the optimal investment allocation:

□

In Sects. 1.4 and 1.5, we will give a general description of the optimal control
problem. The description of the optimal control problem generally includes the state
dynamic equation of the control system, the state constraints, the target set, the
admissible set of the control input, the performance function, and so on.

1.4   Formulation of Continuous-Time Optimal Control Problems

In this part, a class of optimal control problems over a continuous-period of time is
firstly formulated.

- State dynamics equation of the control system

  The dynamics equations of a continuous system can be expressed as a set of
  first-order differential equation, called *the state equations*, and a set of algebraic
  equations called *the output equations*, that is,

$$\dot{x}(t) = f(x, u, t),$$
$$y(t) = \widehat{f}(x, u, t),$$

where $x \in \mathbb{R}^n$ and $u \in \mathbb{R}^r$ are the state vector and the control input of the control
system, respectively, and $y \in \mathbb{R}^m$ is the output vector of the system, the vector-
valued function $f(\cdot) \in \mathbb{R}^n$ satisfies certain conditions, for example, the Lipschitz
condition.

Considering the above situation, there exists a unique solution to the above dynamics equation subject to the piecewise continuous control input u, and the vector-valued function $g(\cdot) \in \mathbb{R}^m$ is the output function. In some cases, the state x is measurable and can be used to design the associated control, while in some cases, only the output y can be measured to construct the control.

- State constraints and target sets

According to the specific situation, the initial state and final state of the dynamics equation of the control system may be required to satisfy various constraints, including equality constraints and inequality constraints. The general forms of equality constraints are

$$h_1(x(t_0), t_0) = 0,$$
$$h_2(x(t_f), t_f) = 0,$$

where $h_1(\cdot)$ and $h_2(\cdot)$ are function vectors. A typical equality constraint is given as

$$x(t_0) = x_0,$$
$$x(t_f) = x_f,$$

where x_0 and x_f are given constant vectors, respectively.

Notice that sometimes the initial state $x(t_0)$ or final state $x(t_f)$ cannot be uniquely determined by the equality constraints.

The general forms of inequality constraints are

$$h_1(x(t_0), t_0) \le 0,$$
$$h_2(x(t_f), t_f) \le 0.$$

The meaning of the above inequalities is that each element of the function vectors h_1 and h_2 is less than or equal to 0. For example, when the final state is required to be located in a circular domain with the origin as the center and 1 as the radius in the state space, it can be described as an inequality constraint on the final state, i.e.,

$$x_1^2(t_f) + x_2^2(t_f) + x_3^2(t_f) + \cdots + x_n^2(t_f) \le 1.$$

In most cases, the initial state $x(t_0)$ is given, and the final state $x(t_f)$ is required to meet certain constraints. The set of all final states satisfying the constraints is called the *target set*, whose general definition is given as

$$\mathcal{M} \triangleq \left\{ x(t_f) \in \mathbb{R}^n, \text{ such that } h_1(x(t_f), t_f) = 0, \ h_2(x(t_f), t_f) \le 0 \right\}.$$

There are also certain requirements or constraints putting on the characteristics
of the system state and the control input in the whole control process. This kind
of constraint is usually described by the integral about the function of the system
state and control input, and there are two forms of integral-type equality constraint
and integral-type inequality constraint, i.e.,

$$\int_{t_0}^{t_f} L_e(x, u, t)dt = 0,$$

$$\int_{t_0}^{t_f} L_i(x, u, t)dt \leq 0,$$

where $L_e(\cdot)$ and $L_i(\cdot)$ are integrable function variables.

- Admissible set of the control input
  Generally, the control input u is restricted by some constraints. The set composed
  of all the input vectors satisfying the control constraints is called the admissible
  control set, which is denoted by $\mathcal{U}$.
  A typical control constraint is the boundary limit, such as

$$\alpha_i \leq u_i \leq \beta_i,$$

for all $i = 1, 2, \ldots, r$.
The total amplitude constant constraint is

$$u_1^2 + u_2^2 + \cdots + u_r^2 = \alpha^2,$$

where $\alpha_1, \beta_i, i = 1, 2, \ldots, r$ and α are all constant valued.

- Performance cost function
  The control performance is a quantitative evaluation of the system state characteristics and control input characteristics in the control process, and a quantitative
  description of the main problems. The performance function can be generally
  described as

$$J(x, u) \triangleq h(x(t_f), t_f) + \int_{t_0}^{t_f} g(x(t), u(t), t)dt, \tag{1.3}$$

where $h(\cdot)$ and $g(\cdot)$ are both scalar value functions, and $g(\cdot)$ is integrable function.
h is called the terminal performance function, which is the evaluation of final state
characteristics, while $\int_{t_0}^{t_f} g(x, u, t)dt$ is called the integral performance function
determining the characteristics of system state and control input within time interval $[t_0, t_f]$. For different control problems, it chooses the terminal performance
cost function and integral performance function properly.
The performance function given in (1.3) is actually a mixed one. In the following,
it is given a few specific quadratic forms of performance functions.

– Finite-time linear state regulation problem.

$$J(x, u) \triangleq \frac{1}{2} x^\top (t_f) F x(t_f) + \frac{1}{2} \int_{t_0}^{t_f} \left[x^\top (t) Q(t) x(t) + u^\top (t) R(t) u(t) \right] dt,$$

where $F = F^\top \geq 0$, $Q(t) = Q^\top (t) \geq 0$, $R(t) = R^\top (t) > 0$ are all coefficient matrices.

– Finite-time linear tracking problem.

$$J(e, u) \triangleq \frac{1}{2} e^\top (t_f) F e(t_f) + \frac{1}{2} \int_{t_0}^{t_f} \left[e^\top (t) Q(t) e(t) + u^\top (t) R(t) u(t) \right] dt,$$

where $F = F^\top \geq 0$, $Q(t) = Q^\top (t) \geq 0$, $R(t) = R^\top (t) > 0$, and

$$e(t) = z(t) - y(t)$$

is the output tracking error, where $z(t)$ is the expected output and $y(t)$ is the output vector.

In the variational method, the optimal control problem of functional with mixed performance function is called Bolza problem.

When it does not concern about the characteristics of the final state, say $h = 0$, the above performance function is reduced to the following:

$$J(x, u) \triangleq \int_{t_0}^{t_f} g(x(t), u(t), t) dt. \tag{1.4}$$

For this integral performance function, there are some specific forms as given below.

– Minimum time problem.

$$J(u) \triangleq \int_{t_0}^{t_f} 1 dt = t_f - t_0,$$

with $g(x(t), u(t), t) = 1$.

– Minimum fuel problem.

$$J(u) \triangleq \int_{t_0}^{t_f} \left[\sum_{j=1}^{m} |u_j(t)| \right] dt.$$

– Minimum energy problem.

$$J(u) \triangleq \int_{t_0}^{t_f} u^\top(t)u(t)dt.$$

– Minimum time-fuel problem.

$$J(u) \triangleq \int_{t_0}^{t_f} \left[\beta + |u_j(t)|\right] dt,$$

where β, with $\beta > 0$, represents a weighting factor.
– Minimum time-energy problem.

$$J(u) \triangleq \int_{t_0}^{t_f} \left[\beta + u^\top(t)u(t)\right] dt.$$

In the variational method, the optimal control problem of functional with integral performance function is called Lagrangian problem.

Sometimes it mainly concerns with the terminal characteristics of the system state. At this time, the following performance function is used

$$J(x, u) \triangleq h(x(t_f), t_f). \tag{1.5}$$

For example, a missile is required to have a minimum miss distance. t_f can be fixed or flexible, which is determined by the nature of the optimal control problem.

With the above discussions, in the following it is formulated a class of optimal control problems.

Problem 1.1 For a given control system,

$$\dot{x}(t) = f(x, u, t),$$
$$y(t) = \widehat{f}(x, u, t),$$

it is required to design an admissible control $u \in \mathcal{U}$ such that the system state reaches the target set

$$x(t_f) \in \mathcal{M}.$$

During the whole control process, the constraints of the state and control input are satisfied,

$$\int_{t_0}^{t_f} g_e(x, u, t)dt = 0,$$

$$\int_{t_0}^{t_f} g_i(x, u, t)dt \le 0,$$

and the performance function is maximized or minimized at the same time.

$$J(x, u) \triangleq h(x(t_f), t_f) + \int_{t_0}^{t_f} g(x(t), u(t), t)dt.$$

□

Notice that in this book without loss of generality consider the performance cost function which is minimized subject to the optimal control.

If an admissible control $u \in \mathscr{U}$ is the solution of the above optimal control problem, it is called the optimal control, and the state of the corresponding control system is called the optimal trajectory.

In the design of the optimal control system, the option of performance function is very important. Improper selection will cause the performance of the control system to fail to meet the expected requirements, or even the optimal control problem has no solution.

1.5   Formulation of Discrete-Time Optimal Control Problems

By following the similar discussions given in Sect. 1.4 for continuous-time state systems, we can formulate a class of optimal control problems over a discrete-period of time below.

Problem 1.2   For a given control system,

$$x(k + 1) = f(x(k), u(k), k),$$
$$y(k) = \widehat{f}(x(k), u(k), k),$$

with $k \in \{k_0, \ldots, k_f\}$.

It is required to design an admissible control $u \in \mathscr{U}$ such that the system state reaches the following target set at the final time k_f:

$$x(k_f) \in \mathscr{M}.$$

During the whole control process, we may consider the following constraints of the state and control input are satisfied:

$$\sum_{k=k_0}^{k_f-1} g_e(x(k), u(k), k) = 0,$$

$$\sum_{k=k_0}^{k_f-1} g_i(x(k), u(k), k) \leq 0,$$

and the following performance cost function

$$J(x, u) \triangleq h(x(k_f), k_f) + \sum_{k=k_0}^{k_f} g(x(k), u(k), k),$$

is minimized as well.

□

1.6 Organization

- In this Chapter, it is given the overall background and the motivation of this text-book, as well as a brief introduction of the main works in this book.
- In Chap. 2, it is specified the condition for the extrema of functionals via the variational method and studies how to implement the extrema problems of functionals considering constraints by the elimination/direct method and the Lagrange method.
- In Chap. 3, it is studied how to solve the optimal control problems by applying the developed results of the extremal of functional via the variational method. It is given the necessary and sufficient conditions for the optimal solution to the optimal control problems with unbounded controls, and the optimal solution to the optimal control problems with different boundary conditions is studied. Moreover, it analyzes specific quadratic regulation and linear-quadratic tracking problems.
- In Chap. 4, it is studied how to implement the optimal control problems considering constrained controls and states by applying the variational method. More specifically, it analyzes the minimum time, minimum fuel, and minimum energy problems respectively.
- Besides, the Pontraygin's minimum principle described in the last chapter, in Chap. 5, introduces another key branch of optimal control methods, say the dynamic programming. Also for the purpose of comparison, it studies the relationship between these two optimal control methods.
- Based on the results developed in previous chapters in this book, Chap. 6 introduces some games, such as noncooperative differential games and two-person zero-sum differential games, where the system is driven by individual players each of which would like to minimize his own performance cost function. The NE strategies are solved and analyzed by applying the variational method.
- In Chap. 7, it is studied the different classes of optimal control problems in discrete-time case.
- In Chap. 8, it is given a brief conclusion of this book.

References

1. E.T. Fortmann, K.L. Hitz, *An Introduction to Linear Control Systems* (CRC Press, Boca Raton, 1977)
2. D.N. Burghes, A. Graham, *Introduction to Control Theory Including Optimal Control* (Wiley, New York, 1980)
3. G.F. Franklin, D.J. Powell, A. Emami-Naeini, *Feedback Control of Dynamic Systems* (Addison-Wesley, Reading, MA, 1994)
4. B.C. Kuo, *Automatic Control Systems*, 7th edn. (Prentice Hall, Englewood Cliffs, NJ, 1995)
5. F.H. Raven, *Automatic Control Engineering* (McGraw-Hill, New York, 1995)
6. R.T. Stefani, B. Shahian, C.J. Savant, G. Hostetter, *Design of Feedback Control Systems* (Oxford University Press, Oxford, 2001)
7. A.D. Lewis, A mathematical approach to classical control—single-input, single-output, time-invariant, continuous time, finite-dimensional, deterministic, linear systems. Lecture notes (2004)
8. C.-T. Chen, *Analog and Digital Control System Design Transfer-Function, State-Space, and Algebraic Methods* (Oxford University Press, Oxford, 2006)
9. J.H. Williams, F. Kahrl, Electricity reform and sustainable development in china. Environ. Res. Lett. **3**(4):044009 (2008)
10. K.J. Åström, R.M. Murray, *Feedback Systems* (Princeton University Press, Princeton, 2010)
11. N.S. Nise, *Control Systems Engineering* (Wiley, Hoboken, 2011)
12. R. Munasinghe. *Classical Control Systems: Design and Implementation* (Alpha Science, Oxford, 2012)
13. R.C. Dorf, *Modern Control Systems* (Addison-Wesley, Reading, MA, 1992)
14. Z. Gajec, M.M. Lelic, *Modern Control Systems Engineering* (Prentice-Hall, London, 1996)
15. K. Ogata, *Modern Control Engineering* (Prentice-Hall, Englewood Cliffs, NJ, 1997)
16. S.M. Shinners, *Modern Control System Theory and Design* (Wiley, New York, 1992)
17. P.N. Paraskevopoulos, *Modern Control Engineering* (CRC Press, Boca Raton 2001)
18. S.H. Zak, *Systems and Control* (Oxford University Press, Oxford, 2002)
19. D.N. Burghes, A. Graham, *Control and Optimal Control Theories with Applications* (Elsevier—Woodhead Publishing, Sawston, 2004)
20. R. Weinstock, *Calculus of Variations: With Applications to Physics and Engineering* (Dover Publications, New York, 1974)
21. G.M. Ewing, *Calculus of Variations with Applications* (Dover Publications, New York, 1985)
22. I.M. Gelfand, S.V. Fomin, R.A. Silverman, *Calculus of Variations* (Dover Publications, New York, 2000)
23. L.A. Pars, *An Introduction to the Calculus of Variations* (Dover Publications, New York, 2010)
24. M.I. Kamien, N.L. Schwartz, *Dynamic Optimization: The Calculus of Variations and Optimal Control in Economics and Management* (Dover Publications, New York, 2012)
25. O. Bolza, *Lectures on the Calculus of Variations* (Dover Publications, New York, 2018)
26. C. Fox, *An Introduction to the Calculus of Variations* (Dover Publications, New York, 1987)
27. H. Kielhöfer, *Calculus of Variations–An Introduction to the One-Dimensional Theory with Examples and Exercises* (Springer, Berlin, 2018)
28. B. van Brunt, *The Calculus of Variations—Universitext* (Springer, Berlin, 2004)
29. H.J. Pesch, R.Z. Bulirsch, Bellman's equation and Caratheodory's work. J. Optim. Theory Appl. **80**(2), 203–229 (1994)
30. H.J. Pesch, M. Plail, The maximum principle of optimal control: A history of ingenious ideas and missed opportunities. Control Cybern. **38**(4A), 973–995 (2009)
31. V.G. Boltyanskii, R.V. Gamkrelidze, L.S. Pontryagin, On the theory of optimal processes. Doklady Akad. Nauk SSSR **110**(1), 7–10 (1956)
32. R.V. Gamkrelidze, The theory of time-optimal processes for linear systems. Izvest. Akad. Nauk SSSR. Ser. Mat. **22**, 449–474 (1958)

33. V.G. Boltyanskii, The maximum principle in the theory of optimal processes. Doklady Akad. Nauk SSSR **119**(6), 1070–1073 (1958)
34. R.V. Gamkrelidze, On the general theory of optimal processes. Doklady Akad. Nazlk SSSR **123**(2), 223–226 (1958)
35. V.G. Boltyanskii, Optimal processes with parameters. Doklady Akad. Nnuk Uzbek. SSR **10**, 9–13 (1959)
36. R.V. Gamkrelidze, Time optimal processes with bounded phase coordinates. Doklady Akad. Nazlk SSSR **126**(3), 475–478 (1959)
37. L.S. Pontryagin, Optimal control processes. Uspekki Mat. Naak **14**(1), 3–20 (1959)
38. L.S. Pontryagin, V.G. Boltyanskii, et al, *The Mathematical Theory of Optimal Processes* (Wiley, New York-London, 1962). Translated by K.N. Trirogoff
39. A.J. Burns, *Introduction to the Calculus of Variations and Control with Modern Applications*(CRC Press, Boca Raton, 2013)
40. D.S. Naidu, *Optimal Control Systems* (CRC Press, Boca Raton, FL, 2002)
41. E.J. McShane, The calculus of variations from the beginning through optimal control. SIAM J. Control Optim. **27**, 916–939 (1989)
42. L.M. Hocking, *Optimal Control—An Introduction to the Theory with Applications* (Clarendon Press, Oxford, 1991)
43. A. Sagan, *Introduction to the Calculus of Variations* (Dover Publishers, Mineola, NY, 1992)
44. R.E. Pinch, *Optimal Control and Calculus of Variations* (Oxford University Press, Oxford, 1993)
45. J.L. Troutman, *Variational Calculus and Optimal Control: Optimization with Elementary Convexity* (Springer, Berlin, 1995)
46. J. Macki, A. Strauss, *Introduction to Optimal Control Theory* (Springer, New York, 1995)
47. D.R. Smith, *Variational Methods in Optimization* (Dover Publications, New York, 1998)
48. S.D. Naidu, *Optimal Control Systems* (CRC Press, New York, 2003)
49. D.E. Kirk, *Optimal Control Theory: An introduction* (Dover Publications, New York, 2012)
50. F. Clarke, *Calculus of Variations and Optimal Control, Functional Analysis* (Springer, Berlin, 2013)
51. Emanuel Todorov. *Optimal Control Theory*. MIT press, 2006
52. P.H. Geering, *Optimal Control with Engineering Applications* (Springer, Berlin, Heidelberg, 2007)
53. T.L. Friesz, *Dynamic Optimization and Differential Games* (Springer Science+Business Media, Berlin, 2010)
54. R. Vinter, *Optimal Control* (Birkhauser, Boston, 2000)
55. E. Miersemann, *Calculus of Variations*. Lecture Notes in Leipzig University (2012)
56. D. Liberzon, *Calculus of Variations and Optimal Control Theory: A Concise Introduction* (Princeton University Press, Princeton, 2012)
57. G. Leitmann, *Optimization Techniques: With Applications to Aerospace Systems* (Academic Press Inc., New York, 1962)
58. J.E. Corban, A.J. Calise, G.A. Flandro, Rapid near-optimal aerospace plane trajectory generation and guidance. J. Guid. **14**(6), 1181–90 (1991)
59. Z.J. Ben-Asher, *Optimal Control Theory with Aerospace Applications* (American Institute of Aeronautics and Astronautics Inc., Reston, 2010)
60. E. Trelat, Optimal control and applications to aerospace: some results and challenges. J. Optim. Theory Appl. **154**, 713–758 (2012)
61. J.M. Longuski, J.J. Guzman, J.E. Prussing, *Optimal Control with Aerospace Applications* (Springer, New York, 2014)
62. H.C. Lee, A generalized minimum principle and its application to the vibration of a wedge with rotatory inertia and shear. J. Appl. Mech. **30**(2), 176–180 (1963)
63. W.S. Howard, V. Kumar, A minimum principle for the dynamic analysis of systems with frictional contacts. In *Proceedings IEEE International Conference on Robotics and Automation, Atlanta, GA, USA*, pp. 437–442 (1993)

64. H. Li, G. Ben, Application of the principle of minimum entropy production to shock wave reflections. i. steady flows. J. Appl. Phys. **80**(4), 2027 (1996)
65. W. Jung, J. Park, S. Lee, K. Kim, S. Kim, Optimal control of passive locking mechanism for battery exchange using pontryagin's minimum principle. In *the 8th Asian Control Conference (ASCC), Kaohsiung, Taiwan*, pp. 1227–1232, 15–18 May (2011)
66. H.K. Moharam, N. Mostafa, R.N. Hamed, Dynamic load carrying capacity of flexible manipulators using finite element method and pontragin's minimum principle. J. Optim. Ind. Eng. **6**(12), 17–24 (2013)
67. H. Andrei, F. Spinei, The minimum energetical principle in electric and magnetic circuits. In *The 18th European Conference on Circuit Theory and Design, Seville*, pp. 906–909, 27–30 Aug. (2007)
68. C. Zheng, S.W. Cha, Real-time application of pontryagin's minimum principle to fuel cell hybrid buses based on driving characteristics of buses. Int. J. Precis. Eng. Manuf. -Green Technol. **4**, 199–209 (2017)
69. S. Zhang, R. Xiong, C. Zhang, Pontryagin's minimum principle-based power management of a dual-motor-driven electric bus. Appl. Energy **159**, 370–380 (2015)
70. H. Andrei, F. Spinei, N. Jula, The principle of minimum dissipated power and the minimum energetic principle—two general theorems of the linear electric and magnetic circuits in stationary regime. In *ISTASC'08: Proceedings of the 8th Conference on Systems Theory and Scientific Computation*, pp. 75–87 (2008)
71. N. Kim, S. Cha, H. Peng, Optimal control of hybrid electric vehicles based on pontryagin's minimum principle. IEEE Trans. Control Syst. Technol. **19**(5), 1279–1287 (2011)
72. N. Kim, A. Rousseau, Sufficient conditions of optimal control based on pontryagin's minimum principle for use in hybrid electric vehicles. J. Automob. Eng. **226**(9), 1160–1170 (2012)
73. S.R. Upreti, *Optimal Control for Chemical Engineers* (CRC Press, Baco Raton, 2017)
74. E. Aydin, D. Bonvin, K. Sundmacher, Dynamic optimization of constrained semi-batch processes using pontryagin's minimum principle—an effective quasi-newton approach. Comput. Chem. Eng. **99**, 135–144 (2017)
75. E. Aydin, D. Bonvin, K. Sundmacher, Nmpc using pontryagin's minimum principle—application to a two-phase semi-batch hydroformylation reactor under uncertainty. Comput. Chem. Eng. **108**, 47–56 (2018)
76. Y.-T. Tseng, J.D. Ward, Comparison of objective functions for batch crystallization using a simple process model and pontryagin's minimum principle. Comput. Chem. Eng. **99**, 271–279 (2017)
77. D.A. Carlson, A. Haurie, *Infinite horizon optimal control theory and applications*. Lecture Notes in Economics and Mathematical Systems (Springer, New York, 1987)
78. G. Erickson, *Dynamic Models of Advertising Competition* (Kluwer, Boston, 1991)
79. D. Leonard, *Optimal Control Theory and Static Optimization in Economics* (Cambridge University Press, Cambridge, 1992)
80. O. Maimon, E. Khmelnitsky, K. Kogan, *Optimal Flow Control in Manufacturing Systems: Production Planning and Scheduling*. (Springer, New York, U.S., 1998)
81. Engelbert J. Dockner, Steffen Jorgensen, Ngo Van Long, and Gerhard Sorger. *Differential Games in Economics and Management Science*. Cambridge University Press, 2000
82. P. Suresh, *Sethi and Gerald L* (Springer, Thompson. Optimal Control Theory - Applications to Management Science and Economics, 2000)
83. Suresh P. Sethi. *Optimal Control Theory: Applications to Management Science and Economics, 3rd Edition*. Springer International Publishing, 2019
84. F. Boselie, E. Leeuwenberg, A test of the minimum principle requires a perceptual coding system. Perception **15**(3), 331–354 (1986)
85. G. Hatfield, W. Epstein, The status of the minimum principle in the theoretical analysis of visual perception. Psychological Bulletin **95**, 155–186 (1985)
86. R.E. Bellman, *Dynamic Programming* (Princeton University Press, 1957)
87. R.E. Bellman, S.E. Dreyfus, *Applied Dynamic Programming* (Princeton University Press, 1962)

88. R.E. Bellman, R.E. Kalaba, *Dynamic Programming and Modem Control Theory* (Academic Press, 1965)
89. Brian Gluss. *An Elementary Introduction to Dynamic Programming: A State Equation Approach*. Allyn and Bacon, Inc., 1972
90. Dimitri P. Bertsekas. *Dynamic Programming and Optimal Control, Vol. I, Edition 4*. Athena Scientific, 2017
91. John Bather and J. A. Bather. *Decision Theory: An Introduction to Dynamic Programming and Sequential Decisions*. Wiley, 2000
92. L. Frank, *Lewis and Derong Liu* (Wiley-IEEE Press, Reinforcement Learning and Approximate Dynamic Programming for Feedback Control, 2012)
93. P. Dimitri, *Bertsekas and John N* (Athena Scientific, Tsitsiklis. Neuro-Dynamic Programming, 1996)
94. Warren B. Powell, *Approximate Dynamic Programming: Solving the Curses of Dimensionality*, 2nd edn. (Wiley, 2011)
95. C.F. Roos, A mathematical theory of competition. American Journal of Mathematics **47**(3), 163–175 (1925)
96. R. Isaacs, *Differential Games: A Mathematical Theory with Applications to Warfare and Pursuit, Control and Optimization* (John Wiley and Sons, London, 1965)
97. Meir Pachter and Y. Yavin. Simple-motion pursuit-evasion differential games. *Journal of Optimization Theory and Applications*, 51:95–159, 1986
98. Tamer Basar and Geert Jan Olsder, *Dynamic Noncooperative Game Theory* (Academic Press, New York, 1995)
99. Tamer Basar and Alain Haurie. *Advances in Dynamic Games and Applications*. Birkhauser Basel, 1994
100. Engelbert Dockner, Steffen Jorgensen, Ngo Van Long, Gerhard Sorger, *Differential Games in Economics and Management Science* (Cambridge University Press, 2001)
101. Leon Petrosyan. *Differential Games of Pursuit*. World Scientific Publishers, 1993
102. C.K. Leong, W. Huang, A stochastic differential game of capitalism. Journal of Mathematical Economics **46**(4), 552 (2010)
103. H. Tembine and Tyrone E. Duncan. Linear-quadratic mean-field-type games: A direct method. *Games*, 9 (1): 7, 2018
104. M. Y. Huang, P. E. Caines, and R. P. Malhame. Individual and mass behaviour in large population stochastic wireless power control problems: centralized and Nash equilibrium solutions. In *Proc. the 42nd IEEE Conference on Decision and Control, Maui, Hawaii*, pages 98–103, December 2003
105. M. Y. Huang, P. E. Caines, and R. P. Malhame. Large-population cost-coupled LQG problems: generalizations to non-uniform individuals. In *Proc. the 43rd IEEE Conference on Decision and Control, Atlantis, Paradise Island, Bahamas*, pages 3453–3458, December 2004
106. M. Y. Huang, R. P. Malhame, and P. E. Caines. On a class of large-scale cost-coupled Markov games with applications to decentralized power control. In *Proc. the 43rd IEEE Conference on Decision and Control, Atlantis, Paradise Island, Bahamas*, pages 2830–2835, December 2004
107. M.Y. Huang, R.P. Malhame, P.E. Caines, Large population stochastic dynamic games: Closed-loop Mckean-Vlasov systems and the Nash certainty equivalence principle. Communications in Information and Systems **6**(3), 221–252 (2006)
108. M.Y. Huang, P.E. Caines, R.P. Malhame, Large-population cost-coupled LQG problems with non-uniform agents: individual-mass behavior and decentralized epsilon-Nash equilibria. IEEE Transactions on Automatic Control **52**(9), 1560–1571 (2007)
109. M. Nourian, P.E. Caines, Epsilon - Nash mean field game theory for nonlinear stochastic dynamical systems with major and minor agents. SIAM Journal on Control and Optimization **51**(4), 3302–3331 (2013)
110. P.-L. Lions, J.-M. Lasry, Large investor trading impacts on volatility. Annales de l'Institut Henri Poincaré C. **24**(2), 311–323 (2007)

111. J.-M. Lasry, P.-L. Lions, Mean field games. Japanese Journal of Mathematics **2**(1), 229–260 (2007)
112. Jean-Michel Lasry and Pierre-Louis Lions. Mean field games. II - finite horizon and optimal control. *Comptes Rendus Mathematique (in French)*, 343 (10):679–684, 2006
113. Pierre Cardaliaguet. *Notes on Mean Field Games*. 2013
114. H. Tembine, Risk-sensitive mean-field-type games with Lp-norm drifts. Automatica **59**, 224–237 (2015)
115. B. Djehiche, A. Tcheukam, H. Tembine, Mean-field-type games in engineering. AIMS Electronics and Electrical Engineering **1**(1), 18–73 (2017)
116. H. Tembine, Mean-field-type games. AIMS Mathematics **2**(4), 706–735 (2017)
117. Tyrone Duncan, Hamidou Tembine, Linear-quadratic mean-field-type games: A direct method. Games **9**(1), 7 (2018)
118. D. Andersson, B. Djehiche, A maximum principle for SDEs of mean-field type. Applied Mathematics & Optimization **63**(3), 341–356 (2010)
119. Alain Bensoussan, Jens Frehse, and Phillip Yam. *Mean Field Games and Mean Field Type Control Theory*. Springer Briefs in Mathematics New York: Springer-Verlag, 2013
120. H. Bock and K. Plitt. A multiple shooting algorithm for direct solution of optimal control problems. In *Proceedings of the 9th world congress of the international federation of automatic control*, volume 9, 1984
121. D. Kraft. *Computational Mathematical Programming*, volume F15 of *in NATO ASI Series*, chapter On converting optimal control problems into nonlinear programming codes. Springer, 1985
122. O. von Stryk, R. Bulirsch, Direct and indirect methods for trajectory optimization. Annals of Operations Research **37**, 357–373 (1992)
123. T.J. Böhme and B. Frank. *in Hybrid Systems, Optimal Control and Hybrid Vehicles — Advances in Industrial Control*, chapter Direct Methods for Optimal Control. Springer, 2017
124. Ismail Bin Mohd, Interval elimination method for stochastic spanning tree problem. Applied Mathematics and Computation **66**(2–3), 325–341 (1994)
125. Stefan Edelkamp and Stefan Schrödl. *Heuristic Search – Theory and Applications*. Elsevier, 2012
126. Elijah Polak, *Optimization: Algorithms and Consistent Approximations* (Springer-Verlag, 1997)
127. Hui Peng, T. Ozaki, V. Haggan-Ozaki, and Y. Toyoda. A parameter optimization method for radial basis function type models. *IEEE Transactions on Neural Networks*, 14(2):432–438, Mar. 2003
128. Tao Zhang, L. Li, Yanluan Lin, Wei Xue F. Xie, H. Xu, and X. Huang. An automatic and effective parameter optimization method for model tuning. *Geoscientific Model Development*, 8(5):3791–3822, 2015
129. G. Madan, *Singh* (Dynamical Hierarchical Control. North-Holl and Pabl. Co., Amsterdam, 1980)
130. Z.C. Shi, W.B. Gao, Dynamic hierarchical control for large-scale systems. Acta Automatica Sinica **13**(2), 111–119 (1987)
131. N.V. Findler, J. Gao, Dynamic hierarchical control for distributed problem solving. Data and Knowledge Engineering **2**(4), 285–301 (1987)
132. W.S. Mcculloch, H.D. Landahl, W. Pitts, A statistical consequence of the logical calculus of nervous nets. Bulletin of Mathematical Biology **5**(4), 135–137 (1943)
133. J.J. Hopfield, Neural networks and physical systems with emergent collective computational abilities. Proceedings of the National Academy of Sciences **79**(8), 2554–2558 (1982)
134. J.J. Hopfield, D.W. Tank, "neural" computation of decisions in optimization problems. Biological Cybernetics **52**(3), 141–152 (1985)
135. J.J. Hopfield, D.W. Tank, Simple neural optimization network: An A/D converter, signal decision circuit and a linear programming circuit. IEEE Transactions on Circuits and Systems **33**(5), 533–541 (1986)

136. M.P. Kennedy, L.O. Chua, Neural networks for nonlinear programming. IEEE Transactions on Circuits and Systems **35**(5), 554–562 (1988)
137. C. Peterson. *Neural optimization*. The handbook of brain theory and neural networks, 1998
138. U. Maulik, S. Bandyopadhyay, Genetic algorithm-based clustering technique. Pattern Recognition **33**(9), 1455–1465 (2000)
139. D.E. Goldberg, *Genetic Algorithm in Search, Optimization, and Machine Learning* (Addison-Wesley, MA Publisher, 1989)
140. R.E. Bellman, L.A. Zadeh, *Local and Fuzzy Logics* (Selected Papers, Fuzzy Sets, Fuzzy Logic, And Fuzzy Systems, 1977)
141. Marco Dorigo, Marco A. Montes de Oca, Sabrina Oliveira, and Thomas Stutzle. *Ant Colony Optimization*. Wiley Encyclopedia of Operations Research and Management Science, 2011
142. J. Kennedy, *Particle Swarm Optimization* (Springer, 2011)
143. S.Z. Selim, K. Alsultan, A simulated annealing algorithm for the clustering problem. Pattern Recognition **24**(10), 1003–1008 (1991)
144. Donald E. Kirk, Optimal Control Theory: An Introduction. American Scientist (1971)
145. G. Oster, E.O. Wilson, *Caste and Ecology in Social Insects* (Princeton University Press, 1978)
146. C. Lawrence, *Evans* (University of Maryland, An Introduction to Mathematical Optimal Control Theory, 1995)
147. S.P. Sethi. *In: Mathematical Models in Marketing. Lecture Notes in Economics and Mathematical Systems (Operations Research)*, volume 132, chapter Optimal Control of the Vidale-Wolfe Advertising Model. Springer, 1976
148. L. Meirovitch, *Dynamics and Control of Structures* (John Wiley & Sons, New York, NY, 1990)
149. W.F. Ramirez, *Process Control and Identification* (Academic Press, San Diego, CA, 1994)
150. Z. Ma, D. Callaway, I. Hiskens, Decentralized charging control of large populations of plug-in electric vehicles. IEEE Transactions on Control Systems Technolgy **21**(1), 67–78 (2013)
151. L. Gan, U. Topcu, S.H. Low, Optimal decentralized protocol for electric vehicle charging. IEEE Transactions on Power Systems **28**(2), 940–951 (2013)
152. Z. Ma, S. Zou, X. Liu, A distributed charging coordination for large-scale plug-in electric vehicles considering battery degradation cost. IEEE Transactions on Control Systems Technology **23**(5), 2044–2052 (2015)

Chapter 2
Extrema of a Functional via the Variational Method

It is widely known that the calculus of variations had been considered as an important mathematical branch after Leonhard Euler published the famous monographs, *Elementa Calculi Variationum*, in 1733, and *A method for finding curved lines enjoying properties of maximum or minimum, or solution of isoperimetric problems in the broadest accepted sense* in 1744. The variational method is a powerful mathematical tool to deal with the implementation of the extrema (maximum or minimum) of a functional. In his book on the calculus of variations, Euler extended known methods of the calculus of variations to form and solve differential equations for the general problem of optimizing single-integral variational quantities.

Nevertheless, it is worth stating, before Euler studied the variational method in a systematic way, quite a few specific optimization problems that had been essentially solved by using the variational principles. Queen Dido was faced with the problem to find the closed curve with a fixed perimeter that encloses the maximum area. Certainly, the extremal solution is a circle which can be obtained by applying the variational method. Isaac Newton designed the shape of a body moving in the air with the least resistance. Another interesting problem proposed by Johann Bernoulli at the end of the seventeenth century is the Brachistochrone problem which was solved by Jacob Bernoulli, Isaac Newton, L'Hospital, and himself.

This chapter is organized as follows. It firstly introduces some fundamental terms related to functions and functionals, respectively, in Sect. 2.1, which will be used later. Then in Sect. 2.2, it specifies the necessary and sufficient condition for the extrema of functionals via the variational method. Based upon the results developed in Sect. 2.2, it gives necessary conditions for the extrema of functionals with respect to multiple functions, which are independent of each other, in Sect. 2.3. In Sects. 2.4 and 2.5, it introduces the extrema problems of functions and functionals, respectively, considering constraints which are solved by the elimination/direct method and the Lagrange method. In Sect. 2.6, it gives a brief summary of the results developed in this chapter. Lastly, in Sect. 2.7, it lists some exercises for readers to assist them to grasp the content covered in this chapter.

© The Author(s), under exclusive license to Springer Nature Singapore Pte Ltd. 2021   39
Z. Ma and S. Zou, *Optimal Control Theory*,
https://doi.org/10.1007/978-981-33-6292-5_2

2.1 Fundamental Notations

In this part, a collection of fundamental terms will be introduced.

2.1.1 Linearity of Function and Functional

Firstly define the linearity of a function as follows:

Definition 2.1 $x(\cdot)$ is a function of a variable t, if there is a single related x to each value of t in a range of domain $\mathcal{T}$; then call $x(t)$ as a *linear function* of t if and only if it satisfies the *principle of homogeneity*

$$x(\alpha t) = \alpha x(t), \tag{2.1}$$

for all $t \in \mathcal{T}$ and for all real numbers α such that $\alpha t \in \mathcal{T}$, and the *principle of additivity*

$$x(t^{(1)} + t^{(2)}) = x(t^{(1)}) + x(t^{(2)}), \tag{2.2}$$

for all $t^{(1)}$, $t^{(2)}$, and $t^{(1)} + t^{(2)}$ in $\mathcal{T}$.   □

Notice that here t may represent time or any scalar or vector variable.

Example 2.1 Consider a function $x(t) = at + b$; then it is straightforward to show that $x(t)$ satisfies (2.1) and (2.2). Hence it is a linear function of t. However, it can be shown that $x(t) = t^2 + 1$ is not a linear function of t.   □

Definition 2.2 $J(x(\cdot))$ is a functional of a function $x(\cdot)$, if there is a single related J to each function $x(\cdot)$ in a class of functions $\mathcal{X}$; then call $J(x)$ as a *linear functional* of function x if and only if it satisfies the *principle of homogeneity*

$$\alpha J(x) = J(\alpha x), \tag{2.3}$$

for all $x \in \mathcal{X}$ and for all real numbers α such that $\alpha x \in \mathcal{X}$, and the *principle of additivity*

$$J(x^{(1)} + x^{(2)}) = J(x^{(1)}) + J(x^{(2)}), \tag{2.4}$$

for all $x^{(1)}$, $x^{(2)}$, and $x^{(1)} + x^{(2)}$ in $\mathcal{X}$.   □

Consider a functional

$$J(x) = \int_{t_0}^{t_f} x(t)dt, \tag{2.5}$$

where x is a continuous function of $t \in [t_0, t_f]$; then $J(x) = \frac{1}{2}a + b$ with $t_0 = 0$, $t_f = 1$, and $x(t) = at + b$ as given in Example 2.1.

Example 2.2 Consider the functional as given in (2.5); then it can be verified that $J(x)$ satisfies (2.1) and (2.4) as follows:

• Additivity:

$$J(x^{(1)} + x^{(2)}) = \int_{t_0}^{t_f} \left[x^{(1)} + x^{(2)} \right] dt$$

which implies that

$$J(x^{(1)} + x^{(2)}) = J(x^{(1)}) + J(x^{(2)}),$$

for all $x^{(1)}$, $x^{(2)}$, and $x^{(1)} + x^{(2)}$ in $\mathscr{X}$;

• Homogeneity:

$$J(\alpha x) = \int_{t_0}^{t_f} \alpha x(t) dt = \alpha \int_{t_0}^{t_f} x(t) dt = \alpha J(x).$$

Consequently, we can claim that it is a linear functional of function x. □

Example 2.3 Consider another functional in the following:

$$J(x) = \int_{t_0}^{t_f} \left[x^2(t) + 2 \right] dt; \tag{2.6}$$

then we obtain

$$J(\alpha x) = \int_{t_0}^{t_f} \left[[\alpha x(t)]^2 + 2 \right] dt$$

$$= \alpha^2 \int_{t_0}^{t_f} \left[x^2(t) + 2 \right] dt + \left[2 - 2\alpha^2 \right] [t_f - t_0]$$

$$= \alpha^2 J(x) + 2 \left[1 - \alpha^2 \right] [t_f - t_0], \tag{2.7}$$

which is not equal to $\alpha J(x)$ in general.

Hence, $J(x)$ specified in (2.6) is a nonlinear functional of function x. □

2.1.2 Norm in Euclidean Space and Functional

Definition 2.3 The norm in n-dimensional Euclidean space is a rule of correspondence that assigns to each point $x \equiv (x_1, \ldots, x_n)$ a real number. The norm of x denoted by $\|x\|$ satisfies the following properties:

$$\|x\| \geq 0, \text{ and } \|x\| = 0 \text{ if and only if } x = 0, \tag{2.8a}$$

$$\|\alpha x\| = |\alpha| \cdot \|x\|, \text{ for all real valued } \alpha, \tag{2.8b}$$

$$\|x^{(1)} + x^{(2)}\| \leq \|x^{(1)}\| + \|x^{(2)}\|. \tag{2.8c}$$

□

Here, some of norms in the Euclidean space are listed as follows:

$$\|x\|_1 \triangleq \sum_{i=1}^{n} |x_i|, \tag{2.9a}$$

$$\|x\|_2 \triangleq \left[\sum_{i=1}^{n} x_i^2\right]^{1/2}, \tag{2.9b}$$

$$\|x\|_\infty \triangleq \max_{i=1,\ldots,n} \{|x_i|\}. \tag{2.9c}$$

It can be verified that each of the terms specified above satisfies each of the properties in (2.8).

Besides, the norm in the finite-dimensional Euclidean space as specified above, the norm in the infinite-dimensional space is defined as follows:

Definition 2.4 The norm of a function is a rule of correspondence that assigns to each function $x(\cdot) \in \mathcal{X}$, defined for $t \in [t_0, t_f]$, a real number. The norm of $x(\cdot)$, denoted by $\|x(\cdot)\|$, satisfies the following properties:

$$\|x(\cdot)\| \geq 0, \text{ and } \|x(\cdot)\| = 0 \text{ if and only if } x(t) = 0, \text{ for all } t \in [t_0, t_f], \tag{2.10a}$$

$$\|\alpha x(\cdot)\| = |\alpha| \cdot \|x(\cdot)\|, \text{ for all real valued } \alpha, \tag{2.10b}$$

$$\|x^{(1)}(\cdot) + x^{(2)}(\cdot)\| \leq \|x^{(1)}(\cdot)\| + \|x^{(2)}(\cdot)\|. \tag{2.10c}$$

□

Besides the norms defined in the n-dimensional Euclidean space above, in the following some norms in the infinite-dimensional space are introduced as well: x is a continuous scalar function of t defined in the interval $[t_0, t_f]$.

Define certain norms for $x(\cdot)$ as follows:

$$\|x(\cdot)\|_1 \triangleq \int_{t_0}^{t_f} |x(t)| dt, \tag{2.11a}$$

$$\|x(\cdot)\|_2 \triangleq \left[\int_{t_0}^{t_f} x^2(t) dt\right]^{1/2}, \tag{2.11b}$$

$$\|x(\cdot)\|_\infty \triangleq \max_{t_0 \leq t \leq t_f} \{|x(t)|\}, \tag{2.11c}$$

each of which is a norm of function $x(\cdot)$ since it can be verified that all of them satisfy the three properties given in (2.10), respectively.

For notational simplicity, the norms defined in (2.11) are denoted by L_1-norm, L_2-norm, and L_∞-norm, respectively.

Example 2.4 Consider a function $x(t) = 2t^2 + 1$ with $t \in [0, 1]$; then the norms for $x(\cdot)$ are given as

$$\|x(\cdot)\|_1 \triangleq \int_0^1 \left| 2t^2 + 1 \right| dt = \frac{5}{3}, \tag{2.12a}$$

$$\|x(\cdot)\|_2 = \left[\int_0^1 \left[2t^2 + 1 \right]^2 dt \right]^{1/2} = \sqrt{\frac{47}{15}}, \tag{2.12b}$$

$$\|x(\cdot)\|_\infty \triangleq \max_{0 \le t \le 1} \{ \left| 2t^2 + 1 \right| \} = 3. \tag{2.12c}$$

□

2.1.3 Increment of Function and Functional

In this part, we consider the increment of a function and a functional.

Definition 2.5 The increment of a function f, denoted by Δf, is defined as the following:

$$\Delta f = f(x + \Delta x) - f(x), \tag{2.13}$$

for all x and $x + \Delta x \in \mathcal{Y}$ the domain of the function f. □

Notice that, as defined in Definition 2.5, Δf depends on both x and Δx. Hence, we may write Δf as $\Delta f(x, \Delta x)$.

Example 2.5 Consider the function

$$f(x) = 2x_1^2 + x_1 x_2 + x_1, \tag{2.14}$$

for all real q_1 and q_2.

The increment of the function f given above is

$$\Delta f = f(x + \Delta x) - f(x)$$
$$= 2[x_1 + \Delta x_1]^2 + [x_1 + \Delta x_1][x_2 + \Delta x_2] + [x_1 + \Delta x_1] - \left[2x_1^2 + x_1 x_2 + x_1 \right]$$
$$= 4x_1 \Delta x_1 + 2(\Delta x_1)^2 + \Delta x_1 x_2 + \Delta x_2 x_1 + \Delta x_1 \Delta x_2 + \Delta x_1. \tag{2.15}$$

□

Parallel to the increment of a function below, the increment of a functional is introduced.

Definition 2.6 The increment of a functional J, denoted by ΔJ, is defined as the following:

$$\Delta J(x, \delta x) = J(x(t) + \delta x(t)) - J(x(t)),\tag{2.16}$$

for all x and $x + \delta x \in \mathcal{X}$ the domain of the functional J, with $\delta x(t)$ representing the variation of the function $x(t)$.   □

Notice that, as defined in Definition 2.6, ΔJ depends on both $x(t)$ and $\delta x(t)$. Hence, we may write ΔJ as $\Delta J(x(t), \delta x(t))$.

Example 2.6 Specify the increment of the following functional:

$$J(x) = \int_{t_0}^{t_f} [2x(t) + 1]^2 dt,\tag{2.17}$$

where x is a continuous function of t.

By Definition 2.6, the increment is

$$\begin{aligned}
\Delta J &= J(x + \delta x) - J(x)\\
&= \int_{t_0}^{t_f} [2x(t) + 2\delta x(t) + 1]^2\, dt - \int_{t_0}^{t_f} [2x(t) + 1]^2 dt\\
&= \int_{t_0}^{t_f} \left[8x(t)\delta x(t) + 4\,[\delta x(t)]^2 + 4\delta x(t)\right] dt.
\end{aligned}\tag{2.18}$$

□

2.1.4   Differential of Function and Variation of Functional

Definition 2.7 By Definition 2.5, the increment of the function f at a variable x is given as

$$\Delta f(x, \Delta x) \triangleq f(x + \Delta x) - f(x, \Delta x);\tag{2.19}$$

then by expanding $f(x + \Delta x)$ in a Taylor series at x, we obtain

$$\begin{aligned}
\Delta f(x, \Delta x) &= f(x) + \frac{df}{dx}\Delta x + \frac{1}{2!}\frac{d^2 f}{dx^2}[\Delta x]^2 + \mathcal{O}\left([\Delta x]^3\right) - f(x)\\
&= \frac{df}{dx}\Delta x + \frac{1}{2!}\frac{d^2 f}{dx^2}[\Delta x]^2 + \mathcal{O}\left([\Delta x]^3\right),
\end{aligned}\tag{2.20}$$

where $\mathcal{O}\left([\Delta x]^3\right)$ represents the total parts with the order three or higher in Δx.

Fig. 2.1 An illustration of the increment Δf, the differential df, and the derivative $\dot{f}$

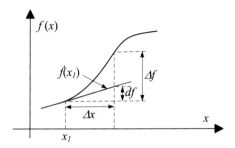

Denote by df the linear part of Δx in the increment of the function f, then $df = \dfrac{df}{dx} \Delta x$. That is to say, df is the linear approximation of the increment of function f, with respect to Δx.

Also denote by $\dot{f}(x)$ the derivative of $f(x)$, then $\dot{f}(x) = \dfrac{df}{dx}$. □

Figure 2.1 displays an illustration of the increment Δf, the differential df, and the derivative $\dot{f}$. As illustrated $\dot{f}(x)$ is the slope of the line that is tangent to f at x; $\dot{f}(x)\Delta x$ is a linear approximation to Δf. Actually, the smaller the Δx, the better the approximation.

In particular, consider that f is a differentiable function of n variables; then the differential df is given by

$$df = \frac{\partial f}{\partial x_1} \Delta x_1 + \frac{\partial f}{\partial x_2} \Delta x_2 + \cdots + \frac{\partial f}{\partial x_n} \Delta x_n. \tag{2.21}$$

Example 2.7 Specify the increment of

$$f(x) = x_1^2 + 2x_1 x_2. \tag{2.22}$$

Firstly, we have the increment as

$$\Delta f(x, \Delta x) = [2x_1 + 2x_2]\Delta x_1 + 2x_1 \Delta x_2 + [\Delta x_1]^2 + 2\Delta x_1 \Delta x_2. \tag{2.23}$$

The first two terms are linear in Δx; then the differential is

$$df(x, \Delta x) = [2x_1 + 2x_2]\,\Delta x_1 + [2x_1]\,\Delta x_2. \tag{2.24}$$

□

Definition 2.8 As given in Definition 2.6, and by expanding $J(x(t) + \delta x(t))$ in a Taylor series at x, the increment of a functional can be written as

$$\Delta J(x, \delta x)$$

$$= J(x(t) + \delta x(t)) - J(x(t))$$

$$= J(x(t)) + \frac{\partial J}{\partial x} \delta x(t) + \frac{1}{2!} \frac{\partial^2 J}{\partial x^2} [\delta x(t)]^2 + \mathcal{O}\left([\delta x(t)]^3\right) - J(x(t))$$

$$= \frac{\partial J}{\partial x} \delta x(t) + \frac{1}{2!} \frac{\partial^2 J}{\partial x^2} [\delta x(t)]^2 + \mathcal{O}\left([\delta x(t)]^3\right), \tag{2.25}$$

where $\mathcal{O}\left([\delta x(t)]^3\right)$ represents the total parts with the order three or higher in $\delta x(t)$, and δJ is the variation of J evaluated for the function x.

Denote by δJ the linear part of $\delta x(t)$ in the increment of the functional J; then $\delta J = \frac{\partial J}{\partial x} \delta x(t)$, and it is called the variation of the functional J. That is to say, the variation δJ is the linear approximation of the increment of functional J, with respect to $\delta x(t)$.

Also denote by $\delta^2 J$ the second variation of the functional J, such that

$$\delta^2 J = \frac{1}{2!} \frac{\partial^2 J}{\partial x^2} [\delta x(t)]^2, \tag{2.26}$$

where $\delta^2 J$ is the second part of $\delta x(t)$ in the increment of the functional J.   □

Example 2.8 Specify the variation of the functional studied in Example 2.6, where $J(x) = \int_{t_0}^{t_f} [2x(t) + 1]^2 dt$; then the increment of the functional is

$$\Delta J(x, \delta x) = 4[2x(t) + 1]\delta x(t) + 4[\delta x(t)]^2, \tag{2.27}$$

and hence the variation of the functional

$$\delta J(x, \delta x) = 4[2x(t) + 1]\delta x(t),$$

and the second variation of the functional

$$\delta^2 J(x, \delta x) = 4[\delta x(t)]^2.$$ □

2.2   Extrema of Functional

In this part, some definitions are given related to the extrema (maximum or minimum) of the function and functional, respectively.

Definition 2.9 A function f with domain $\mathscr{D}$ reaches a local extremal value at x^* if there exists an $\varepsilon > 0$ such that for all points x in $\mathscr{D}$ that satisfy $\|x - x^*\| < \varepsilon$, the increment of f has the same sign. If

Fig. 2.2 A function f with several extrema

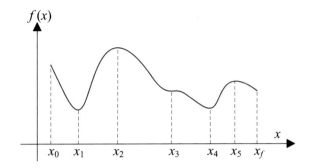

$$\Delta f = f(x) - f(x^*) \geq 0, \tag{2.28}$$

$f(x^*)$ reaches a local minimum value at this point; otherwise, if

$$\Delta f = f(x) - f(x^*) \leq 0, \tag{2.29}$$

$f(x^*)$ reaches a local maximum value at this point.

Furthermore, if (2.28) is satisfied for arbitrarily large ε, then $f(x^*)$ reaches a global minimum value at x^* whereas, if (2.29) holds for arbitrarily large ε, then $f(x^*)$ reaches a global maximum value at x^*. □

Example 2.9 Consider the function of a single variable $x \in [x_0, x_f]$, as illustrated in Fig. 2.2.

Since the domain is bounded and closed, the extrema may be located at those points where the differential vanishes and at the end points as well.

As illustrated, the differential vanishes at x_1, x_2, x_3, x_4, and x_5. The function reaches a maximum value at x_2, x_5, and a minimum value at x_1 and x_4, respectively.

More specially, x_2 is the global maximum and x_1 is the global minimum. However, different from other stationary points, x_3 is not an extremal point. Concerning the end points x_0 and x_f, x_0 is a local maximum and x_f is a local minimum, respectively. □

Definition 2.10 A functional J with domain Ω reaches a local maximum value at x^* if there is an $\varepsilon > 0$ such that for all functions x in Ω which satisfy $\|x - x^*\| < \varepsilon$, the increment of J has the same sign. Moreover, if

$$\Delta J = J(x) - J(x^*) \geq 0, \tag{2.30}$$

$J(x^*)$ reaches a local minimum value at this point; otherwise, if

$$\Delta J = J(x) - J(x^*) \leq 0, \tag{2.31}$$

$J(x^*)$ reaches a local maximum value at this point.

Furthermore, if (2.30) is satisfied for arbitrarily large ε, then $J(x^*)$ reaches a global minimum value at x^*, and x^* is called a global minimum whereas, if (2.31) holds for arbitrarily large ε, then $J(x^*)$ reaches a global maximum value at x^*, and x^* is called a global maximum.   □

Theorem 2.1 (Fundamental Theorem of the Calculus of Variations) *Suppose that* x^* *is an extremum; then*

$$\delta J\left(x^*, \delta x\right) = 0, \tag{2.32}$$

for all admissible δx, *i.e., the variation of* J *vanishes at* x^*.

Furthermore, x^* *is a minimum in case* $\delta^2 J\left(x^*, \delta x\right) > 0$; *and* x^* *is a maximum in case* $\delta^2 J\left(x^*, \delta x\right) < 0$.

Proof The necessary condition for the extrema will be shown by contradiction.

Assume that x^* is an extremum and $\delta J\left(x^*, \delta x\right) \neq 0$. Let us show that these assumptions imply that the increment ΔJ can be made to change sign in an arbitrarily small neighborhood of x^*. The increment is

$$\Delta J(x^*, \delta x) = \delta J(x^*, \delta x) + \mathcal{O}\left([\delta x(t)]^2\right), \tag{2.33}$$

with $\delta J(x^*, \delta x) = \dfrac{\partial J}{\partial x}\delta x(t)$.

Hence, there exists an ε, such that δJ dominates the value of ΔJ, for all $\|\delta x\| < \varepsilon$. Now let us select the variation

$$\delta x = \alpha \delta x^{(1)}, \tag{2.34}$$

as shown in Fig. 2.3 (for a scalar function), where $\alpha > 0$ and $\|\alpha \delta x^{(1)}\| < \varepsilon$.

Suppose that

$$\delta J(x^*, \alpha \delta x^{(1)}) < 0. \tag{2.35}$$

Fig. 2.3 An illustration of a function x^* and its variation

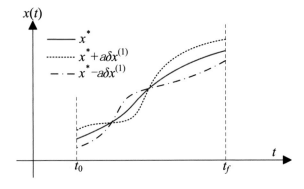

Since δJ is a linear functional of δx, the principle of homogeneity gives

$$\delta J(x^*, \alpha \delta x^{(1)}) = \alpha \delta J(x^*, \delta x^{(1)}) < 0. \tag{2.36}$$

The signs of ΔJ and δJ are the same for $\|\delta x\| < \varepsilon$; thus

$$\Delta J(x^*, \alpha \delta x^{(1)}) < 0. \tag{2.37}$$

Next, consider a variation, such that

$$\delta x = -\alpha \delta x^{(1)},$$

as shown in Fig. 2.3.

Clearly, $\|\alpha \delta x^{(1)}\| < \varepsilon$ implies that $\| - \alpha \delta x^{(1)}\| < \varepsilon$; then the sign of $\Delta J(x^*, -\alpha \delta x^{(1)})$ is the same as the sign of $\delta J(x^*, -\alpha \delta x^{(1)})$.

Again using the principle of homogeneity, we obtain

$$\delta J(x^*, -\alpha \delta x^{(1)}) = -\alpha \delta J(x^*, \delta x^{(1)}), \tag{2.38}$$

therefore, since $\delta J(x^*, \alpha \delta x^{(1)}) < 0$, $\delta J(x^*, -\alpha \delta x^{(1)}) > 0$, and this implies

$$\Delta J(x^*, -\alpha \delta x^{(1)}) > 0. \tag{2.39}$$

Thus, we have shown that if $\delta J(x^*, \delta x) \neq 0$, there exists $\delta x^{(1)}$ in an arbitrarily small neighborhood of x^*

$$\Delta J(x^*, \alpha \delta x^{(1)}) < 0, \tag{2.40}$$

and

$$\Delta J(x^*, -\alpha \delta x^{(1)}) > 0. \tag{2.41}$$

This is contradicted by the assumption that x^* is an extremum as defined in Definition 2.10. Therefore, if x^* is an extremum it is necessary that

$$\delta J(x^*, \delta x) = 0, \text{ for arbitrary } \delta x. \tag{2.42}$$

The assumption that the functions in Ω are not bounded guarantees that $\alpha \delta x^{(1)}$ and $-\alpha \delta x^{(1)}$ are both admissible variations.

Following a similar technique applied above, we can show the sufficient condition for the extremum (minimum or maximum) as well. $\qquad\square$

Based upon the fundamental theorem of the calculus of variations, in Sect. 2.2.1, we will study the necessary and sufficient conditions for the extrema for the functional with fixed final time and fixed final state.

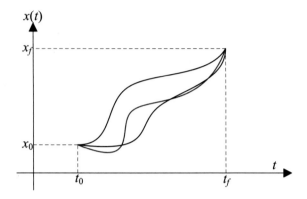

Fig. 2.4 Some functions with fixed final time and fixed final state

2.2.1 Extrema with Fixed Final Time and Fixed Final State

Firstly study *the simplest variational problem* such that t_0 and t_f are fixed and the initial and final states are fixed as well with

$$x(t_0) = x_0, \quad x(t_f) = x_f, \tag{2.43}$$

for some given values x_0 and x_f; see Fig. 2.4 for an illustration.

Problem 2.1 (*Optimization Problems with Fixed Final Time and Fixed Final State*)
Let x be a scalar function in the case of functions with continuous first derivatives.
It is desired to find the function x^* for which the functional

$$J(x) = \int_{t_0}^{t_f} g\left(x(t), \dot{x}(t), t\right) dt \tag{2.44}$$

has a local extremum.   □

Consider the following assumption for the integrand of the functional g.

Assumption 2.1 Suppose that g has continuous first and second partial derivatives with respect to all of its arguments.   □

In Theorem 2.2, we will develop a necessary condition for the extremal solution to Problem 2.1. Before that, we firstly give an interesting lemma below.

Lemma 2.1 *It can be shown that if a function h is continuous and*

$$\int_{t_0}^{t_f} h(t)\delta x(t)dt = 0, \tag{2.45}$$

for every function δx that is continuous in the interval $[t_0, t_f]$, then the function h must be zero everywhere in the interval $[t_0, t_f]$, i.e., $h(t) = 0$, for all $t \in [t_0, t_f]$.

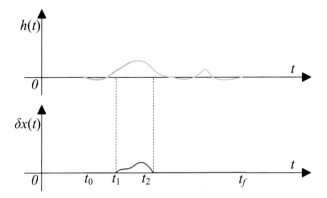

Fig. 2.5 A nonzero function h and a specific variation δx

Proof of contradiction Suppose that the function h is nonzero valued in subintervals over $[t_0, t_f]$; then due to the continuity property of h and without loss of generality, we can further suppose that h is positive valued in a subinterval $[t_1, t_2] \subset [t_0, t_f]$. Set a function of δx such that it is positive valued during $[t_1, t_2]$ and zero valued otherwise. See an illustration in Fig. 2.5.

Thus, we can have

$$\int_{t_0}^{t_f} h(t)\delta x(t)dt = \int_{t_1}^{t_2} h(t)\delta x(t)dt > 0,$$

which is contradicted by the assumed $\int_{t_0}^{t_f} h(t)\delta x(t)dt = 0$ for all continuously valued functions $\delta x(t)$. □

Theorem 2.2 *Suppose that x^* is an extremal solution to Problem 2.1; then x^* satisfies the following equation:*

$$\frac{\partial g(x^*(t), \dot{x}^*(t), t)}{\partial x} - \frac{d}{dt}\left(\frac{\partial g(x^*(t), \dot{x}^*(t), t)}{\partial \dot{x}}\right) = 0, \qquad (2.46)$$

for all $t \in [t_0, t_f]$.

Proof The increment of functional at x is given as

$$\Delta J(x, \delta x) = J(x + \delta x) - J(x)$$

$$= \int_{t_0}^{t_f} g(x(t) + \delta x(t), \dot{x}(t) + \delta \dot{x}(t), t)\, dt - \int_{t_0}^{t_f} g(x(t), \dot{x}(t), t)dt; \qquad (2.47)$$

then combining the integrals gives

$$\Delta J(x, \delta x) = \int_{t_0}^{t_f} [g\,(x(t) + \delta x(t), \dot{x}(t) + \delta \dot{x}(t), t) - g(x(t), \dot{x}(t), t)]\,dt. \quad (2.48)$$

Notice that the dependence on $\dot{x}$ and $\delta \dot{x}$ is not indicated in the argument of ΔJ, since $\dot{x}$ and $\delta \dot{x}$ are determined by x and δx, respectively, as

$$\dot{x} = \frac{dx(t)}{dt}, \quad \delta \dot{x} = \frac{d\delta x(t)}{dt}.$$

By expanding the integrand of (2.47) in a Taylor series at $x(t)$ and $\dot{x}(t)$ gives the following:

$$\Delta J(x, \delta x) = \int_{t_0}^{t_f} \left\{ g(x(t), \dot{x}(t), t) + \frac{\partial g(x(t), \dot{x}(t), t)}{\partial x} \delta x(t) \right.$$

$$+ \frac{\partial g(x(t), \dot{x}(t), t)}{\partial \dot{x}} \delta \dot{x}(t) + \frac{1}{2} \left[\frac{\partial^2 g(x(t), \dot{x}(t), t)}{\partial x^2} [\delta x(t)]^2 \right.$$

$$+ 2 \frac{\partial^2 g(x(t), \dot{x}(t), t)}{\partial x \partial \dot{x}} \delta x(t) \delta \dot{x}(t) + \frac{\partial^2 g(x(t), \dot{x}(t), t)}{\partial \dot{x}^2} [\delta \dot{x}(t)]^2 \right]$$

$$\left. + o\left([\delta x(t)]^2, [\delta \dot{x}(t)]^2 \right) - g(x(t), \dot{x}(t), t) \right\} dt. \quad (2.49)$$

The notation $o\left([\delta x(t)]^2, [\delta \dot{x}(t)]^2 \right)$ denotes terms in the expansion of order three or higher in $\delta x(t)$ and $\delta \dot{x}(t)$. These terms are smaller in magnitude than the $\delta x(t)$, $\delta \dot{x}(t)$, $\delta x(t)\delta \dot{x}(t)$, $[\delta x(t)]^2$, and $[\delta \dot{x}(t)]^2$ as $\delta x(t)$ and $\delta \dot{x}(t)$ go to zero. As indicated, the partial derivatives above are evaluated on the trajectory x, $\dot{x}$.

Next, extract the terms in ΔJ which are linear in $\delta x(t)$ and $\delta \dot{x}(t)$ to obtain the variation of the functional J

$$\delta J(x, \delta x) = \int_{t_0}^{t_f} \left[\frac{\partial g(x(t), \dot{x}(t), t)}{\partial x} \delta x(t) + \frac{\partial g(x(t), \dot{x}(t), t)}{\partial \dot{x}} \delta \dot{x}(t) \right] dt, \quad (2.50)$$

where $\delta x(t)$ and $\delta \dot{x}(t)$ are related by

$$\delta x(t) = \int_{t_0}^{t_t} \delta \dot{x}(t)dt + \delta x(t_0). \quad (2.51)$$

Thus, selecting δx uniquely determines $\delta \dot{x}$, and we shall regard δx as the function that can be varied independently.

To express $\delta J(x, \delta x)$ just in terms of δx, we integrate by parts the term involving $\delta \dot{x}$ to obtain

$$\delta J(x, \delta x) = \frac{\partial g(x(t), \dot{x}(t), t)}{\partial \dot{x}} \delta x(t) \Big|_{t_0}^{t_f}$$

$$+ \int_{t_0}^{t_f} \left[\frac{\partial g(x(t), \dot{x}(t), t)}{\partial x} - \frac{d}{dt} \left(\frac{\partial g(x(t), \dot{x}(t), t)}{\partial \dot{x}} \right) \right] \delta x(t) dt. \tag{2.52}$$

Since $x(t_0)$ and $x(t_f)$ are specified, all admissible curves must pass through these points; therefore, $\delta x(t_0)$, $\delta x(t_f)$, and the terms outside the integral vanish.
By applying the fundamental theorem, we can obtain

$$\delta J(x^*, \delta x) = 0$$

$$= \int_{t_0}^{t_f} \left[\frac{\partial g(x^*(t), \dot{x}^*(t), t)}{\partial x} - \frac{d}{dt} \left(\frac{\partial g(x^*(t), \dot{x}^*(t), t)}{\partial \dot{x}} \right) \right] \delta x(t) dt. \tag{2.53}$$

Applying Lemma 2.1 to the above theorem, we obtain that (2.46) holds.   □

As observed, the Euler equation (2.46) for Problem 2.1 is a nonlinear time-variant ordinary differential equation with a pair of two-point boundary values. Hence, it usually has a nonlinear two-point boundary value problem which is in general challenging to deal.

2.2.2 Specific Forms of Euler Equation in Different Cases

Theorem 2.2 gives the Euler equation for Problem 2.1 in general cases. It is worth studying the specific forms for several special cases.
Firstly, by the chain rule, we have

$$\frac{d}{dt} \left(\frac{\partial g}{\partial \dot{x}} \right) = \frac{\partial^2 g}{\partial x \partial \dot{x}} \dot{x}(t) + \frac{\partial^2 g}{\partial \dot{x} \partial \dot{x}} \ddot{x}(t) + \frac{\partial^2 g}{\partial t \partial \dot{x}}, \tag{2.54}$$

with $g \equiv g(x(t), \dot{x}(t), t)$ for notational simplicity.

- Case I: Suppose that $g(\dot{x}(t), t)$, say $g(\cdot)$ is only with $\dot{x}(t)$ and t, and is independent of $x(t)$; then the Euler equation (2.46) can degenerate into the following form:

$$\frac{d}{dt} \left(\frac{\partial g(\dot{x}^*(t), t)}{\partial \dot{x}} \right) = 0, \tag{2.55}$$

for all $t \in [t_0, t_f]$, since $\dfrac{\partial g(\dot{x}^*(t), t)}{\partial x} = 0$. Hence, we have

$$\frac{\partial g(\dot{x}^*(t), t)}{\partial \dot{x}} = \xi, \ \forall t \in [t_0, t_f], \quad \text{for some constant valued } \xi. \tag{2.56}$$

- Case II: Suppose that $g(\dot{x}(t))$, say $g(\cdot)$ is only with $\dot{x}(t)$, and is independent of $x(t)$ and t; then the Euler equation (2.46) degenerates into (2.55) as well. Hence, in this case we have that (2.56) is still a necessary condition, say

$$\frac{\partial g(\dot{x}^*(t))}{\partial \dot{x}} = \xi, \ \forall t \in [t_0, t_f], \quad \text{for some constant valued } \xi. \tag{2.57}$$

- Case III: Suppose that $g(x(t), \dot{x}(t))$, say $g(\cdot)$ is with $x(t)$ and $\dot{x}(t)$, and is independent of t; then, by (2.54), the Euler equation (2.46) degenerates into the following form:

$$\frac{\partial g(x^*(t), \dot{x}^*(t))}{\partial x} - \frac{\partial^2 g(x^*(t), \dot{x}^*(t))}{\partial x \partial \dot{x}} \dot{x}^*(t) - \frac{\partial^2 g(x^*(t), \dot{x}^*(t))}{\partial \dot{x} \partial \dot{x}} \ddot{x}^*(t) = 0, \tag{2.58}$$

for all $t \in [t_0, t_f]$.
- Case IV: Suppose that $g(x(t))$ or $g(x(t), t)$, say $g(\cdot)$ is with $x(t)$ or is with $x(t)$ and t; then, the Euler equation becomes

$$\frac{\partial g(x^*(t))}{\partial x} = 0, \tag{2.59a}$$

$$\frac{\partial g(x^*(t), t)}{\partial x} = 0. \tag{2.59b}$$

As observed, the above Euler equation in this case is an algebra equation; hence, there is no arbitrary constant involved as that in differential equations. Thus, the solution to the Euler equation usually does not fit with the given fixed boundary conditions on states $x(t_0)$ and $x(t_f)$. As a consequence, there may not exist any solutions for the extremal problems in most cases. □

In a few of the examples given below, we will specify a necessary condition for the extrema of Problem 2.1 by applying Theorem 2.2.

Example 2.10 Specify the shortest-length curve between a given pair of two points (t_0, x_0) and (t_f, x_f) in a two-dimensional plane.

Solution. Firstly, as illustrated in Fig. 2.6, we have $[d\Gamma]^2 = [dt]^2 + [dx]^2$. It implies that

$$d\Gamma = \sqrt{1 + \dot{x}^2(t)} dt. \tag{2.60}$$

Hence, the length of a curve between (t_0, x_0) and (t_f, x_f) is specified as the following:

$$J(x) = \int_{t_0}^{t_f} g(\dot{x}(t)) dt, \quad \text{with } g(\dot{x}(t)) \triangleq \sqrt{1 + \dot{x}^2(t)}, \tag{2.61}$$

which implies that g is a function of $\dot{x}(t)$ only and is independent of $x(t)$ and t.

Fig. 2.6 An illustration of
the length of a curve

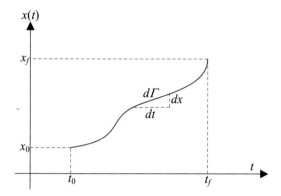

By applying the Euler equation specified in (2.56), we have

$$\frac{\partial g(\dot{x}^*(t))}{\partial \dot{x}} = \frac{\dot{x}^*(t)}{\sqrt{1 + [\dot{x}^*(t)]^2}} = \xi, \tag{2.62}$$

for all $t \in [t_0, t_f]$, with some constant-valued ξ.

We can obtain that $x^*(t) = \xi_1 t + \xi_2$, say $x^*(t)$ is a straight line, with some constant-valued coefficients ξ_1 and ξ_2 which can be specified by the boundary conditions $x(t_0) = x_0$ and $x(t_f) = x_f$.

For example, suppose that $t_0 = 0$, $t_f = 5$, $x_0 = 5$, and $x_f = 20$; then we obtain that $x^*(t) = 3t + 5$.   □

Next, in Example 2.11 below, we will study how to apply the variational method to solve the well-known Brachistochrone problem which was firstly proposed by Johann Bernoulli in 1696 and successively solved by Jacob Bernoulli (his brother), Isaac Newton, L' Hosptial, and himself.

Example 2.11 (The Brachistochrone Problem)

Solution. Denote by $v(t)$ the velocity of the mass, which is determined by the vertical position of the mass $y(x)$, say

$$v(t) = \sqrt{2gy(x)}, \tag{2.63}$$

where g represents the gravitational acceleration. Also denote by $\Gamma(t)$ the length of the curve; then as illustrated in Fig. 2.7 and by (2.60),

$$v(t) = \frac{d\Gamma(t)}{dt} = \sqrt{1 + [\dot{y}(x)]^2}\frac{dx}{dt}; \tag{2.64}$$

Fig. 2.7 An illustration of
the Brachistochrone problem

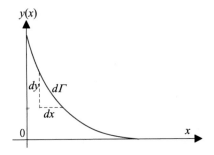

then we have

$$dt = \frac{\sqrt{1 + [\dot{y}(x)]^2}}{v} dx = \frac{\sqrt{1 + [\dot{y}(x)]^2}}{\sqrt{2gy(x)}} dx, \qquad (2.65)$$

which implies that

$$J = \int_0^P dt = \int_0^P g(\dot{y}(x), y(x)) dx, \qquad (2.66)$$

with $g \equiv g(\dot{y}(x), y(x)) \triangleq \dfrac{\sqrt{1 + [\dot{y}(x)]^2}}{\sqrt{2gy(x)}}$; then by the Euler equation developed ear-

lier

$$\frac{\partial g^*}{\partial y(x)} - \frac{d}{dx}\left(\frac{\partial g^*}{\partial \dot{y}(x)}\right) = 0, \qquad (2.67)$$

with $g^* \triangleq \dfrac{\sqrt{1 + [\dot{y}^*(x)]^2}}{\sqrt{2gy^*(x)}}$.

Also by the specification of $g(\dot{y}(x), y(x))$, we have

$$\frac{d}{dx}\left(g^* - \dot{y}^*(x)\frac{\partial g^*}{\partial \dot{y}(x)}\right)$$

$$= \dot{y}^*(x)\frac{\partial g^*}{\partial y} + \ddot{y}^*(x)\frac{\partial g^*}{\partial \dot{y}(x)} - \ddot{y}^*(x)\frac{\partial g^*}{\partial \dot{y}(x)} - \dot{y}^*(x)\frac{d}{dx}\left(\frac{\partial g^*}{\partial \dot{y}(x)}\right)$$

$$= \dot{y}^*(x)\left[\frac{\partial g^*}{\partial y(x)} - \frac{d}{dx}\left(\frac{\partial g^*}{\partial \dot{y}(x)}\right)\right] = 0; \qquad (2.68)$$

then we have

$$g^* - \dot{y}^*(x)\frac{\partial g^*}{\partial \dot{y}(x)} = \xi, \quad \text{for some constanted valued } \xi. \qquad (2.69)$$

So we can obtain

$$\frac{\sqrt{1+[\dot{y}^*(x)]^2}}{\sqrt{2gy^*(x)}} - \dot{y}^*(x)\frac{\dot{y}^*(x)}{\sqrt{2gy^*(x)}\sqrt{1+[\dot{y}^*(x)]^2}} = \frac{1}{\sqrt{2gy^*(x)}\sqrt{1+[\dot{y}^*(x)]^2}} = \xi. \quad (2.70)$$

This implies that

$$y^*(x)\left[1+[\dot{y}^*(x)]^2\right] = \frac{1}{2g\xi^2}. \quad (2.71)$$

To solve the curve of $y^*(x)$, here we introduce another variable θ, such that

$$x = x(\theta), \text{ and } \dot{y}^*(x) = \cot(\theta/2), \quad (2.72)$$

with $\cot(\theta/2) \triangleq \frac{\cos(\theta/2)}{\sin(\theta/2)}$; then we have

$$y^*(\theta) = \frac{1}{2g\xi^2}\sin^2(\theta/2) = \frac{1}{4g\xi^2}[1-\cos(\theta)]. \quad (2.73)$$

By implementing the derivative of both sides of the above equality, we have

$$\dot{y}^*(x)\frac{dx(\theta)}{d\theta} = \cot(\theta/2)\frac{dx(\theta)}{d\theta} = \frac{1}{4g\xi^2}\sin(\theta); \quad (2.74)$$

then

$$\frac{dx(\theta)}{d\theta} = \frac{1}{2g\xi^2}\sin^2(\theta/2) = \frac{1}{4g\xi^2}[1-\cos(\theta)], \quad (2.75)$$

by which we obtain $x(\theta) = \frac{1}{4g\xi^2}[\theta - \sin(\theta)] + x_0$.

In summary, together with the boundary condition, we finally get the extremal solution to the Brachistochrone problem as follows:

$$x(\theta) = \frac{1}{4g\xi^2}[\theta - \sin(\theta)], \quad (2.76a)$$

$$y^*(\theta) = \frac{1}{4g\xi^2}[1-\cos(\theta)]. \quad (2.76b)$$

$\square$

Example 2.12 (Electric Vehicle Charging Problems)
Suppose that an electric vehicle (EV) is charged in a power system, such that

Fig. 2.8 An illustration of
an EV charging coordination
problem

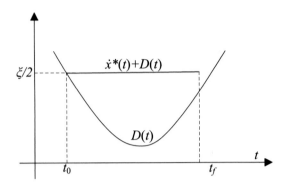

$$J(x) = \int_{t_0}^{t_f} [\dot{x}(t) + D(t)]^2 \, dt, \qquad (2.77)$$

where $x(t)$ and $D(t)$ represent the state of charge (SoC) of the EV and the given fixed base demand in the system, respectively.

We also consider that $x(t_0) = x_0$ and $x(t_f) = x_{max}$.

Solution. Firstly by applying the Euler equation, we obtain

$$\frac{\partial g(\dot{x}^*(t))}{\partial \dot{x}} = 2[\dot{x}^*(t) + D(t)] = \xi, \qquad (2.78)$$

for all $t \in [t_0, t_f]$, with some constant-valued ξ; see an illustration in Fig. 2.8.

As $\dot{x}(t)$ actually represents the charging power at time t, by the Euler equation (2.78), we obtain that the optimal charging power is to fill the valley of the base demand.

Moreover by (2.78) and the boundary condition, the SoC of EV is given as

$$x^*(t) = x_0 + \frac{\xi}{2}[t - t_0] - \int_{t_0}^{t_f} D(t) dt, \qquad (2.79)$$

for all $t \in [t_0, t_f]$, with $\xi = \dfrac{2}{t_f - t_0} \left[x_{max} - x_0 + \int_{t_0}^{t_f} D(t) dt \right]$.   □

Example 2.13 Give a necessary condition for the extrema of the functional

$$J(x) = \int_0^2 \left[\dot{x}^2(t) + t^2 x(t) \right] dt, \qquad (2.80)$$

with the boundary conditions as $x(t_0) = 2$ and $x(t_f) = 10$.

Solution. By applying the Euler equation, we obtain

$$t^2 - \frac{d}{dt} \left(2\dot{x}^*(t) \right) = 0; \qquad (2.81)$$

then we have $\ddot{x}^*(t) = \frac{1}{2}t^2$, and hence

$$x^*(t) = \frac{1}{24}t^4 + \xi_1 t + \xi_2, \tag{2.82}$$

with some constant-valued ξ_1 and ξ_2 which are equal to $\frac{11}{3}$ and 2, respectively by the boundary conditions $x(t_0) = 2$ and $x(t_f) = 10$. □

Example 2.14 Specify an extremum for the functional

$$J(x) = \int_0^{\pi/2} [\dot{x}^2(t) - x^2(t)]dt, \tag{2.83}$$

which satisfies the boundary conditions $x(0) = 0$ and $x(\pi/2) = 1$.
   Solution. By applying Theorem 2.2, the Euler equation is

$$0 = \frac{\partial g(x^*(t), \dot{x}^*(t), t)}{\partial x} - \frac{d}{dt}\left(\frac{\partial g(x^*(t), \dot{x}^*(t), t)}{\partial \dot{x}}\right)$$

$$= -2x^*(t) - 2\frac{d\dot{x}^*(t)}{dt}. \tag{2.84}$$

It is equivalent to the following:

$$\ddot{x}^*(t) + x^*(t) = 0, \tag{2.85}$$

with a solution of the form $x^*(t) = \kappa \exp(st)$, for some constant-valued κ and s; then by substituting this into the above differential equation, the following should hold:

$$\kappa s^2 \exp(st) + \kappa \exp(st) = 0, \text{ for all } t, \tag{2.86}$$

which implies that

$$s^2 + 1 = 0, \tag{2.87}$$

with roots of $\pm j$.
   So the solution has the form

$$x^*(t) = \xi_1 \exp(-jt) + \xi_2 \exp(jt), \tag{2.88}$$

or equivalently

$$x^*(t) = \xi_3 \cos(t) + \xi_4 \sin(t), \tag{2.89}$$

where ξ_i, with $i = 1, \ldots, 4$, are constant values that can be uniquely determined by applying the boundary conditions.

The constants can be specified by satisfying the boundary conditions $x(0) = 0$ and $x(\pi/2) = 1$, such that

$$x^*(0) = \xi_3 \cos(0) + \xi_4 \sin(0) = 0, \tag{2.90a}$$

$$x^*\left(\frac{\pi}{2}\right) = \xi_3 \cos\left(\frac{\pi}{2}\right) + \xi_4 \sin\left(\frac{\pi}{2}\right) = 1, \tag{2.90b}$$

which implies that $\xi_3 = 0$ and $\xi_4 = 1$. Thus, the solution to the Euler equation is

$$x^*(t) = \sin(t). \tag{2.91}$$

$\square$

It is interesting to check the optimality of x^* by exploring the increment with some neighboring curves of x^* implemented in the last example. Consider a class of curves with the following form

$$x(t) = \sin(t) + \alpha \sin(2t) = x^*(t) + \delta x(t), \tag{2.92}$$

with α real constant valued.

It can be verified that each δx curve, defined above, goes through zero at $t = 0$ and at $t = \pi/2$; thus $x^* + \delta x$ satisfies the required boundary conditions.

$x^*(t) = \sin(t)$, hence $\dot{x}^*(t) = \cos(t)$ and $J(x^*) = 0$; while considering another function $x(t) = \sin(t) + \alpha \sin(2t)$, the corresponding performance cost is given as follows:

$$J(x) = \frac{3\pi}{4}\alpha^2$$

which is strictly larger than $J(x^*) = 0$, for any nonzero-valued α.

2.2.3 Sufficient Condition for Extrema

In the last section, we have developed the necessary condition, the Euler equation, for the extremal solution to a class of optimization problems. However, to establish the extremal property (maximum or minimum) of a solution, we need to study the sufficient condition as stated in Theorem 2.1, say x^* is a minimum in case $\delta^2 J(x^*, \delta x) > 0$, with $\delta^2 J$ denoting the second variation of the functional J as defined in (2.26); and x^* is a maximum in case $\delta^2 J(x^*, \delta x) < 0$.

In the following part, as before, for notational simplicity, consider $g \equiv g(x(t), \dot{x}(t), t)$.

Considering the integral performance cost function specified in Problem 2.1, we can obtain

$$\delta^2 J = \frac{1}{2!} \frac{\partial^2 J}{\partial x^2} [\delta x(t)]^2$$

$$= \frac{1}{2} \int_{t_0}^{t_f} \left[\frac{\partial^2 g}{\partial x^2} [\delta x(t)]^2 + 2 \frac{\partial^2 g}{\partial x \partial \dot{x}} \delta x(t) \delta \dot{x}(t) + \frac{\partial^2 g}{\partial \dot{x}^2} [\delta \dot{x}(t)]^2 \right] dt$$

$$= \frac{1}{2} \int_{t_0}^{t_f} [\delta x(t) \delta \dot{x}(t)] \mathbb{M} \begin{bmatrix} \delta x(t) \\ \delta \dot{x}(t) \end{bmatrix} dt, \tag{2.93}$$

with $\mathbb{M} \equiv \begin{bmatrix} \dfrac{\partial^2 g}{\partial x^2} & \dfrac{\partial^2 g}{\partial x \partial \dot{x}} \\[2mm] \dfrac{\partial^2 g}{\partial x \partial \dot{x}} & \dfrac{\partial^2 g}{\partial \dot{x}^2} \end{bmatrix}$.

As a consequence, by Theorem 2.1, say x^* is a minimum in case $\delta^2 J (x^*, \delta x) > 0$; we obtain that x^* is a minimum if $\mathbb{M}$ is a positive-definite matrix.

Similarly, by Theorem 2.1, say x^* is a maximum in case $\delta^2 J (x^*, \delta x) < 0$; we obtain that x^* is a maximum if $\mathbb{M}$ is a negative-definite matrix.

Besides the above analysis, another sufficient condition for the extrema of x^* is specified.

Firstly, we reorganize a part of integral in the second variation of functional J as follows:

$$\int_{t_0}^{t_f} 2 \frac{\partial^2 g}{\partial x \partial \dot{x}} \delta x(t) \delta \dot{x}(t) \, dt$$

$$= \int_{t_0}^{t_f} 2 \frac{\partial^2 g}{\partial x \partial \dot{x}} \delta x(t) \, d\delta x(t)$$

$$= \int_{t_0}^{t_f} \frac{\partial^2 g}{\partial x \partial \dot{x}} d[\delta x(t)]^2$$

$$= \frac{\partial^2 g}{\partial x \partial \dot{x}} [\delta x(t)]^2 \Big|_{t_0}^{t_f} - \int_{t_0}^{t_f} [\delta x(t)]^2 \, d \frac{\partial^2 g}{\partial x \partial \dot{x}}$$

$$= \frac{\partial^2 g}{\partial x \partial \dot{x}} [\delta x(t)]^2 \Big|_{t_0}^{t_f} - \int_{t_0}^{t_f} [\delta x(t)]^2 \frac{d}{dt} \left(\frac{\partial^2 g}{\partial x \partial \dot{x}} \right) dt, \tag{2.94}$$

where the last equality from the second holds by applying integration by parts, say $\int \mu d\nu = \mu \nu - \int \nu d\mu$ with $\mu \equiv \dfrac{\partial^2 g}{\partial x \partial \dot{x}}$ and $\nu \equiv [\delta x(t)]^2$, respectively.

Hence by (2.94) together with (2.93), we have

$$\delta^2 J = \frac{\partial^2 g}{\partial x \partial \dot{x}} [\delta x(t)]^2 \Big|_{t_0}^{t_f} + \frac{1}{2} \int_{t_0}^{t_f} \left\{ \left[\frac{\partial^2 g}{\partial x^2} - \frac{d}{dt} \left(\frac{\partial^2 g}{\partial x \partial \dot{x}} \right) \right] [\delta x(t)]^2 + \frac{\partial^2 g}{\partial \dot{x}^2} [\delta \dot{x}(t)]^2 \right\} dt. \tag{2.95}$$

Since $\delta x(t_0)$ and $\delta x(t_f)$ are both equal to zero for Problem 2.1, we have

$$\delta^2 J = \frac{1}{2} \int_{t_0}^{t_f} \left\{ \left[\frac{\partial^2 g}{\partial x^2} - \frac{d}{dt} \left(\frac{\partial^2 g}{\partial x \partial \dot{x}} \right) \right] [\delta x(t)]^2 + \frac{\partial^2 g}{\partial \dot{x}^2} [\delta \dot{x}(t)]^2 \right\} dt. \qquad (2.96)$$

Thus, by Theorem 2.1, say x^* is a minimum in case $\delta^2 J(x^*, \delta x) > 0$, we obtain that x^* is a minimum solution if the following inequalities are satisfied:

$$\frac{\partial^2 g(x^*(t), \dot{x}^*(t), t)}{\partial x^2} - \frac{d}{dt} \left(\frac{\partial^2 g(x^*(t), \dot{x}^*(t), t)}{\partial x \partial \dot{x}} \right) > 0, \qquad (2.97a)$$

$$\frac{\partial^2 g(x^*(t), \dot{x}^*(t), t)}{\partial \dot{x}^2} > 0. \qquad (2.97b)$$

Similarly, by Theorem 2.1, say x^* is a maximum in case $\delta^2 J(x^*, \delta x) < 0$, we obtain that x^* is a maximum solution if the following inequalities are satisfied:

$$\frac{\partial^2 g(x^*(t), \dot{x}^*(t), t)}{\partial x^2} - \frac{d}{dt} \left(\frac{\partial^2 g(x^*(t), \dot{x}^*(t), t)}{\partial x \partial \dot{x}} \right) < 0, \qquad (2.98a)$$

$$\frac{\partial^2 g(x^*(t), \dot{x}^*(t), t)}{\partial \dot{x}^2} < 0. \qquad (2.98b)$$

In this section, we have developed sufficient conditions for the extrema (maximum or minimum) of functionals with fixed boundary conditions. Here we will verify the results with a few examples as follows.

Firstly revisit Example 2.10, where it gave the necessary condition for the shortest curve between a given pair of points in a two-dimensional space. It turned out to be a straight line. In the following, we will claim with the analysis that (2.97) is also the sufficient condition to be a shortest curve.

Example 2.15 Verify that the straight line is the shortest-length curve between a given pair of two points (t_0, x_0) and (t_f, x_f) in a plane.

Proof Firstly by following the modeling issue specified in Example 2.10, we have

$$\frac{\partial^2 g(\dot{x}^*(t))}{\partial \dot{x}^2} = \frac{\partial^2}{\partial \dot{x}^2} \left(\sqrt{1 + [\dot{x}^*(t)]^2} \right) = \frac{1}{\left[1 + [\dot{x}^*(t)]^2 \right]^{3/2}}, \qquad (2.99)$$

which is always strictly positive for all t.

Hence, we can make the conclusion that the straight line, obtained in Example 2.10, is the shortest path between a pair of points in the plane. ☐

Example 2.16 Verify that the extreme function, given in (2.82) for the problem in Example 2.13, is the minimum solution.

Proof By Example 2.13, we have

$$\frac{\partial^2 g(\dot{x}^*(t))}{\partial \dot{x}^2} = \frac{\partial^2}{\partial \dot{x}^2} \left(\left[\dot{x}^*(t) \right]^2 + t^2 x^*(t) \right) = 2, \qquad (2.100)$$

which is strictly positive for all t.

Hence we can make the conclusion. ☐

Example 2.17 Verify that the solution implemented in Example 2.12 for EV charging problems is a minimum solution.

Proof By Example 2.12, we have $g(\dot{x}(t)) = [\dot{x}(t) + D(t)]^2$; then

$$\frac{\partial^2 g(\dot{x}^*(t))}{\partial \dot{x}^2} = \frac{\partial^2}{\partial \dot{x}^2} \left(\left[\dot{x}^*(t) + D(t) \right]^2 \right) = 2, \qquad (2.101)$$

which is positive for all t.

Hence we can make the conclusion as well. ☐

2.2.4 Extrema with Fixed Final Time and Free Final State

In the last section, we studied the optimal problem with fixed final time and final states. Here, we study another case such that the final time t_f is fixed and, however, the final state at t_f, say $x(t_f) \equiv x_f$, is free.

See an illustration of those admissible curves as displayed in Fig. 2.9.

Problem 2.2 (*Optimization Problems with Fixed Final Time and Free Final State*) Specify a necessary condition for a function to be an extremum for the functional

$$J(x) = \int_{t_0}^{t_f} g\left(x(t), \dot{x}(t), t\right) dt, \qquad (2.102)$$

where t_0, $x(t_0)$, and t_f are specified, and $x(t_f)$ is free. ☐

Fig. 2.9 Some functions with fixed final time and free final states

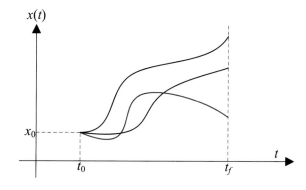

Theorem 2.3 *Suppose that x^* is an extremal solution to Problem 2.2; then x^* satisfies the Euler equation (2.46) with the following boundary condition:*

$$\frac{\partial g\left(x^*(t_f), \dot{x}^*(t_f), t_f\right)}{\partial \dot{x}} = 0. \tag{2.103}$$

Proof In the proof of Theorem 2.2, we have obtained

$$
\begin{aligned}
\delta J(x, \delta x) &= \frac{\partial g(x(t), \dot{x}(t), t)}{\partial \dot{x}} \delta x(t) \Big|_{t_0}^{t_f} \\
&\quad + \int_{t_0}^{t_f} \left[\frac{\partial g(x(t), \dot{x}(t), t)}{\partial x} - \frac{d}{dt} \left(\frac{\partial g(x(t), \dot{x}(t), t)}{\partial \dot{x}} \right) \right] \delta x(t) dt \\
&= \frac{\partial g(x(t_f), \dot{x}(t_f), t_f)}{\partial \dot{x}} \delta x(t_f) \\
&\quad + \int_{t_0}^{t_f} \left[\frac{\partial g(x(t), \dot{x}(t), t)}{\partial x} - \frac{d}{dt} \left(\frac{\partial g(x(t), \dot{x}(t), t)}{\partial \dot{x}} \right) \right] \delta x(t) dt,
\end{aligned}
\tag{2.104}
$$

where the last equality holds since $\delta x(t_0) = 0$.

Consider an extremum curve x^*; then by applying Theorem 2.1, we obtain

$$
\begin{aligned}
\delta J(x^*, \delta x) &= \frac{\partial g(x^*(t_f), \dot{x}^*(t_f), t_f)}{\partial \dot{x}} \delta x(t_f) \\
&\quad + \int_{t_0}^{t_f} \left[\frac{\partial g(x^*(t), \dot{x}^*(t), t)}{\partial x} - \frac{d}{dt} \left(\frac{\partial g(x^*(t), \dot{x}^*(t), t)}{\partial \dot{x}} \right) \right] \delta x(t) dt \\
&= 0.
\end{aligned}
\tag{2.105}
$$

Applying Lemma 2.1 to the last equality, and since $x(t_f)$ is free, $\delta x(t_f)$ is arbitrary; therefore, it is necessary that the Euler equation (2.46) and the boundary condition (2.103) as well should be held.   □

Notice that, as verified in Theorem 2.3, the Euler equation for Problem 2.2 is a collection of differential equations which is the same as that for Problem 2.1. Different from Problem 2.1, besides the initial boundary condition, $x(t_0) = x_0$, it also provides the second required end boundary condition in (2.103) which is considered as the natural boundary condition.

Example 2.18 Give a smooth curve with the shortest length connecting the point $x(t_0) = 3$, with $t_0 = 4$, to the line $t = 20$.

Solution. Firstly, by Example 2.10, the solution to the Euler equation is a straight line such that

$$x^*(t) = \xi_1 t + \xi_2, \tag{2.106}$$

Fig. 2.10 The shortest curve between a fixed point and a free final state

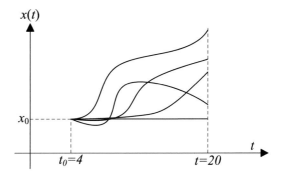

with ξ_1 and ξ_2 as constant valued and to be determined by boundary conditions, respectively.

Firstly $\xi_2 = 2$ by the initial boundary condition $x^*(4) = 3$, and by the necessary condition for the end point as given in (2.103), we obtain

$$\frac{\partial g(x^*(t_f), \dot{x}^*(t_f), t_f)}{\partial \dot{x}} = \frac{\dot{x}^*(t_f)}{\sqrt{1 + [\dot{x}^*(t_f)]^2}} = 0. \qquad (2.107)$$

Thus $\dot{x}^*(t_f) = 0$, with $t_f = 20$. Substituting $\dot{x}^*(t_f) = 0$ into the equation of $\dot{x}^*(t) = \xi_1$, we get $\xi_1 = 0$.

The solution is $x^*(t) = 3$ a straight line parallel to the t-axis; see Fig. 2.10 for an illustration. □

Example 2.19 Give a necessary condition for the extrema of the functional from Example 2.13 with the boundary conditions as $t_0 = 0$, $t_f = 2$, $x(t_0) = 2$, but $x(t_f)$ arbitrarily given.

Solution. Firstly, from Example 2.13, we already have

$$x^*(t) = \frac{1}{24}t^4 + \xi_1 t + \xi_2, \qquad (2.108)$$

with some constant-valued ξ_1 and ξ_2.

Also by Theorem 2.3, x^* shall satisfy the following boundary conditions:

$$x^*(t_0) = 2, \quad \text{with } t_0 = 0, \qquad (2.109a)$$

$$\frac{\partial g(x^*(t_f), \dot{x}^*(t_f), t_f)}{\partial \dot{x}} = 0, \quad \text{with } t_f = 2; \qquad (2.109b)$$

then we have $\xi_2 = 2$ and

$$\frac{\partial g(x^*(t_f), \dot{x}^*(t_f), t_f)}{\partial \dot{x}} = 2\dot{x}^*(t_f) = 0, \quad \text{with } t_f = 2, \qquad (2.110)$$

which implies that $\frac{1}{6}t_f^3 + \xi_1 = 0$.

Hence $\xi_1 = -\frac{4}{3}$.

As a conclusion, we have

$$x^*(t) = \frac{1}{24}t^4 - \frac{4}{3}t + 2. \tag{2.111}$$

$\square$

As it is stated that the final time is fixed for Problems 2.1 and 2.2, as a consequence, the variations of the functionals involved two integrals having the same limits of integration.

However, in case the final time is free, the above result does not hold any longer. Therefore, we now generalize the results of the previous discussion. This is accomplished by separating the total variation of a functional into two partial variations: the variation resulting from the different $\delta x(t)$ in the interval $[t_0, t_f]$ and the variation resulting from the difference in the end points of two curves.

The sum of these two variations is called the *general variation* of a functional. First, let us consider the case where $x(t_f)$ is specified.

2.2.5 Extrema with Free Final Time and Fixed Final State

In the last sections, we have studied the optimal problem with a fixed final time. In the next two sections, we will analyze the cases such that the final time t_f is free.

Firstly in this part, we consider that the final state at t_f, say $x(t_f) \equiv x_f$, is fixed. See an illustration of those admissible curves as displayed in Fig. 2.11 which all start at a fixed point (x_0, t_0) and terminate on a horizontal line with a value x_f.

From Fig. 2.12, it can be observed that $\delta x(t) \triangleq x(t) - x^*(t)$ is only meaningful during $[t_0, t_f]$, since x^* is not defined for any t beyond t_f.

Fig. 2.11 Some functions with free final time and fixed final state

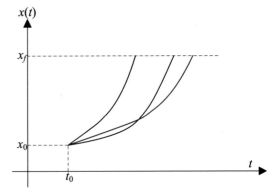

Fig. 2.12 An illustration of the extremal solution x^* and a variation x

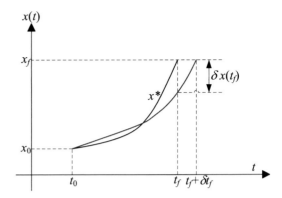

Problem 2.3 (*Extrema Problems with Free Final Time and Fixed Final State*) Specify a necessary condition for a function to be an extremum for the functional

$$J(x) = \int_{t_0}^{t_f} g(x(t), \dot{x}(t), t) dt, \tag{2.112}$$

where t_0, $x(t_0)$, and $x(t_f)$ are specified, but t_f is free. □

Theorem 2.4 *Suppose that x^* is an extremal solution to Problem 2.3; then x^* satisfies the Euler equation (2.46) with the following boundary condition:*

$$g\left(x^*(t_f), \dot{x}^*(t_f), t_f\right) - \frac{\partial g(x^*(t_f), \dot{x}^*(t_f), t_f)}{\partial \dot{x}} \dot{x}^*(t_f) = 0. \tag{2.113}$$

Proof Firstly as earlier, define the increment of the functional:

$$\Delta J(x^*, \delta x) \triangleq J(x^* + \delta x) - J(x^*)$$

$$= \int_{t_0}^{t_f + \delta t_f} g\left(x^*(t) + \delta x(t), \dot{x}^*(t) + \delta \dot{x}(t), t\right) dt - \int_{t_0}^{t_f} g(x^*(t), \dot{x}^*(t), t) dt$$

$$= \int_{t_0}^{t_f} \left[g\left(x^*(t) + \delta x(t), \dot{x}^*(t) + \delta \dot{x}(t), t\right) - g(x^*(t), \dot{x}^*(t), t)\right] dt$$

$$+ \int_{t_f}^{t_f + \delta t_f} g\left(x^*(t) + \delta x(t), \dot{x}^*(t) + \delta \dot{x}(t), t\right) dt \quad \equiv (I). \tag{2.114}$$

We can deal with (I) in the above proof as follows:

$$(I) = g\left(x^*(t) + \delta x(t), \dot{x}^*(t) + \delta \dot{x}(t), t\right) \delta t_f + o(\delta t_f), \tag{2.115}$$

by which, together with (2.52), we have

$$\Delta J(x^*, \delta x) = \frac{\partial g(x^*(t), \dot{x}^*(t), t)}{\partial \dot{x}} \delta x(t) \Big|_{t_0}^{t_f} + g\left(x^*(t) + \delta x(t), \dot{x}^*(t) + \delta \dot{x}(t), t\right) \delta t_f$$

$$+ \int_{t_0}^{t_f} \left[\frac{\partial g\left(x^*(t), \dot{x}^*(t), t\right)}{\partial x} - \frac{d}{dt} \left(\frac{\partial g\left(x^*(t), \dot{x}^*(t), t\right)}{\partial \dot{x}} \right) \right] \delta x(t) dt + o(\cdot).$$

Moreover, by applying the Taylor series expansion, we have

$$g\left(x^*(t) + \delta x(t), \dot{x}^*(t) + \delta \dot{x}(t), t\right)$$

$$= g\left(x^*(t_f), \dot{x}^*(t_f), t_f\right) + \frac{\partial g\left(x^*(t_f), \dot{x}^*(t_f), t_f\right)}{\partial x} \delta x(t_f)$$

$$+ \frac{\partial g\left(x^*(t_f), \dot{x}^*(t_f), t_f\right)}{\partial \dot{x}} \delta \dot{x}(t_f) + o(\cdot).$$

Hence, by putting the above analysis together and the fact that $\delta x(t_0) = 0$, we obtain

$$\Delta J(x^*, \delta x) = \frac{\partial g\left(x^*(t_f), \dot{x}^*(t_f), t_f\right)}{\partial \dot{x}} \delta x(t_f) + g\left(x^*(t_f), \dot{x}^*(t_f), t_f\right) \delta t_f$$

$$+ \int_{t_0}^{t_f} \left[\frac{\partial g\left(x^*(t), \dot{x}^*(t), t\right)}{\partial x} - \frac{d}{dt} \left(\frac{\partial g\left(x^*(t), \dot{x}^*(t), t\right)}{\partial \dot{x}} \right) \right] \delta x(t) dt + o(\cdot).$$

Different from previous optimization problems with fixed t_f, where $\delta x(t_f)$ is either zero or free, in Problem 2.3, $\delta x(t_f)$ is dependent upon the value of δt_f and is neither zero nor arbitrary.

Also since we shall specify the variation of functional, say δJ, which is composed of linear parts in ΔJ, it is useful to give a linear approximation of $\delta x(t_f)$ on the value of δt_f. As demonstrated by Fig. 2.12, we have

$$\delta x(t_f) + \dot{x}^*(t_f)\delta t_f \doteq 0, \tag{2.116}$$

where $y \doteq 0$ denotes that y is approximately equal to zero up to the first order.

Thus we obtain

$$\delta J(x^*, \delta x) = 0, \qquad \text{which holds by Theorem 2.1}$$

$$= \left[-\frac{\partial g\left(x^*(t_f), \dot{x}^*(t_f), t_f\right)}{\partial \dot{x}} \dot{x}^*(t_f) + g\left(x^*(t_f), \dot{x}^*(t_f), t_f\right) \right] \delta t_f$$

$$+ \int_{t_0}^{t_f} \left[\frac{\partial g}{\partial x} \left(x^*(t), \dot{x}^*(t), t\right) - \frac{d}{dt} \left(\frac{\partial g\left(x^*(t), \dot{x}^*(t), t\right)}{\partial \dot{x}} \right) \right] \delta x(t) dt.$$

$$\tag{2.117}$$

As observed from above equation, the variation of J is composed of two parts: the integral one, which is common for all the problems which has been studied so far, related to $\delta x(t)$, $t \in [t_0, t_f]$, and the part involving δt_f.

Similarly, as stated earlier, the extremum for this class of free final time problems is also an extremum for a particular fixed final time problem; therefore, the Euler equation of (2.46) is satisfied as well.

Moreover, as claimed, δt_f is arbitrary, so its coefficient must be zero to make $\delta J(x^*, \delta x) = 0$; then the boundary condition (2.113) holds. ☐

In the following, we will apply the developed result to specify the necessary conditions for the extremal solution with certain problems.

Example 2.20 Specify an extremum for the functional

$$J(x) = \int_{t_0}^{t_f} \left[2x(t) + \dot{x}^2(t) \right] dt, \tag{2.118}$$

with the boundary conditions as $t_0 = 0$, $x(t_0) = 1$, $x(t_f) = 0$, and t_f being free.
Solution. The Euler equation is specified as $\ddot{x}^*(t) = 2$ with its solution

$$x^*(t) = t^2 + \xi_1 t + \xi_2, \tag{2.119}$$

for some constant-valued ξ_1 and ξ_2.

Since t_f is free, by (2.113), we have the following boundary condition:

$$0 = g\left(x^*(t_f), \dot{x}^*(t_f), t_f\right) - \frac{\partial g\left(x^*(t_f), \dot{x}^*(t_f), t_f\right)}{\partial \dot{x}} \dot{x}^*(t_f) \tag{2.120}$$

$$= 2x^*(t) + [\dot{x}^*(t_f)]^2 - 2[\dot{x}^*(t_f)]^2 = 2x^*(t_f) - [\dot{x}^*(t_f)]^2. \tag{2.121}$$

Hence by the specified Euler equation and the boundary condition, we have

$$x^*(0) = 1 = \xi_2, \tag{2.122a}$$

$$x^*(t_f) = 0 = t_f^2 + \xi_1 t_f + \xi_2, \tag{2.122b}$$

$$2x^*(t_f) - [\dot{x}^*(t_f)]^2 = 0 - [2t_f + \xi_1]^2 = 0, \tag{2.122c}$$

by solving which, we can get the values $\xi_1 = -2$ and $\xi_2 = 1$. Hence

$$x^*(t) = t^2 - 2t + 1, \tag{2.123}$$

and $t_f = 1$. ☐

In Example 2.12, we have studied the extremal solution to EV charging problems with fixed final t_f and a fixed SoC state $x(t_f) = x_{max}$, with x_{max} representing the maximum allowed SoC state.

Here in Example 2.21 below, we will revisit this problem by considering a free final t_f, but still with a fixed SoC state $x(t_f) = x_{max}$.

Example 2.21 (*Electric Vehicle Charging Problems with Free Final Time and Fixed Final SoC State*)

Implement the optimal EV charging solution to the problem given in Example 2.12 with fixed $x(t_0) = x_0$ and $x(t_f) = x_{max}$, and but t_f free.

Solution. Firstly by Example 2.12, it has $\dot{x}^*(t) + D(t) = \xi$, for all $t \in [t_0, t_f]$, with some constant-valued ξ.

That is to say, the optimal charging power is yet to fill the valley of the base demand.

By (2.113) stated in Theorem 2.4, the following boundary condition should hold:

$$g\left(x^*(t_f), \dot{x}^*(t_f), t_f\right) - \frac{\partial g\left(x^*(t_f), \dot{x}^*(t_f), t_f\right)}{\partial \dot{x}} \dot{x}^*(t_f)$$

$$= D^2(t_f) - \left[\dot{x}^*(t_f)\right]^2 = 0, \tag{2.124}$$

which implies that at final time t_f,

$$\dot{x}^*(t_f) = D(t_f), \text{ or } \dot{x}^*(t_f) = -D(t_f). \tag{2.125}$$

$\square$

2.2.6 Extrema with Free Final Time and Free Final State

In this section, we continue to study the problem with free final time t_f, however different from Sect. 2.2.5, in this part, suppose that the final boundary state is no longer fixed, say $x(t_f)$ is given free. See an illustration of candidates for x in Fig. 2.13.

Fig. 2.13 Some functions with free final time and free final states

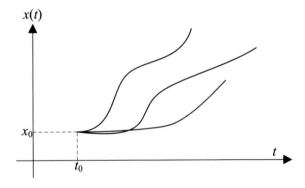

Fig. 2.14 The extreme function and another admissible function in case final time and state are free

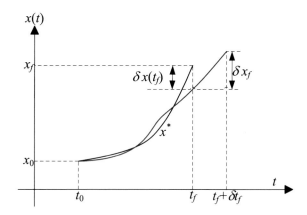

Problem 2.4 (*Extrema Problems with Free Final Time and Final State*) Specify a necessary condition for a function to be an extremum for the functional

$$J(x) = \int_{t_0}^{t_f} g\,(x(t), \dot{x}(t), t)\, dt, \tag{2.126}$$

where t_0 and $x(t_0) = x_0$ are specified, and t_f and $x(t_f)$ are free. $\square$

Figure 2.14 displays an extremum x^* and an admissible comparison curve x.

To use the fundamental theorem, we must first determine the variation by forming the increment. This is accomplished in exactly the same manner as in Problem 2.3 with the increment of the functional given as

$$\Delta J = \frac{\partial g\,(x^*(t_f), \dot{x}^*(t_f), t_f)}{\partial \dot{x}} \delta x(t_f) + g\,(x^*(t_f), \dot{x}^*(t_f), t_f)\,\delta t_f$$
$$+ \int_{t_0}^{t_f} \left[\frac{\partial g\,(x^*(t), \dot{x}^*(t), t)}{\partial x} - \frac{d}{dt} \left(\frac{\partial g\,(x^*(t), \dot{x}^*(t), t)}{\partial \dot{x}} \right) \right] \delta x(t)\,dt + o(\cdot).$$

As demonstrated in Fig. 2.14, we have

$$\delta x_f \doteq \delta x(t_f) + \dot{x}^*(t_f)\delta t_f, \tag{2.127}$$

by which, together with the specification of the increment of the functional J given above and Theorem 2.1, we can obtain the variation of J in the following:

$$\delta J(x^*, \delta x) = 0$$

$$= \frac{\partial g\left(x^*(t_f), \dot{x}^*(t_f), t_f\right)}{\partial \dot{x}} \delta x_f$$

$$+ \left[g\left(x^*(t_f), \dot{x}^*(t_f), t_f\right) - \frac{\partial g\left(x^*(t_f), \dot{x}^*(t_f), t_f\right)}{\partial \dot{x}} \dot{x}^*(t_f) \right] \delta t_f$$

$$+ \int_{t_0}^{t_f} \left[\frac{\partial g\left(x^*(t), \dot{x}^*(t), t\right)}{\partial x} - \frac{d}{dt}\left(\frac{\partial g\left(x^*(t), \dot{x}^*(t), t\right)}{\partial \dot{x}}\right) \right] \delta x(t) dt. \quad (2.128)$$

As before, the Euler equation must be satisfied; therefore, the integral is zero.

As stated in this section, consider that t_f and $x(t_f)$ are not specified. And there are many specific cases concerning this boundary conditions. In Theorems 2.5 and 2.6, we will specify the necessary conditions for extrema problems with different boundary situations, respectively.

Theorem 2.5 *Suppose that x^* is an extremal solution to Problem 2.4 with t_f and $x(t_f)$ independent of each other; then x^* satisfies the Euler equation (2.46) with the following boundary conditions:*

$$\frac{\partial g(x^*(t_f), \dot{x}^*(t_f), t_f)}{\partial \dot{x}} = 0, \quad (2.129a)$$

$$g(x^*(t_f), \dot{x}^*(t_f), t_f) = 0. \quad (2.129b)$$

Proof Consider that t_f and $x(t_f)$ are unrelated to each other. In this case, δx_f and δt_f are independent of one another and arbitrary, so their coefficient must each be zero. From (2.128), then

$$\frac{\partial g\left(x^*(t_f), \dot{x}^*(t_f), t_f\right)}{\partial \dot{x}} = 0, \quad (2.130a)$$

$$g\left(x^*(t_f), \dot{x}^*(t_f), t_f\right) - \frac{\partial g\left(x^*(t_f), \dot{x}^*(t_f), t_f\right)}{\partial \dot{x}} \dot{x}^*(t_f) = 0, \quad (2.130b)$$

which together imply that

$$g(x^*(t_f), \dot{x}^*(t_f), t_f) = 0. \quad (2.131)$$

$\square$

Theorem 2.6 *Suppose that x^* is an extremal solution to Problem 2.4 with t_f and $x(t_f)$ dependent on each other, such that*

$$x(t_f) = \phi(t_f), \quad (2.132)$$

with $\phi(t_f)$ representing a function of t_f; then x^* satisfies the Euler equation (2.46) with the following boundary condition:

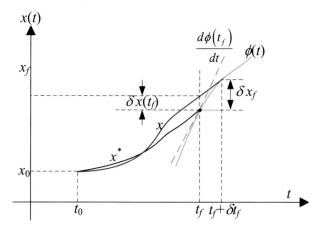

Fig. 2.15 An illustration of the relationship of δx_f, $\delta x(t_f)$, $\phi(t)$, and δt_f

$$\frac{\partial g\left(x^*(t_f), \dot{x}^*(t_f), t_f\right)}{\partial \dot{x}}\left[\frac{d\phi(t_f)}{dt} - \dot{x}^*(t_f)\right] + g\left(x^*(t_f), \dot{x}^*(t_f), t_f\right) = 0. \quad (2.133)$$

Proof As illustrated in Fig. 2.15, the distance η is given as

$$\eta = \frac{d\phi(t_f)}{dt}\delta t_f, \quad (2.134)$$

which is a linear approximation to δx_f; then the following linear approximation between δx_f and δt_f holds:

$$\delta x_f \doteq \frac{d\phi(t_f)}{dt}\delta t_f. \quad (2.135)$$

Hence, we can obtain the conclusion of (2.133) by replacing δx_f with $\frac{d\phi(t_f)}{dt}\delta t_f$ in (2.128) and because δt_f is free. This equation may be called _the transversality condition_. $\qquad\square$

In the following, we will study a few examples to demonstrate the application of the results proposed in Theorems 2.5 and 2.6 for specific cases, respectively.

Firstly in Examples 2.22 and 2.23, we will revisit the shortest-length problems, which have been studied earlier, with free final time and final state, respectively.

Example 2.22 Give an extremum curve for the functional (2.61) specified in Example 2.10 with $t_0 = 0$, $x(t_0) = 2$, and t_f and $x(t_f)$ unspecified such that $x(t_f)$ is required to lie on the straight line

$$\phi(t) = t - 3. \tag{2.136}$$

Solution. As studied in Example 2.10, the Euler equation for the underlying problem is $x^*(t) = \xi_1 t + \xi_2$.

By the boundary condition $x^*(0) = 2$, we have $\xi_2 = 2$. Thus $x^*(t) = \xi_1 t + 2$ for all t.

Since $x(t_f)$ and t_f are related, by applying Theorem 2.6, we have the following boundary condition:

$$\frac{\dot{x}^*(t_f)}{\sqrt{1 + [\dot{x}^*(t_f)]^2}} \left[1 - \dot{x}^*(t_f)\right] + \sqrt{1 + [\dot{x}^*(t_f)]^2} = 0. \tag{2.137}$$

Thus we get

$$\dot{x}^*(t_f) + 1 = 0, \tag{2.138}$$

by which, together with $x^*(t) = \xi_1 t + 2$ for all t, we obtain that $\xi_1 = -1$.

Consequently, we have

$$x^*(t_f) = -t_f + 2 = \phi(t_f) = t_f - 3,$$

which has the solution $t_f = \frac{5}{2}$.

Thus, the extremal solution is $x^*(t) = -t + 2$, with $t_f = \frac{5}{2}$.

As illustrated in Fig. 2.16, the implemented curve with the shortest length is a straight line perpendicular to $\phi(t)$. □

Fig. 2.16 The shortest curve from a fixed point to a point on a straight line

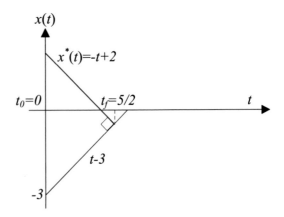

Example 2.23 Give an extremum for the functional defined in (2.61) which begins at the origin and terminates on the curve below:

$$\phi(t) = [t - 4]^2. \tag{2.139}$$

Solution. The Euler equation and its solution are the same as in the previous example, and since $x^*(0) = 0$, we have $\xi_2 = 0$. Same as Example 2.22, since $x(t_f)$ and t_f are related as well, by applying Theorem 2.6, we have the following boundary condition:

$$\frac{\dot{x}^*(t_f)}{\sqrt{1 + [\dot{x}^*(t_f)]^2}} \left[2t_f - 8 - \dot{x}^*(t_f)\right] + \sqrt{1 + [\dot{x}^*(t_f)]^2} = 0, \tag{2.140}$$

by which, together with $\dot{x}^*(t_f) = \xi_1$, we have

$$\xi_1[2t_f - 8] + 1 = 0. \tag{2.141}$$

Also since the boundary condition $x^*(t_f) = \phi(t_f)$, we get

$$x^*(t_f) = \xi_1 t_f = \phi(t_f) = [t_f - 4]^2. \tag{2.142}$$

Thus by solving the above two equations, we can implement that $\xi_1 = 0.44$ and $t_f = 3$, so the solution is $x^*(t) = 0.44t$.

As illustrated in Fig. 2.17, the implemented curve is a straight line perpendicular to $\phi(t) = [t - 4]^2$. □

In summary, we have studied how to implement the extrema of a functional with different boundary conditions. The key equation is (2.128), since from it we can deduce all of the results we have obtained so far.

As developed in this section, the Euler equation is always satisfied with different boundary conditions. Thus, the integral term of (2.128) will be zero. If, as stated

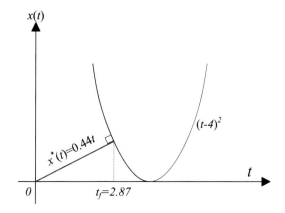

Fig. 2.17 The shortest curve from a fixed point to a point on a nonlinear curve $\phi(t)$

in Problem 2.1, t_f and $x(t_f)$ are specified, then $\delta t_f = 0$ and $\delta x_f = \delta x(t_f) = 0$ in (2.128).

To obtain the boundary condition equations for Problem 2.2, say t_f is specified and $x(t_f)$ is free, simply let $\delta t_f = 0$ and $\delta x_f = \delta x(t_f)$ in (2.128). Similarly, to obtain the boundary equations for Problem 2.3, set $\delta x_f = 0$ in (2.128).

Certainly, we can claim that the necessary conditions for Problems 2.2 and 2.3 could be considered as special cases.

2.3 Extrema of Functional with Multiple Independent Functions

Up to now, we have explored the extrema of a functional with respect to a single scalar-valued function $x(\cdot)$. In this section, we may extend it to the extrema of a functional on a collection of scalar-valued functions x_i, with $i = 1, \cdot, n$.

As before, here we still start from the extrema problems with fixed end points.

Problem 2.5 Consider a functional J such that

$$J(x_i; i = 1, \ldots, n) = \int_{t_0}^{t_f} g\left(x_1(t), \ldots, x_n(t), \dot{x}_1(t), \ldots, \dot{x}_n(t), t\right) dt, \quad (2.143)$$

where $x_1, x_2, \ldots, x_n$ are independent of each other, and t_0, t_f, and boundary values are all specified, such that $x_i(t_0) = x_{i0}$ and $x_i(t_f) = x_{if}$, for all $i = 1, \ldots, n$. □

For notational simplicity, we may consider

$$x \equiv (x_1, \ldots, x_n)^\top \quad \text{and} \quad \delta x \equiv (\delta x_1, \ldots, \delta x_n)^\top,$$

and also $\dot{x} \equiv (\dot{x}_1, \ldots, \dot{x}_n)^\top$ and $\delta \dot{x} \equiv (\delta \dot{x}_1, \ldots, \delta \dot{x}_n)^\top$.

In Theorem 2.7 below, we will specify a necessary condition for the extremal solution $x^* \equiv (x_1^*, \ldots, x_n^*)^\top$ for Problem 2.5.

Theorem 2.7 *A necessary condition for the extremal solution x^* to Problem 2.5 is specified as the following:*

$$\frac{\partial g(x^*(t), \dot{x}^*(t), t)}{\partial x} - \frac{d}{dt}\left(\frac{\partial g(x^*(t), \dot{x}^*(t), t)}{\partial \dot{x}}\right) = 0, \quad (2.144)$$

for all $t \in [t_0, t_f]$.

Proof As usual, we firstly give the increment of J and then its variation δJ and finally get the Euler equation for the underlying extremal problems with a collection of functions by applying the fundamental result of Theorem 2.1.

By definition, the increment of J is given as follows:

$$\Delta J = \int_{t_0}^{t_f} [g(x(t) + \delta x(t), \dot{x}(t) + \delta \dot{x}(t), t) - g(x(t), \dot{x}, t)] \, dt;$$

then by expanding it in a Taylor series about $x(t)$ and $\dot{x}$, we get

$$\Delta J = \int_{t_0}^{t_f} \left[\sum_{i=1}^{n} \frac{\partial g(x(t), \dot{x}(t), t)}{\partial x_i} \delta x_i(t) + \sum_{i=1}^{n} \frac{\partial g(x(t), \dot{x}(t), t)}{\partial \dot{x}_i} \delta \dot{x}_i(t) \right.$$

$$\left. + o(\delta x_i(t), \delta \dot{x}_i(t); i = 1, \ldots, n) \right] dt.$$

In the following, we still apply the integration by parts to eliminate the dependence of δJ on each $\delta \dot{x}_i$.

$$\delta J = 0 = \sum_{i=1}^{n} \frac{\partial g(x^*(t), \dot{x}^*(t), t)}{\partial \dot{x}_i} \delta x_i^*(t) \Big|_{t_0}^{t_f}$$

$$+ \int_{t_0}^{t_f} \sum_{i=1}^{n} \left[\frac{\partial g(x^*(t), \dot{x}^*(t), t)}{\partial x_i} - \frac{d}{dt} \left(\frac{\partial g(x^*(t), \dot{x}^*(t), t)}{\partial \dot{x}_i} \right) \right] \delta x_i^*(t) dt;$$

then we have

$$\delta J = 0 = \int_{t_0}^{t_f} \sum_{i=1}^{n} \left[\frac{\partial g(x^*(t), \dot{x}^*(t), t)}{\partial x_i} - \frac{d}{dt} \left(\frac{\partial g(x^*(t), \dot{x}^*(t), t)}{\partial \dot{x}_i} \right) \right] \delta x_i^*(t) dt,$$

since $\delta x^*(t_0) = 0$ and $\delta x^*(t_f) = 0$, due to the fact that all of the x_i's are fixed at t_0 and t_f, respectively.

The δx_i, with $i = 1, \ldots, n$, are independent of each other; then, except δx_1, we can select $\delta x_i = 0$, for all $i = 2, \ldots, n$. Thus, we have

$$\delta J = 0 = \int_{t_0}^{t_f} \left[\frac{\partial g(x^*(t), \dot{x}^*(t), t)}{\partial x_i} - \frac{d}{dt} \left(\frac{\partial g(x^*(t), \dot{x}^*(t), t)}{\partial \dot{x}_i} \right) \right] \delta x_i(t) dt.$$

$$(2.145)$$

But δx_1 can assume arbitrary values as long as it is zero at the end points t_0 and t_f; therefore, by applying Lemma 2.1, the coefficient of $\delta x_1(t)$ must be zero everywhere in the interval $[t_0, t_f]$.

By repeating the above analysis for each of δx_i, with $i = 2, \ldots, n$, we can obtain the conclusion of (2.144). ☐

Example 2.24 Verify that the straight line is the shortest-length curve between a given pair of two points (t_0, x_0) and (t_f, x_f), with $x_0 \equiv (x_{10}, x_{20})^\top$ and $x_f \equiv (x_{1f}, x_{2f})^\top$.

Proof Firstly, by the modeling issue discussed in Example 2.10, we have $[d\Gamma]^2 = [dt]^2 + [dx_1]^2 + [dx_2]^2$. It implies that

$$d\Gamma = \sqrt{1 + \dot{x}_1^2 + \dot{x}_2^2}dt. \tag{2.146}$$

Hence, the length of a curve between (t_0, x_0) and (t_f, x_f) is specified as the following:

$$J(x) = \int_{t_0}^{t_f} g(\dot{x}(t))dt, \quad \text{with } g(\dot{x}(t)) \triangleq \sqrt{1 + \dot{x}_1^2 + \dot{x}_2^2}, \tag{2.147}$$

which implies that g is a function of $\dot{x}(t)$ only and is independent of $x(t)$ and t.

By applying the Euler equation specified in (2.144), we have

$$\frac{\partial g\left(\dot{x}_i^*(t)\right)}{\partial \dot{x}} = \frac{\dot{x}_i^*(t)}{\sqrt{1 + \dot{x}_1^2 + \dot{x}_2^2}} = \xi_i, \quad \text{with } i = 1, 2, \tag{2.148a}$$

for all $t \in [t_0, t_f]$, with some constant-valued $\xi 1$ and ξ_2.

We can obtain $x^*(t) = \begin{bmatrix} \xi_1 \\ \xi_2 \end{bmatrix} t + \begin{bmatrix} \eta_1 \\ \eta_2 \end{bmatrix}$, say $x^*(t)$ is a straight line, with some constant-valued coefficients ξ_i and η_i, $i = 1, 2$, which can be specified by the boundary conditions $x(t_0) = x_0$ and $x(t_f) = x_f$; see an illustration in Fig. 2.18.

For example, suppose that $t_0 = 0$, $t_f = 6$, $x_0 = \begin{bmatrix} 0 \\ 0 \end{bmatrix}$, and $x_f = \begin{bmatrix} 3 \\ 4 \end{bmatrix}$; then we

obtain $x^*(t) = \begin{bmatrix} \frac{1}{2} \\ \frac{2}{3} \end{bmatrix} t$.   □

Example 2.25 Suppose that a collection of EVs is charged in a power system, such that

$$J(x) = \int_{t_0}^{t_f} \left[\sum_{i=1}^{n} \dot{x}_i(t) + D(t) \right]^2 dt, \tag{2.149}$$

where $x_i(t)$ and $D(t)$ represent the state of charge (SoC) of the EV i and the given fixed base demand in the system, respectively.

We also consider that $x_i(t_0) = x_{i0}$ and $x_i(t_f) = x_{i,max}$.

Solution. Firstly by applying the Euler equation, we obtain

$$\frac{\partial g\left(\dot{x}^*(t)\right)}{\partial \dot{x}_i} = 2 \left[\sum_{i=1}^{n} \dot{x}_i^*(t) + D(t) \right] = \xi, \tag{2.150}$$

for each $i = 1, \ldots, n$, and all $t \in [t_0, t_f]$, with some constant-valued ξ.

Fig. 2.18 The shortest curve
from a fixed point to a point

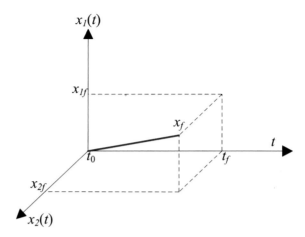

As $\dot{x}_i(t)$ actually represents the charging power of EV i at time t, by the Euler
equation (2.150), we obtain that the optimal charging power is to fill the valley of
the base demand.

By (2.150) and the boundary condition, we can get

$$\xi = \frac{2}{t_f - t_0} \left[\sum_{i=1}^{n} \left[x_{i,max} - x_{i0} \right] + \int_{t_0}^{t_f} D(t)dt \right]. \tag{2.151}$$

However, different from Example 2.12, in this charging problem involving several
EVs, the extrema solution of $x_i^*(\cdot)$ is not unique.   □

Example 2.26 Give a necessary condition for the extrema of the functional

$$J(x) = \int_0^2 \left[\dot{x}_1^2(t) + \dot{x}_2^2(t) + 2t^2 \dot{x}_2(t) \right] dt, \tag{2.152}$$

with the boundary conditions as $x(t_0) = \begin{bmatrix} 2 \\ 1 \end{bmatrix}$ and $x(t_f) = \begin{bmatrix} 4 \\ 7 \end{bmatrix}$.

Solution. By applying the Euler equation, we obtain

$$- 2\ddot{x}_1^*(t) = 0, \tag{2.153a}$$

$$t^2 - \ddot{x}_2^*(t) = 0; \tag{2.153b}$$

then we have $x_1^*(t) = \xi_1 t + \xi_2$, and $x_2^*(t) = \frac{1}{12}t^4 + \xi_3 t + \xi_4$.

By applying the boundary conditions, we get

$$x_1^*(t) = t + 2, \tag{2.154a}$$

$$x_2^*(t) = -\frac{1}{3}t^3 + \frac{13}{3}t + 1. \tag{2.154b}$$

□

Example 2.27 Determine the Euler equations for the functional

$$J(x) = \int_{t_0}^{t_f} \left[x_1^2(t)x_2(t) + \frac{1}{2}t^2\dot{x}_1^2(t)\dot{x}_2^2(t) \right] dt, \tag{2.155}$$

where all of the boundary points are specified, respectively.
Solution. By the Euler equations, we get

$$0 = 2x_1^*(t)x_2^*(t) - \frac{d}{dt} \left[t^2\dot{x}_1^*(t)[\dot{x}_2^*(t)]^2 \right]$$
$$= 2x_1^*(t)x_2^*(t) - 2t\dot{x}_1^*(t)[\dot{x}_2^*(t)]^2 - t^2\ddot{x}_1^*(t)[\dot{x}_2^*(t)]^2 - 2t^2\dot{x}_1^*(t)\dot{x}_2^*(t)\ddot{x}_2^*(t), \tag{2.156a}$$

$$0 = x_1^{*2}(t) - \frac{d}{dt} \left[t^2[\dot{x}_1^*(t)]^2\dot{x}_2^*(t) \right]$$
$$= x_1^{*2}(t) - 2t[\dot{x}_1^*(t)]^2\dot{x}_2^*(t) - 2t^2\dot{x}_1^*(t)\ddot{x}_1^*(t)\dot{x}_2^*(t) - t^2[\dot{x}_1^*(t)]^2\ddot{x}_2^*(t). \tag{2.156b}$$

□

Example 2.28 Give an extremum curve for the functional

$$J(x) = \int_0^{3\pi/4} \left[x_1^2(t) + 4x_2^2(t) + \dot{x}_1(t)\dot{x}_2(t) \right] dt, \tag{2.157}$$

which satisfies the boundary conditions $x(0) = \begin{bmatrix} 0 \\ 1 \end{bmatrix}$ and $x(\frac{3\pi}{4}) = \begin{bmatrix} 1 \\ -1 \end{bmatrix}$.
Solution. By the Euler equations of (2.144), we get

$$2x_1^*(t) - \ddot{x}_2^*(t) = 0 \tag{2.158a}$$

$$8x_2^*(t) - \ddot{x}_1^*(t) = 0 \tag{2.158b}$$

are linear, time-invariant, and homogeneous.
    The Euler equations can be solved by applying the classical methods or Laplace transform method

$$x_1^*(t) = \xi_1 \exp(2t) + \xi_2 \exp(-2t) + \xi_3 \cos(2t) + \xi_4 \sin(2t), \tag{2.159}$$

where ξ_1, ξ_2, ξ_3, and ξ_4 are constants of integration which will be determined with given end points.

Further differentiate $x_1^*(t)$ twice and substitute it into the second Euler equation; then we can get the solution of x_2^* as

$$x_2^*(t) = \frac{1}{2}\left[\xi_1 \exp(2t) + \xi_2 \exp(-2t) - \xi_3 \cos(2t) - \xi_4 \sin(2t)\right]. \qquad (2.160)$$

Considering the boundary condition for the extrema x^*,

$$x_1^*(0) = 0, \quad x_2^*(0) = 1, \quad x_1^*\left(\frac{3\pi}{4}\right) = 1, \quad x_2^*\left(\frac{3\pi}{4}\right) = -1, \qquad (2.161)$$

together with the expression of $x_1^*(t)$ and $x_2^*(t)$ specified above, we can obtain

$$\xi_1 = -\frac{\exp(3\pi/2) + 2}{2[\exp(3\pi) - 1]}, \quad \xi_2 = \frac{2\exp(3\pi) + \exp(3\pi/2)}{2[\exp(3\pi) - 1]}, \quad \xi_3 = -1, \quad \xi_4 = -\frac{3}{2}.$$

$\square$

In Theorem 2.8 below, we will give a necessary condition for the extrema of the functional on vector-valued functions with free or unspecified end points.

Theorem 2.8 *Consider the functional*

$$J(x) = \int_{t_0}^{t_f} g(x(t), \dot{x}(t), t)dt,$$

where x and g satisfy the continuity and differentiability requirements stated earlier, and both the final time t_f and the final state $x(t_f)$ are free; then a necessary condition for an extremum of the above functional is the Euler equation (2.144) together with the following boundary condition:

$$\left[\frac{\partial g(x^*(t_f), \dot{x}^*(t_f), t_f)}{\partial \dot{x}}\right]^{\mathsf{T}} \delta x_f$$

$$+ \left[g(x^*(t_f), \dot{x}^*(t_f), t_f) - \left[\frac{\partial g(x^*(t_f), \dot{x}^*(t_f), t_f)}{\partial \dot{x}}\right]^{\mathsf{T}} \dot{x}^*(t_f)\right]\delta t_f = 0. \quad (2.162)$$

Proof By applying the exact same way to implement the variation of the functional with scalar-valued functions considered in Problem 2.4, we can get the variation of the functional with the free end points. The difference is that it deals with the vector functions in this case. More specially, by forming the increment, integrating by parts the term involving $\delta\dot{x}(t)$, retaining terms of first order, considering the relation of $\delta x(t_f)$ with δx_f and δt_f given as

$$\delta x(t_f) = \delta x_f - \dot{x}^*(t_f)\delta t_f, \qquad (2.163)$$

and by applying Theorem 2.1, we can obtain the following result for the variation of the underlying functional with free end points:

$$\delta J(x^*, \delta x) = 0$$
$$= \left[\frac{\partial g(x^*(t_f), \dot{x}^*(t_f), t_f)}{\partial \dot{x}} \right]^{\mathrm{T}} \delta x_f$$
$$+ \left[g(x^*(t_f), \dot{x}^*(t_f), t_f) - \left[\frac{\partial g(x^*(t_f), \dot{x}^*(t_f), t_f)}{\partial \dot{x}} \right]^{\mathrm{T}} \dot{x}^*(t_f) \right] \delta t_f$$
$$+ \int_{t_0}^{t_f} \left[\frac{\partial g(x^*(t), \dot{x}^*(t), t)}{\partial x} - \frac{d}{dt} \left(\frac{\partial g(x^*(t), \dot{x}^*(t), t)}{\partial \dot{x}} \right) \right]^{\mathrm{T}} \delta x^*(t) dt. \quad (2.164)$$

As before, we can claim that an extremum for this free end point problem should be an extremum for a given fixed end point problem as well; therefore, by Theorem 2.7, x^* for the functional with free end points is also a solution to the Euler equations (2.144).

The boundary conditions at the final time are then specified by the relationship of (2.162).   □

Notice that, Theorem 2.8 only gives a necessary condition for the extrema of a functional with multiple independent functions in a pretty general case. Certainly, we can obtain the necessary condition for each of the specific cases following the same technique applied in the last section to develop necessary conditions for the extrema of a functional with a single function.

For example, suppose that the final time t_f is fixed, $x_i(t_f)$, with $i = 1, 2, \ldots, m$, for some $m \le n$ are specified, and the rest of $x_j(t_f)$, with $j = m + 1, \ldots, n$, are free. In this case, the boundary condition specified in (2.162) becomes the following specific form:

$$g(x^*(t_f), \dot{x}^*(t_f), t_f) - \frac{\partial g(x^*(t_f), \dot{x}^*(t_f), t_f)}{\partial \dot{x}} \dot{x}^*(t_f) = 0, \qquad (2.165a)$$
$$x_i(t_f) = x_{if}, \quad \text{for each of } i = 1, \ldots, m, \qquad (2.165b)$$
$$\frac{\partial g(x^*(t_f), \dot{x}^*(t_f), t_f)}{\partial \dot{x}_j} = 0, \quad \text{for each of } j = m + 1, \ldots, n, \qquad (2.165c)$$

due to the following:

$$\delta t_f = 0,$$
$$\delta x_i(t_f) = 0, \quad \text{for each of } i = 1, \ldots, m,$$
$$\delta x_j(t_f) \text{ is arbirary}, \quad \text{for each of } j = m + 1, \ldots, n.$$

Example 2.29 Give a necessary condition for the extrema of the functional given in Example 2.26 with a revised boundary condition such that $x(0) = \begin{bmatrix} 2 \\ 1 \end{bmatrix}$, $x_1(2) = 4$, and $x_2(2)$ is free.

Solution. Firstly, by the result obtained in Example 2.26,

$$x_1^*(t) = \xi_1 t + \xi_2, \tag{2.167a}$$

$$x_2^*(t) = -\frac{1}{3}t^3 + \xi_3 t + \xi_4; \tag{2.167b}$$

then by Theorem 2.8, we have the boundary condition as follows:

$$x_1^*(0) = \xi_2 = 2, \tag{2.168a}$$

$$x_2^*(0) = \xi_4 = 1, \tag{2.168b}$$

$$x_1^*(2) = 2\xi_1 + \xi_2 = 4, \tag{2.168c}$$

$$\frac{\partial g(x^*(t_f), \dot{x}^*(t_f), t_f)}{\partial \dot{x}_2} = 2\dot{x}_2^*(t_f) + 2t_f^2 = 2\xi_3 = 0. \tag{2.168d}$$

Thus we get the resulting extrema as follows:

$$x_1^*(t) = t + 2, \tag{2.169a}$$

$$x_2^*(t) = -\frac{1}{3}t^3 + 1. \tag{2.169b}$$

As observed, $x_2^*(t)$ is different from that for the same functional but with a distinct boundary condition. □

2.4 Extrema of Function with Constraints

In this section, we will specify the extremal solution to the system with certain constraints.

Problem 2.6 Find the necessary condition for the extremal solution to a system with $(n + m)$ variables, $x_1, \ldots, x_{n+m}$, denoted by $f(x_1, \ldots, x_{n+m})$ with the constraints below

$$a_1(x_1, \ldots, x_{n+m}) = 0,$$

$$\cdots$$

$$a_m(x_1, \ldots, x_{n+m}) = 0.$$

□

By the specification of Problem 2.6, though there are $n + m$ variables, there are only $[n + m] - m = n$ independent variables due to the m constrained equations.

Here, we will solve the constrained optimization problems by applying the elimination (or direct) method and the Lagrange multiplier method in Sects. 2.4.1 and 2.4.2, respectively.

2.4.1 Elimination/Direct Method

Concerning the constraints given in Problem 2.6, there exist m functions such that

$$x_{n+1} = b_1(x_1, \ldots, x_n), \tag{2.170a}$$

$$\ldots$$

$$x_{n+m} = b_m(x_1, \ldots, x_n). \tag{2.170b}$$

By putting these relations into the function $f(\cdot)$, we can obtain a function with n independent variables, say $f(x_1, \ldots, x_n)$.

To find the extremal value of this function, we solve $x_1^*, \ldots, x_n^*$ from the following equations:

$$\frac{\partial f(x_1^*, \ldots, x_n^*)}{\partial x_1} = 0, \tag{2.171a}$$

$$\ldots$$

$$\frac{\partial f(x_1^*, \ldots, x_n^*)}{\partial x_n} = 0, \tag{2.171b}$$

and put the implemented $x_1^*, \ldots, x_n^*$ in the equations of (2.170) to obtain the values of $x_1^*, \ldots, x_n^*$.

Next, we will verify the elimination method via an example.

Example 2.30 Give a necessary condition for the extremal solution to the system with a performance cost function given as

$$f(x_1, x_2) = x_1^2 + 2x_2^2, \tag{2.172}$$

subject to the constraint of $ax_1 + bx_2 = 1$.

Solution. By the equality constraint of $ax_1 + bx_2 = 1$, we can get $x_1 = \frac{1}{a} - \frac{b}{a}x_2$; then we have the problem becomes

$$f(x_2) = \left[\frac{1}{a} - \frac{b}{a}x_2\right]^2 + 2x_2^2 = \frac{1}{a^2}\left[[2a^2 + b^2]x_2^2 - 2bx_2 + 1\right], \tag{2.173}$$

by which we get that the extremum of x_2^* satisfies the following:

$$df(x_2^*) = \frac{1}{a^2}\left[-2b + 2[2a^2 + b^2]x_2^*\right]\Delta x_2 = 0, \tag{2.174}$$

so $x_2^* = \dfrac{b}{2a^2 + b^2}$. Thus

$$x_1^* = \frac{1}{a} - \frac{b}{a}x_2^* = \frac{2a}{2a^2 + b^2}, \tag{2.175}$$

and the minimum value of the function $f(x_1^*, x_2^*)$ is $\dfrac{2}{2a^2 + b^2}$. $\qquad\square$

However, it may be challenging to determine the relation equations (2.170), and the implementation of the solution to (2.171) may be difficult as well.

Alternatively, in the next section, we will present another method, say the Lagrange multiplier method, to solve Problem 2.6.

2.4.2 Lagrange Multiplier Method

Now we introduce the method of Lagrange multipliers to solve the extremal solution to the constrained problem. First, we form the augmented function:

$$\begin{aligned}
&f_a(x_1, \ldots, x_{n+m}, \lambda_1, \ldots, \lambda_n)\\
&\triangleq f(x_1, \ldots, x_{n+m}) + \lambda_1[a_1(x_1, \ldots, x_{n+m})] + \cdots + \lambda_n[a_m(x_1, \ldots, x_{n+m})];
\end{aligned} \tag{2.176}$$

then, the necessary condition for the extremal solution x^* of the function f_a is that

$$\begin{aligned}
df_a(x^*, \lambda) = 0 &= \frac{\partial f_a^*}{\partial x_1}\Delta x_1 + \cdots + \frac{\partial f_a^*}{\partial x_{n+m}}\Delta x_{n+m} + \frac{\partial f_a^*}{\partial \lambda_1}\Delta \lambda_1 + \cdots + \frac{\partial f_a^*}{\partial \lambda_m}\Delta \lambda_m\\
&= \frac{\partial f_a^*}{\partial x_1}\Delta x_1 + \cdots + \frac{\partial f_a^*}{\partial x_{n+m}}\Delta x_{n+m} + a_1^*\Delta \lambda_1 + \cdots + a_m^*\Delta \lambda_m,
\end{aligned} \tag{2.177}$$

where for notational simplicity, consider $f_a^* \equiv f_a(x^*, \lambda)$ and $a_i^* \equiv a_i(x^*)$.

If the constraints are satisfied, the coefficients of $\Delta\lambda_1, \ldots, \Delta\lambda_m$ are all zero. It then selects the m λ_is so that the coefficients of Δx_i, with $i = 1, \ldots, n$, are zero. The remaining n Δx_is are independent of each other, and for df_a to equal zero their coefficients must vanish. The result is that the extreme point $x_1^*, \ldots, x_{n+m}^*$ are found by solving the equations

$$a_i(x_1^*, \ldots, x_{n+m}^*) = 0, \qquad i = 1, 2, \ldots, m, \tag{2.178a}$$

$$\frac{\partial f_a(x_1^*, \ldots, x_{n+m}^*, \lambda_1^*, \ldots, \lambda_m^*)}{\partial x_j} = 0, \quad j = 1, 2 \ldots, n + m. \tag{2.178b}$$

In the following, we will illustrate the above procedure to determine the extremal solution to the problem stated in Example 2.30.

Firstly define the augmented function as

$$f_a(x_1, x_2, \lambda) \triangleq x_1^2 + 2x_2^2 + \lambda[ax_1 + bx_2 - 1]. \tag{2.179}$$

For values of x_1 and x_2 that satisfy the constraining relation defined in Example 2.30, the augmented function f_a equals f, and regardless of the value of λ we have simply added zero to f to obtain f_a.

By satisfying the constraint and minimizing f_a, the constrained extreme point of f can be found.

To find an extreme point of f_a, we use the necessary condition

$$df_a(x_1^*, x_2^*, \lambda) = 0 = [2x_1^* + a\lambda]\Delta x_1 + [4x_2^* + b\lambda]\Delta x_2 + [ax_1^* + bx_2^* - 1]\Delta\lambda.$$

Since only those points that satisfy the constraints are admissible, we have $ax_1^* + bx_2^* - 1 = 0$.

By (2.177), $[2x_1^* + a\lambda]\Delta x_1 + [4x_2^* + b\lambda]\Delta x_2 = 0$, but Δx_1 and Δx_2 are not independent.

Since the constraint must be satisfied, λ can be any valued, so λ can be set such that $4x_2^* + b\lambda^* = 0$.

Δx_1 can be arbitrary valued; for each Δx_1, there is an associated valued Δx_2.

By (2.177), and Δx_1 being arbitrary, the coefficient of Δx_1 must be zero; then $2x_1^* + a\lambda^* = 0$.

By the above analysis, we can obtain the extremal solution as

$$x_1^* = \frac{2a}{2a^2 + b^2}, \quad x_2^* = \frac{b}{2a^2 + b^2}, \quad \lambda^* = -\frac{4}{2a^2 + b^2}. \tag{2.180}$$

Example 2.31 Implement the extremal solution to the problem stated in Example 2.30 via the Lagrange multipliers method.

Solution. By (2.178), we can directly obtain

$$ax_1^* + bx_2^* - 1 = 0, \tag{2.181a}$$

$$2x_1^* + a\lambda^* = 0, \tag{2.181b}$$

$$4x_2^* + b\lambda^* = 0, \tag{2.181c}$$

with the solution as given in (2.180).   □

2.5 Extrema of Functional with Constraints

Now we are ready to consider the presence of constraints in variational problems. To simplify the variational equations, it will be assumed that the admissible curves are smooth.

2.5.1 Extrema of Functional with Differential Constraints

Problem 2.7 Let us determine a set of necessary conditions for a function x^* to be an extremum for a functional of the form

$$J(x) = \int_{t_0}^{t_f} g(x(t), \dot{x}(t), t) dt, \tag{2.182}$$

with $x(t) \equiv (x_1(t), \ldots, x_{n+m}(t))^\top$ which satisfies the following constraints:

$$f_i(x(t), \dot{x}(t), t) = 0, \tag{2.183}$$

with $i = 1, \ldots, n$. □

As an alternative approach, we can use Lagrange multipliers. The first step is to form the augmented functional by adjoining the constraining relations to J, which yields

$$L_a(x, \lambda) = \int_{t_0}^{t_f} \left[g(x(t), \dot{x}(t), t) + \sum_{i=1}^{n} \lambda_i(t) f_i(x(t), \dot{x}(t), t) \right] dt$$

$$= \int_{t_0}^{t_f} \left[g(x(t), \dot{x}(t), t) + \lambda^\top(t) f(x(t), \dot{x}(t), t) \right] dt, \tag{2.184}$$

with $\lambda(t) \equiv (\lambda_1(t), \ldots, \lambda_n(t))^\top$, and

$$f(x(t), \dot{x}(t), t) \equiv [f_1(x(t), \dot{x}(t), t), \ldots, f_n(x(t), \dot{x}(t), t)]^\top.$$

Notice if the constraints are satisfied; then we have

$$L_a(x, \lambda) = J(x, \lambda), \tag{2.185}$$

for any x and λ.

For notational simplicity, we define an augmented integrand function for Problem 2.7 as

$$g_a(x(t), \dot{x}(t), \lambda(t), t) \triangleq g(x(t), \dot{x}(t), \lambda(t), t) + \lambda^\top(t) f(x(t), \dot{x}(t), t). \tag{2.186}$$

Theorem 2.9 *The Euler equation for the extremal solution to Problem 2.7 with constraints is given as follows:*

$$\frac{\partial g_a(x^*(t), \dot{x}^*(t), \lambda^*(t), t)}{\partial x} - \frac{d}{dt}\left(\frac{\partial g_a(x^*(t), \dot{x}^*(t), \lambda^*(t), t)}{\partial \dot{x}}\right) = 0, \quad (2.187)$$

for all $t \in [t_0, t_f]$.

Proof By (2.184), the variation of the functional L_a can be specified as

$$\delta L_a(x, \delta x, \lambda, \delta \lambda) = \int_{t_0}^{t_f} \left\{ \left[\left[\frac{\partial g(x(t), \dot{x}(t), t)}{\partial x}\right]^\top + [\lambda(t)]^\top \frac{\partial f(x(t), \dot{x}(t), t)}{\partial x} \right] \delta x(t) \right.$$

$$+ \left[\left[\frac{\partial g(x(t), \dot{x}(t), t)}{\partial \dot{x}}\right]^\top + [\lambda(t)]^\top \frac{\partial f(x(t), \dot{x}(t), t)}{\partial \dot{x}} \right] \delta \dot{x}(t)$$

$$\left. + [f(x(t), \dot{x}(t), t)]^\top \delta \lambda(t) \right\} dt, \quad (2.188)$$

where $\frac{\partial f}{\partial x}$ is given as $\begin{bmatrix} \frac{\partial f_1}{\partial x_1} & \cdots & \frac{\partial f_1}{\partial x_{n+m}} \\ \vdots & \cdots & \vdots \\ \frac{\partial f_n}{\partial x_1} & \cdots & \frac{\partial f_n}{\partial x_{n+m}} \end{bmatrix}$ which is an $n \times [n + m]$ matrix.

Integrating by parts the term containing $\delta \dot{x}$ and retaining only the terms inside the integral, we obtain

$$\delta L_a(x, \delta x, \lambda, \delta \lambda) = \int_{t_0}^{t_f} \left\{ \left[\left[\frac{\partial g(x(t), \dot{x}(t), t)}{\partial x}\right]^\top + [\lambda(t)]^\top \frac{\partial f(x(t), \dot{x}(t), t)}{\partial x} \right. \right.$$

$$\left. - \frac{d}{dt}\left(\left[\frac{\partial g(x(t), \dot{x}(t), t)}{\partial \dot{x}}\right]^\top + [\lambda(t)]^\top \frac{\partial f(x(t), \dot{x}(t), t)}{\partial \dot{x}} \right) \right] \delta x(t)$$

$$\left. + [f(x(t), \dot{x}(t), t)]^\top \delta \lambda(t) \right\} dt.$$

On an extremum, the variation must be zero; that is, $\delta L_a(x^*, \lambda^*) = 0$. Also, the constraints should be satisfied by an extremum; thus

$$f(x^*(t), \dot{x}^*(t), t) = 0, \quad \forall t \in [t_0, t_f]. \quad (2.189)$$

Hence, we can set these costate functions in a way, such that the coefficients of the n components of $\delta x(t)$ are zero valued over the whole interval $[t_0, t_f]$.

The remaining m components of δx are then independent of each other; then, by Lemma 2.1, the coefficients of these $\delta x(t)$ must be zero over $[t_0, t_f]$.

Finally, by the above analysis, we can obtain

$$\left[\frac{\partial g(x^*(t), \dot{x}^*(t), t)}{\partial x}\right]^\top + [\lambda^*(t)]^\top \frac{\partial f(x^*(t), \dot{x}^*(t), t)}{\partial x}$$

$$-\frac{d}{dt}\left(\left[\frac{\partial g(x^*(t), \dot{x}^*(t), t)}{\partial \dot{x}}\right]^\top + [\lambda^*(t)]^\top \frac{\partial f(x^*(t), \dot{x}^*(t), t)}{\partial \dot{x}}\right) = 0,$$

$$(2.190)$$

for all $t \in [t_0, t_f]$.

Hence we can get the conclusion of (2.187). $\qquad\square$

It can be observed that (2.187) are a set of differential equations, and the constraining relations $f(x(t), \dot{x}(t), t) = 0$ are a set of n equations which may have specific forms which we may study further later in detail. Together, these 2n+m equations constitute a set of necessary conditions for x^* to be an extremum.

The reader may have already noticed that (2.187) together with the collection of constraints $f(x(t), \dot{x}(t), t) = 0$ are the same as if the results from Problem 2.1 had been applied to the functional

$$L_a(x, \lambda) = \int_{t_0}^{t_f} g_a(x(t), \dot{x}(t), \lambda(t), t)dt, \qquad (2.191)$$

with the assumption that the functions x and λ are independent. It should be emphasized that, although the results are the same, the reasoning used is quite different.

Example 2.32 Give a necessary condition for an extremum of the following functional:

$$J(x) = \int_{t_0}^{t_f} \left[x_1^2(t) + x_2^2(t)\right] dt, \qquad (2.192)$$

with given fixed boundary conditions $(t_0, x(t_0))$ and $(t_f, x(t_f))$, and subject to the following relation: $\dot{x}_1(t) = -ax_1(t) + x_2(t)$.

Solution. Firstly, by applying the Lagrange multiplier method, define the augmented function g_a as

$$g_a(x(t), \dot{x}(t), \lambda(t)) \triangleq x_1^2(t) + x_2^2(t) + \lambda(t)[\dot{x}_1(t) + ax_1(t) - x_2(t)]; \qquad (2.193)$$

then by the Euler equation (2.187) specified in Theorem 2.9, we can obtain

$$2x_1^*(t) + a\lambda^*(t) - \dot{\lambda}^*(t) = 0, \qquad (2.194a)$$
$$2x_2^*(t) - \lambda^*(t) = 0, \qquad (2.194b)$$
$$\dot{x}_1^*(t) + ax_1^*(t) - x_2^*(t) = 0. \qquad (2.194c)$$

By the above equations, x_1^* is the solution to the following second differential equation:

$$\ddot{x}_1^*(t) - \left[1 + a^2\right] x_1^*(t) = 0, \tag{2.195}$$

and

$$x_2^*(t) = \dot{x}_1^*(t) + a x_1^*(t). \tag{2.196}$$

□

Considering a specific form of constraints $f_i(x(t), t)$, say $f_i(x(t), t)$ is a function of $x(t)$ and t and independent of $\dot{x}(t)$, the Euler equation (2.187) specified in Theorem 2.9 degenerates to

$$\left[\frac{\partial g(x^*(t), \dot{x}^*(t), t)}{\partial x}\right]^{\mathrm{T}} + [\lambda^*(t)]^{\mathrm{T}} \frac{\partial f(x^*(t), \dot{x}^*(t), t)}{\partial x}$$
$$- \frac{d}{dt} \left(\left[\frac{\partial g(x^*(t), \dot{x}^*(t), t)}{\partial \dot{x}}\right]^{\mathrm{T}}\right) = 0, \tag{2.197}$$

for all $t \in [t_0, t_f]$.

Example 2.33 Give a necessary condition that must be satisfied by the curve with the shortest length which lies on the plane $\eta_1 x_1(t) + \eta_2 x_2(t) + t = c$, for $t \in [t_0, t_f]$, and satisfies the fixed boundary conditions of t_0, t_f, x_0 and x_f.

Solution. As modeled in Example 2.24, the functional to be minimized is

$$J(x) = \int_{t_0}^{t_f} \sqrt{1 + \dot{x}_1^2(t) + \dot{x}_2^2(t)} dt. \tag{2.198}$$

So the augmented integrand function is

$$g_a(x(t), \dot{x}(t), \lambda(t), t)$$
$$= \sqrt{1 + \dot{x}_1^2(t) + \dot{x}_2^2(t)} + \lambda(t) \left[\eta_1 x_1(t) + \eta_2 x_2(t) + t - c\right]. \tag{2.199}$$

By applying the Euler equation of (2.197) the extremal solution (x^*, λ^*) satisfies the following necessary condition:

$$\eta_i \lambda^*(t) - \frac{d}{dt} \left(\frac{\dot{x}_i^*(t)}{\sqrt{1 + [\dot{x}_1^*(t)]^2 + [\dot{x}_2^*(t)]^2}}\right) = 0, \tag{2.200}$$

with $i = 1, 2$, together with the constraint of $\eta_1 x_1^*(t) + \eta_2 x_2^*(t) + t = c$.

Hence we have

$$\frac{d}{dt} \left(\frac{\eta_1 \dot{x}_2^*(t) - \eta_2 \dot{x}_1^*(t)}{\sqrt{1 + [\dot{x}_1^*(t)]^2 + [\dot{x}_2^*(t)]^2}}\right) = 0; \tag{2.201}$$

then $\eta_1 \dot{x}_2^*(t) - \eta_2 \dot{x}_1^*(t) = \xi$ for some constant-valued ξ. Thus

$$\eta_1 x_2^*(t) - \eta_2 x_1^*(t) = \xi t + \hat{\xi}, \tag{2.202}$$

by which, and by the given constraint $\eta_1 x_1(t) + \eta_2 x_2(t) + t = c$, we get that the shortest curve is a straight line such that

$$x_1^*(t) = \frac{1}{\eta_2^2 [\eta_1 + \eta_2]} [at + b], \tag{2.203a}$$

$$x_2^*(t) = \frac{1}{\eta_2 [\eta_1 + \eta_2]} \left[[\xi_1 \eta_1 - \eta_2] t + \xi_2 \eta_1 + c \eta_2 \right], \tag{2.203b}$$

with $a \equiv \xi_1 (\eta_1^2 - \eta_1 \eta_2 - \eta_2^2) - \eta_1 \eta_2$ and $b \equiv \xi_2 (\eta_1^2 - \eta_1 \eta_2 - \eta_2^2) + c \eta_1 \eta_2$, where the coefficients can be determined by the boundary conditions. □

Example 2.34 Give a necessary condition that must be satisfied by the curve with the shortest length which lies on the sphere $x_1^2(t) + x_2^2(t) + t^2 = R^2$, for $t \in [t_0, t_f]$, and joins the specified points (t_0, x_0) and (t_f, x_f).

Solution. The functional to be minimized is

$$J(x) = \int_{t_0}^{t_f} \sqrt{1 + \dot{x}_1^2(t) + \dot{x}_2^2(t)} dt. \tag{2.204}$$

So the augmented integrand function is

$$g_a(x(t), \dot{x}(t), \lambda(t), t) = \sqrt{1 + \dot{x}_1^2(t) + \dot{x}_2^2(t)} + \lambda(t) \left[x_1^2(t) + x_2^2(t) + t^2 - R^2 \right]. \tag{2.205}$$

Thus by applying the Euler equation (2.187), the necessary condition of extremal solution is given as below:

$$2x_1^*(t)\lambda^*(t) - \frac{d}{dt} \left(\frac{\dot{x}_1^*(t)}{\sqrt{1 + [\dot{x}_1^*(t)]^2 + [\dot{x}_2^*(t)]^2}} \right) = 0, \tag{2.206a}$$

$$2x_2^*(t)\lambda^*(t) - \frac{d}{dt} \left(\frac{\dot{x}_2^*(t)}{\sqrt{1 + [\dot{x}_1^*(t)]^2 + [\dot{x}_2^*(t)]^2}} \right) = 0, \tag{2.206b}$$

$$[x_1^*(t)]^2 + [x_2^*(t)]^2 + t^2 = R^2. \tag{2.206c}$$

□

2.5.2 Extrema of Functional with Isoperimetric Constraints

Queen Dido's land transaction that occurred during the seventeenth century has been considered as the original extremal problem with an isoperimetric constraint, such that she tried to determine a curve that possesses a given fixed length and encloses maximum area.

In Problem 2.8 below, we will formulate a class of problems to seek for the extrema of a functional with the isoperimetric constraint, say a fixed integration of the integrand on x, $\dot{x}$ and t.

Problem 2.8 Determine necessary conditions for a function x^* to be an extremum for a functional of the form

$$J(x) = \int_{t_0}^{t_f} g(x(t), \dot{x}(t), t)dt, \tag{2.207}$$

with $x(t) \equiv (x_1(t), \ldots, x_{n+m}(t))^\top$ which satisfies the following fixed integral constraints:

$$\int_{t_0}^{t_f} e_i(x(t), \dot{x}(t), t)dt = \xi_i, \tag{2.208}$$

with ξ_i, $i = 1, \ldots, n$, given as specified constants, and the boundary conditions $x(t_0) = x_0$ and $x(t_f) = x_f$. □

Theorem 2.10 gives the necessary condition for the extrema of Problem 2.8 Before that, define the following notions below.

Firstly, the collection of constraints (2.208) can be equivalently converted into the form of differential equation constraints by defining new variables $z_i(t)$, with $i = 1, \ldots, n$, such that

$$z_i(t) \triangleq \int_{t_0}^{t} e_i(x(t), \dot{x}(t), t)dt, \tag{2.209}$$

or in a more cleaner vector form of $\dot{z}(t) = e(x(t), \dot{x}(t), t)$, with the boundary conditions $z_i(t_0) = 0$ and $z_i(t_f) = \xi_i$, for all i. Thus, furthermore, as before, form the associated augmented function

$$g_a(x(t), \dot{x}(t), \lambda(t), \dot{z}(t), t) \triangleq g(x(t), \dot{x}(t), t) + \lambda^\top(t)[e(x(t), \dot{x}(t), t) - \dot{z}(t)]. \tag{2.210}$$

Theorem 2.10 *The Euler equation for the extremal solution to Problem 2.8 with constraints is given as follows:*

$$\frac{\partial g_a^*}{\partial x} - \frac{d}{dt}\left(\frac{\partial g_a^*}{\partial \dot{x}}\right) = 0, \tag{2.211a}$$

$$\frac{\partial g_a^*}{\partial z} - \frac{d}{dt}\left(\frac{\partial g_a^*}{\partial \dot{z}}\right) = 0, \tag{2.211b}$$

with $g_a^* \equiv g_a(x^*(t), \dot{x}^*(t), \lambda^*(t), \dot{z}_i^*(t), t)$, for all $t \in [t_0, t_f]$, with the boundary conditions $z^*(t_0) = 0$, $z^*(t_f) = \xi$, $x^*(t_0) = x_0$, and $x^*(t_f) = x_f$. $\square$

By Theorem 2.10, we can have the following:

- Since g_a does not contain $z(t)$, we have $\frac{\partial g_a^*}{\partial z} = 0$.
- By (2.211), $\frac{\partial g_a^*}{\partial \dot{z}} = -\lambda^*(t)$; thus, we obtain $\dot{\lambda}^*(t) = 0$ which implies that the Lagrange multipliers $\lambda^*(t)$ are constants.

In practical control problems, such isoperimetric constraints defined in (2.208) may appear in the form of total fuel or energy available to perform a mission.

Example 2.35 Give a necessary condition for x^* to be an extremum of the functional

$$J(x) = \frac{1}{2}\int_{t_0}^{t_f}\left[x_1^2(t) + 2\dot{x}_1(t)\dot{x}_2(t)\right]dt, \tag{2.212}$$

subject to the following constraint: $\int_{t_0}^{t_f} x_2^2(t)dt = \xi$, with ξ a specific constant, and the boundary conditions $x(t_0) = x_0$ and $x(t_f) = x_f$.
Solution. Let $\dot{z}(t) \triangleq x_2^2(t)$; then

$$g_a(x(t), \dot{x}(t), \lambda(t), \dot{z}_i(t)) \triangleq \frac{1}{2}x_1^2(t) + \dot{x}_1(t)\dot{x}_2(t) + \lambda(t)\left[x_2^2(t) - \dot{z}(t)\right].$$

Thus by (2.211), we can obtain

$$x_1^*(t) - \ddot{x}_2^*(t) = 0, \tag{2.213a}$$
$$2\ddot{x}_2^*(t)\lambda^*(t) - \ddot{x}_1^*(t) = 0, \tag{2.213b}$$
$$\dot{\lambda}^*(t) = 0. \tag{2.213c}$$

In addition, the solution to the differential equation

$$\dot{z}^*(t) = [x_2^*(t)]^2, \tag{2.214}$$

with the boundary conditions $z^*(t_0) = 0$ and $z^*(t_f) = \xi$. $\square$

In optimal control problems, the system trajectory is driven by the control input, say, the optimization problem is constrained by the controlled dynamics of the process.

2.6 Summary

This chapter develops the specific conditions to implement the extrema of functionals via the variational method in different boundary conditions, respectively. It further introduces the extrema problems of functionals with constraints which are solved by the elimination/direct method and the Lagrange method. The developed results in this part will be applied as a fundamental base to solve the optimal control problems introduced in the rest of the book.

2.7 Exercises

Exercise 2.1 Implement the increments and differentials of the following functions:

$$f(x) = 2x + \dot{x}^2,$$
$$f(x) = x^3 - x\dot{x}^2,$$
$$f(x_1, x_2) = x_1^2 + 3x_2^2 + \dot{x}_2^2.$$

Exercise 2.2 Implement the increments, variations, and second variations of the following functionals:

$$J(x) = \int_{t_0}^{t_f} \left[x(t) + 2\dot{x}^2(t) \right] dt,$$

$$J(x) = \int_{t_0}^{t_f} \left[x^3(t) - t\dot{x}^2(t) \right] dt,$$

$$J(x) = \int_{t_0}^{t_f} \left[tx_1(t) + x_2^2(t) + 3\dot{x}_2^3(t) \right] dt.$$

Exercise 2.3 Consider a functional as follows:

$$J(x) = \int_0^1 \left[x^2(t) + 2tx(t) + t^2\dot{x}^2(t) \right] dt.$$

Give the expression of the variation of J, and specify the value of δJ in case $x(t) = t$ and $\delta x = 0.02t$.

Exercise 2.4 Give necessary and sufficient conditions for the extrema of the functions specified in Exercise 2.1, respectively.

Exercise 2.5 Give necessary and sufficient conditions for the extrema of the functional

$$J(x) = \int_{t_0}^{t_f} \left[2x^2(t) + \dot{x}^2(t) \right] dt,$$

with the boundary conditions as $x(t_0) = 0$, with $t_0 = 0$, and $x(t_f) = 2$, with $t_f = 3$.

Exercise 2.6 Give necessary conditions for the extrema of the functional

$$J(x) = \int_0^2 \left[2tx(t) + \dot{x}^2(t) \right] dt,$$

with the boundary conditions as $x(0) = 2$ and $x(2) = 8$.

Exercise 2.7 Give necessary conditions for the extrema of the functional given in Exercise 2.6 with the boundary conditions as $x(0) = 0$ and $x(t_f)$ is free, with $t_f = 3$.

Exercise 2.8 Give necessary conditions for the extrema of the functional given in Exercise 2.6 with the boundary conditions as $x(0) = 2$, and the final state $x(t_f)$ lies on a curve $\phi = 2t + 3$.

Exercise 2.9 Give necessary conditions for the extrema of the functional

$$J(x) = \int_0^5 \left[\dot{x}_1^2(t) + \dot{x}_1(t)\dot{x}_2(t) - tx_2^2(t) \right] dt,$$

with the boundary conditions as $x(0) = \begin{bmatrix} 0 \\ 0 \end{bmatrix}$ and $x(5) = \begin{bmatrix} -1 \\ 1 \end{bmatrix}$.

Exercise 2.10 Specify the shortest curve from $(0, 0)$ to a point on the curve of $\phi(t) = \dfrac{1}{t}$.

Exercise 2.11 Specify the shortest curve connecting the point $x(0) = \begin{bmatrix} 0 \\ 1 \end{bmatrix}$ to the line $t = 6$.

Exercise 2.12 Specify the extrema for the functional

$$J(x) = \int_0^2 \left[2tx(t) + \dot{x}^2(t) \right] dt,$$

with the boundary conditions as $x(0) = 2$ and $x(2) = 8$.

Exercise 2.13 Specify the extremal solution to the electric vehicle charging problems studied in Example 2.25 with a revised functional given as

$$J(x) = \int_{t_0}^{t_f} \left[\sum_{i=1}^n \dot{x}_i^2(t) + \left[\sum_{i=1}^n \dot{x}_i(t) + D(t) \right]^2 \right] dt.$$

Notice that $\int_{t_0}^{t_f} \sum_{i=1}^{n} \dot{x}_i^2(t)dt$ represents the degradation cost with respect to the charging behaviors $\dot{x}_i(\cdot)$ over the period of $[t_0, t_f]$.

Exercise 2.14 Find an extremum curve for the functional given in Example 2.28 which satisfies a revised boundary condition such that $x(0) = \begin{bmatrix} 0 \\ 1 \end{bmatrix}$, $x_1(\frac{3\pi}{4})$ is free, and $x_2(\frac{3\pi}{4}) = -1$.

Exercise 2.15 Specify the extrema solution x^* to the system with the performance function

$$f(x) = x_1^2 + 2x_2^2 + x_3^2$$

satisfying the constraints of $x_1 + x_2 = 5$ and $2x_2 + x_3 = 10$ by applying the elimination/direct method.

Exercise 2.16 Specify the extrema solution x^* to the system with the performance function given in Exercise 2.15 by applying the Lagrange multiplier method.

Exercise 2.17 Specify the extrema solution x^*, by applying the elimination/direct method, such that the performance cost function, as given below, is minimized as

$$J(x) = \int_0^1 x_2^2(t)dt,$$

with boundary conditions $x_1(0) = 1$ and $x_1(1) = 8$ and satisfying the constraint of $\dot{x}_1(t) = x_1(t) + x_2(t)$.

Exercise 2.18 Specify the extrema solution x^* for the system defined in Exercise 2.17 again by applying the Lagrange multiplier method.

Exercise 2.19 Specify the extrema solution x^* to minimize the performance function

$$J(x) = \int_0^2 \ddot{x}^2(t)(t),$$

with boundary conditions $x(0) = 0$, $\dot{x}(0) = 1$, $x(2) = 1$, and $\dot{x}(2) = 2$.

Hint: Considering $x_1(t) = x(t)$ and by introducing a new variable $x_2(t)$, such that $x_2(t) = \dot{x}(t)$, the extremal solution can be solved by applying the elimination/direct or the Lagrange multiplier method.

Exercise 2.20 Give necessary conditions for x^* to be an extremum of the functional

$$J(x) = \int_0^1 \left[1 + \dot{x}_1^2(t) + \dot{x}_2^2(t)\right] dt,$$

subject to the following constraint: $\int_0^1 x_1(t)dt = 10$, and the boundary conditions $x(0) = \begin{bmatrix} 1 \\ -1 \end{bmatrix}$ and $x(1) = \begin{bmatrix} 5 \\ 8 \end{bmatrix}$.

Chapter 3
Optimal Control via Variational Method

In this chapter, we will study the optimal control problems by applying the developed results of the extremal of functional via the variational method. More specially, in Sect. 3.1, gives the necessary and sufficient conditions for the optimal solution to the optimal control problems with unbounded controls. Based upon the analysis developed in last section, the optimal solution to the optimal control problems with different boundary conditions on the final time and final state, respectively, is studied in Sect. 3.2. A specific class of linear-quadratic regulation (LQR) problems is analyzed in Sect. 3.3. Parallel to Sect. 3.3, in Sect. 3.4, a class of linear-quadratic tracking (LQT) problems which is a generalized version of the LQR problems is further studied. Section 3.5, gives a brief summary of the results developed in this chapter. Lastly, some problems for readers using this book and exercises on the stuffs covered in this chapter in Sect. 3.6 are listed.

3.1 Necessary and Sufficient Condition for Optimal Control

Problem 3.1 specifies a class of optimal control problems, which will be studied via the method of calculus of variations developed in Chap. 2.

Problem 3.1 Consider a system such that the dynamics of the state process $x(\cdot)$ is driven by control strategy u, such that

$$\dot{x}(t) = f(x(t), u(t), t). \tag{3.1}$$

The objective is to determine a control subject to which a performance cost function $J(u)$ as given below

$$J(u) \triangleq h(x(t_f), t_f) + \int_{t_0}^{t_f} g(x(t), u(t), t)dt, \tag{3.2}$$

© The Author(s), under exclusive license to Springer Nature Singapore Pte Ltd. 2021
Z. Ma and S. Zou, *Optimal Control Theory*,
https://doi.org/10.1007/978-981-33-6292-5_3

is minimized, where h represents the performance cost with respect to the final time t_f and the final state at that time $x(t_f)$.   ☐

Denote by x^* the state process with respect to the optimal control u^* subject to which the performance cost function J is minimized.

It shall be initially assumed that the admissible state and control regions are not bounded, and that the initial condition $x(t_0) = x_0$ and the initial time t_0 are specified, respectively. As usual, x is the $n \times 1$ state vector and u is the $m \times 1$ vector of control inputs.

Assume that the performance cost h is differentiable, and hence can be written as follows:

$$h(x(t_f), t_f) = \int_{t_0}^{t_f} \left[\frac{dh(x(t), t)}{dt} \right] dt + h(x(t_0), t_0). \tag{3.3}$$

So that the performance cost function can be expressed as

$$J(u) = \int_{t_0}^{t_f} \left[g(x(t), u(t), t) + \frac{dh(x(t), t)}{dt} \right] dt + h(x(t_0), t_0). \tag{3.4}$$

Since $x(t_0)$ and t_0 are fixed, the minimization does not affect the $h(x(t_0), t_0)$ term, so it needs to consider only the functional

$$J(u) = \int_{t_0}^{t_f} \left[g(x(t), u(t), t) + \frac{dh(x(t), t)}{dt} \right] dt. \tag{3.5}$$

Using the chain rule of differentiation, it becomes

$$J(u) = \int_{t_0}^{t_f} \left\{ g(x(t), u(t), t) + \left[\frac{\partial h(x(t), t)}{\partial x} \right]^\top \dot{x}(t) + \frac{\partial h(x(t), t)}{\partial t} \right\} dt. \tag{3.6}$$

To include the differential equation constraints, the augmented functional is formed

$$J_a(u) = \int_{t_0}^{t_f} \left\{ g(x(t), u(t), t) + \left[\frac{\partial h(x(t), t)}{\partial x} \right]^\top \dot{x}(t) \right.$$
$$\left. + \frac{\partial h(x(t), t)}{\partial t} + [\lambda(t)]^\top [f(x(t), u(t), t) - \dot{x}(t)] \right\} dt, \tag{3.7}$$

by introducing the Lagrange multipliers $\lambda_1(t), \ldots, \lambda_n(t)$.

It further defines

$$g_a(x(t), \dot{x}(t), u(t), \lambda(t), t) \triangleq g(x(t), u(t), t) + [\lambda(t)]^\top [f(x(t), u(t), t) - \dot{x}(t)]$$
$$+ \left[\frac{\partial h(x(t), t)}{\partial x}\right]^\top \dot{x}(t) + \frac{\partial h(x(t), t)}{\partial t}, \qquad (3.8)$$

so that

$$J_a(u) = \int_{t_0}^{t_f} g_a(x(t), \dot{x}(t), u(t), \lambda(t), t)dt. \qquad (3.9)$$

Theorem 3.1 *Suppose that u^* is an optimal control for Problem 3.1, then u^* satisfies the following conditions:*

$$\dot{x}^*(t) = \frac{\partial \mathcal{H}(x^*(t), u^*(t), \lambda^*(t), t)}{\partial \lambda}, \qquad (3.10a)$$

$$\dot{\lambda}^*(t) = -\frac{\partial \mathcal{H}(x^*(t), u^*(t), \lambda^*(t), t)}{\partial x}, \qquad (3.10b)$$

$$0 = \frac{\partial \mathcal{H}(x^*(t), u^*(t), \lambda^*(t), t)}{\partial u}, \qquad (3.10c)$$

for all $t \in [t_0, t_f]$, and with the following boundary condition:

$$\left[\frac{\partial h(x^*(t_f), t_f)}{\partial x} - \lambda^*(t_f)\right]^\top \delta x_f$$
$$+ \left[\mathcal{H}(x^*(t_f), u^*(t_f), \lambda^*(t), t_f) + \frac{\partial h(x^*(t_f), t_f)}{\partial t}\right]\delta t_f = 0, \qquad (3.11)$$

where $\mathcal{H}(x(t), u(t), \lambda(t), t)$, which is called the Hamiltonian, *is defined as*

$$\mathcal{H}(x(t), u(t), \lambda(t), t) \triangleq g(x(t), u(t), t) + [\lambda(t)]^\top f(x(t), u(t), t). \qquad (3.12)$$

Proof Following the same procedure applied in Chap. 2, the variation of the performance cost function J_a subject to the control u^* can be specified as the follows:

$$\delta J_a(u^*)$$
$$= \left[\frac{\partial g_a^*(t_f)}{\partial \dot{x}}\right]^\top \delta x_f + \left[g_a^*(t_f) - \left[\frac{\partial g_a^*(t_f)}{\partial \dot{x}}\right]^\top \dot{x}^*(t_f)\right]\delta t_f \qquad \equiv (I)$$
$$+ \int_{t_0}^{t_f} \left\{\left[\left[\frac{\partial g_a^*(t)}{\partial x}\right]^\top - \frac{d}{dt}\left[\frac{\partial g_a^*(t)}{\partial \dot{x}}\right]^\top\right]\delta x(t)\right.$$
$$\left. + \left[\frac{\partial g_a^*(t)}{\partial u}\right]^\top \delta u(t) + \left[\frac{\partial g_a^*(t)}{\partial \lambda}\right]^\top \delta\lambda(t)\right\} dt \qquad \equiv (II) \qquad (3.13)$$

where, for notational simplicity, it may be considered

$$g_a^*(t) \equiv g_a(x^*(t), \dot{x}^*(t), u^*(t), \lambda^*(t), t), \text{ for all } t. \tag{3.14}$$

We can obtain (II), say the integral part of $\delta J_a(u^*)$

$$(II) = \int_{t_0}^{t_f} \left\{ \left[\left[\frac{\partial g^*(t)}{\partial x} \right]^\mathsf{T} + [\lambda^*(t)]^\mathsf{T} \frac{\partial f^*(t)}{\partial x} + \frac{\partial}{\partial x} \left[\frac{\partial h^*(t)}{\partial x} \right]^\mathsf{T} \dot{x}^*(t) \right. \right.$$
$$+ \frac{\partial}{\partial x} \left[\frac{\partial h^*(t)}{\partial t} \right]^\mathsf{T} - \frac{d}{dt} \left(-[\lambda^*(t)]^\mathsf{T} + \left[\frac{\partial h^*(t)}{\partial x} \right]^\mathsf{T} \right) \right] \delta x(t)$$
$$+ \left[\frac{\partial g^*(t)}{\partial u} + [\lambda^*(t)]^\mathsf{T} \frac{\partial f^*(t)}{\partial u} \right]^\mathsf{T} \delta u(t) + \left[f^*(t) - \dot{x}^*(t) \right]^\mathsf{T} \delta \lambda(t) \right\} dt \tag{3.15}$$

where for notational simplicity, consider

$$g^*(t) \equiv g(x^*(t), u^*(t), t), \ f^*(t) \equiv f(x^*(t), u^*(t), t), \ h^*(t) \equiv h(x^*(t), t).$$

It gives the following by dealing with the parts in (II) involved with $h^*(t)$:

$$\frac{\partial}{\partial x} \left[\left[\frac{\partial h^*(t)}{\partial x} \right]^\mathsf{T} \dot{x}^*(t) + \frac{\partial h^*(t)}{\partial t} \right] - \frac{d}{dt} \left[\frac{\partial h^*(t)}{\partial x} \right]^\mathsf{T}$$
$$= \frac{\partial^2 h^*(t)}{\partial x^2} \dot{x}^*(t) + \frac{\partial^2 h^*(t)}{\partial t \partial x} - \frac{\partial^2 h^*(t)}{\partial x^2} \dot{x}^*(t) - \frac{\partial^2 h^*(t)}{\partial x \partial t} = 0, \tag{3.16}$$

where the first equality holds by working out the indicated partial derivatives and by applying the chain rule for $\frac{d}{dt} \left(\frac{\partial h^*(t)}{\partial x} \right)$.

Thus, it has

$$(II) = \int_{t_0}^{t_f} \left\{ \left[\left[\frac{\partial g^*(t)}{\partial x} \right]^\mathsf{T} + [\lambda^*(t)]^\mathsf{T} \frac{\partial f^*(t)}{\partial x} + \frac{d[\lambda^*(t)]^\mathsf{T}}{dt} \right] \delta x(t) \right.$$
$$+ \left[\frac{\partial g^*(t)}{\partial u} + [\lambda^*(t)]^\mathsf{T} \frac{\partial f^*(t)}{\partial u} \right]^\mathsf{T} \delta u(t) + \left[f^*(t) - \dot{x}^*(t) \right]^\mathsf{T} \delta \lambda(t) \right\} dt. \tag{3.17}$$

First observe that the constraints

$$\dot{x}^*(t) = f(x^*(t), u^*(t), t) \tag{3.18}$$

must be satisfied by an extremum so that the coefficient of $\delta \lambda(t)$ is zero.

The Lagrange multipliers are arbitrary, and by the fundamental result of Theorem 2.1, $\delta J(u^*) = 0$, so it is necessary to make the coefficient of $\delta x(t)$ equal to zero, that is,

$$\dot{\lambda}^*(t) = -\left[\frac{\partial f(x^*(t), u^*(t), t)}{\partial x}\right]^\top \lambda^*(t) - \frac{\partial g(x^*(t), u^*(t), t)}{\partial x}, \tag{3.19}$$

as compared with the state equation, it may be called *the costate equation* and $\lambda(t)$ *the costate*.

The remaining variation $\delta u(t)$ is independent, so its coefficient must be zero; then

$$\frac{\partial g(x^*(t), u^*(t), t)}{\partial u} + \left[\lambda^*(t)\right]^\top \frac{\partial f(x^*(t), u^*(t), t)}{\partial u} = 0. \tag{3.20}$$

Thus, it is straightforward to get the conclusion of (3.10) by (3.18), (3.19) and (3.20).

Moreover, as the variation of J is equal to zero, it can be obtained the boundary condition (3.11) as well.   □

In the following, we will study a specific example to demonstrate the developed result in Theorem 3.1.

Example 3.1 Consider a two-dimensional system

$$\dot{x}_1(t) = x_2(t), \tag{3.21a}$$
$$\dot{x}_2(t) = -u(t), \tag{3.21b}$$

with $t_0 = 0, t_f = 2$, the boundary condition as $x(t_0) = \begin{bmatrix} 2 \\ 2 \end{bmatrix}$, and a given performance cost function

$$J(u) = [x_1(t_f) - 1]^2 + [x_2(t_f) + 1]^2 + \int_{t_0}^{t_f} u^2(t)dt. \tag{3.22}$$

Give a necessary condition for the optimal control u^* and the associated state processes x^*.

Solution. For the given optimal problem, the Hamiltonian $\mathcal{H}(\cdot)$ is given as

$$\begin{aligned}
&\mathcal{H}(x_1(t), x_2(t), u(t), \lambda_1(t), \lambda_2(t))\\
&\triangleq g(u(t)) + [\lambda(t)]^\top f(x(t), u(t))\\
&= u^2(t) + \lambda_1(t)x_2(t) - \lambda_2(t)u(t);
\end{aligned} \tag{3.23}$$

then by applying Theorem 3.1, we can obtain

$$\frac{\partial \mathcal{H}(x^*(t), u^*(t), \lambda^*(t))}{\partial u} = 2u^*(t) - \lambda_2^*(t) = 0, \tag{3.24}$$

by which $u^*(t) = \frac{1}{2}\lambda_2^*(t)$; then

$$\mathcal{H}^* \equiv \mathcal{H}(x_1^*(t), x_2^*(t), u^*(t), \lambda_1^*(t), \lambda_2^*(t))$$

$$= \frac{1}{4}[\lambda_2^*(t)]^2 + \lambda_1^*(t)x_2^*(t) - \frac{1}{2}[\lambda_2^*(t)]^2$$

$$= \lambda_1^*(t)x_2^*(t) - \frac{1}{4}[\lambda_2^*(t)]^2. \tag{3.25}$$

Hence by (3.10),

$$\dot{\lambda}_1^*(t) = -\frac{\partial \mathcal{H}^*}{\partial x_1} = 0, \tag{3.26a}$$

$$\dot{\lambda}_2^*(t) = -\frac{\partial \mathcal{H}^*}{\partial x_2} = -\lambda_1^*(t), \tag{3.26b}$$

$$\dot{x}_1^*(t) = \frac{\partial \mathcal{H}^*}{\partial \lambda_1} = x_2^*(t), \tag{3.26c}$$

$$\dot{x}_2^*(t) = \frac{\partial \mathcal{H}^*}{\partial \lambda_2} = -\frac{1}{2}\lambda_2^*(t). \tag{3.26d}$$

The solution to the above collection of equations is

$$\lambda_1^*(t) = \xi_1, \tag{3.27a}$$
$$\lambda_2^*(t) = -\xi_1 t + \xi_2, \tag{3.27b}$$
$$x_1^*(t) = \frac{1}{12}\xi_1 t^3 - \frac{1}{4}\xi_2 t^2 + \xi_3 t + \xi_4, \tag{3.27c}$$
$$x_2^*(t) = \frac{1}{4}\xi_1 t^2 - \frac{1}{2}\xi_2 t + \xi_3. \tag{3.27d}$$

As a consequence, the optimal control u^* is

$$u^*(t) = \frac{1}{2}\lambda_2^*(t) = -\frac{1}{2}\xi_1 t + \frac{1}{2}\xi_2. \tag{3.28}$$

By (3.11), the boundary condition is

$$\left[\frac{\partial h(x^*(t_f), t_f)}{\partial x} - \lambda^*(t_f)\right]^T \delta x_f$$

$$+ \left[\mathcal{H}(x^*(t_f), u^*(t_f), \lambda^*(t_f), t_f) + \frac{\partial h(x^*(t_f), t_f)}{\partial t}\right] \delta t_f$$

$$= 2\begin{bmatrix} x_1^*(t_f) - 1 \\ x_2^*(t_f) + 1 \end{bmatrix}^T \begin{bmatrix} \delta x_{f1} \\ \delta x_{f2} \end{bmatrix} + \left[\lambda_1^*(t)x_2^*(t) - \frac{1}{4}[\lambda_2^*(t)]^2\right]\delta t_f = 0. \tag{3.29}$$

$\square$

In the next, we will give a sufficient condition for the optimal control of Problem 3.1, and establish the optimum property (maximum or minimum) of the solution in Theorem 3.2. Before that, firstly define a matrix $\mathbb{N}(x, u, \lambda)$ as follows:

$$\mathbb{N}(x, u, \lambda) \equiv \begin{bmatrix} \frac{\partial^2 \mathscr{H}}{\partial x^2} & \frac{\partial^2 \mathscr{H}}{\partial x \partial u} \\ \frac{\partial^2 \mathscr{H}}{\partial x \partial u} & \frac{\partial^2 \mathscr{H}}{\partial u^2} \end{bmatrix}, \text{ with } \mathscr{H} \equiv \mathscr{H}(x, u, \lambda). \tag{3.30}$$

Theorem 3.2 *Suppose that u^* is a control given in Theorem 3.1, and x^* and λ^* are associated state and costate trajectories; then, if $\mathbb{N}(x^*, u^*, \lambda^*)$ given in (3.30) is definitively positive (or definitively negative), u^* is an optimal control subject to which the performance cost function $J(u)$ is minimized.*

Proof By the previous analysis, it can be obtained

$$\delta^2 J = \frac{1}{2} \int_{t_0}^{t_f} \left[\frac{\partial^2 \mathscr{H}^*}{\partial x^2} [\delta x(t)]^2 + 2 \frac{\partial^2 \mathscr{H}^*}{\partial x \partial u} \delta x(t) \delta u(t) + \frac{\partial^2 \mathscr{H}^*}{\partial u^2} [\delta u(t)]^2 \right] dt$$

$$= \frac{1}{2} \int_{t_0}^{t_f} \begin{bmatrix} \delta x(t) \\ \delta u(t) \end{bmatrix}^\top \mathbb{N}(x^*, u^*, \lambda^*) \begin{bmatrix} \delta x(t) \\ \delta u(t) \end{bmatrix} dt. \tag{3.31}$$

Thus by Theorem 2.1, the conclusion can be obtained. □

Notice that, in most cases, it is sufficient to determine the maximum (or minimum) property of u^* in case $\frac{\partial^2 \mathscr{H}}{\partial u^2}$ is definitely positive (or negative).

It will be demonstrated with a simple example below to determine the sufficient condition developed in Theorems 3.2 for optimal control problems.

Example 3.2 (*Necessary and Sufficient Condition for Optimal Control*) Consider a one-dimensional system

$$\dot{x}(t) = -ax(t) + u(t), \tag{3.32}$$

with $t_0 = 0$, $t_f = 1$, the boundary condition as $x(t_0) = 0$, $x(t_f) = 2$, and a given performance cost function

$$J(u) = \frac{1}{2} \int_{t_0}^{t_f} u^2(t) dt. \tag{3.33}$$

Give a necessary condition for the optimal control u^* and further verify that u^* is an optimal control such that the system is minimized.

Solution. For the given optimal problem, the Hamiltonian $\mathscr{H}(\cdot)$ is given as

$$\mathscr{H} \equiv \mathscr{H}(x(t), u(t), \lambda(t)) = \frac{1}{2} u^2(t) + \lambda(t)[-ax(t) + u(t)]; \tag{3.34}$$

then by applying Theorem 3.1, it can be obtained that

$$\frac{\partial \mathcal{H}}{\partial u} = u^*(t) + \lambda^*(t) = 0, \tag{3.35}$$

by which $u^*(t) = -\lambda^*(t)$. Thus

$$\begin{aligned}
\mathcal{H}^* &\equiv \mathcal{H}(x^*(t), u^*(t), \lambda^*(t)) \\
&= \frac{1}{2}[\lambda^*(t)]^2 + \lambda^*(t)[-ax^*(t) - \lambda^*(t)] \\
&= -\frac{1}{2}[\lambda^*(t)]^2 - a\lambda^*(t)x^*(t).
\end{aligned} \tag{3.36}$$

Hence by (3.10),

$$\dot{\lambda}^*(t) = -\frac{\partial \mathcal{H}(x^*(t), u^*(t), \lambda^*(t))}{\partial x} = a\lambda^*(t), \tag{3.37a}$$

$$\dot{x}^*(t) = \frac{\partial \mathcal{H}(x^*(t), u^*(t), \lambda^*(t))}{\partial \lambda} = -\lambda^*(t) - ax^*(t). \tag{3.37b}$$

The solution to the above collection of equations is

$$\lambda^*(t) = \exp(at)\lambda_0, \tag{3.38a}$$

$$x^*(t) = -\frac{\lambda_0}{2a}\exp(at) + \left[\frac{\lambda_0}{2a} + x_0\right]\exp(-at). \tag{3.38b}$$

By considering the boundary condition of $x^*(0) = 0$ and $x^*(1) = 2$, we can have
$$\lambda_0 = \frac{4a\exp(a)}{1 - \exp(2a)}.$$
Thus it has u^* is

$$u^*(t) = -\lambda^*(t) = \frac{4a\exp(2at)}{\exp(2a) - 1}. \tag{3.39}$$

And by Theorem 3.2, it is straightforward to claim that the control $u^*(t)$ is an optimal control.   □

3.2 Optimal Control Problems with Different Boundary Conditions

Based upon the result developed in Theorem 3.1, it will give the specific boundary conditions for different cases, respectively.

3.2.1 Optimal Control with Fixed Final Time and Fixed Final State

Consider that both t_f and $x(t_f)$ are specified, then $\delta t_f = 0$ and $\delta x_f = 0$. Thus by (3.11), the boundary condition always holds. The boundary condition is simply

$$x^*(t_f) = x_f. \tag{3.40}$$

Example 3.3 Consider the system given in Example 3.1 with the boundary condition as

$$x(t_0) = \begin{bmatrix} 2 \\ 2 \end{bmatrix}, \text{ and } x(t_f) = \begin{bmatrix} 1 \\ -1 \end{bmatrix}, \text{ with } t_0 = 0, t_f = 2, \tag{3.41}$$

and a revised performance cost function

$$J(u) = \int_{t_0}^{t_f} u^2(t)dt. \tag{3.42}$$

Implement the optimal control and the corresponding state trajectory.

Solution. By Theorem 3.1, the necessary condition is the same as that for Example 3.1 except that the boundary condition is replaced by

$$x(0) = \begin{bmatrix} 2 \\ 2 \end{bmatrix}, \text{ and } x(2) = \begin{bmatrix} 1 \\ -1 \end{bmatrix}; \tag{3.43}$$

then it is straightforward to get that the optimal solution, including states and costates, is

$$x_1^*(t) = \frac{1}{2}t^3 - \frac{9}{4}t^2 + 2t + 2, \tag{3.44a}$$

$$x_2^*(t) = \frac{3}{2}t^2 - \frac{9}{2}t + 2, \tag{3.44b}$$

$$\lambda_1^*(t) = 6, \tag{3.44c}$$

$$\lambda_2^*(t) = -6t + 9, \tag{3.44d}$$

and as a consequence, the optimal control u^* is

$$u^*(t) = \frac{1}{2}\lambda_2^*(t) = -3t + \frac{9}{2}. \tag{3.45}$$

See Fig. 3.1 for an illustration of optimal control and its associated state and costate trajectories, respectively.

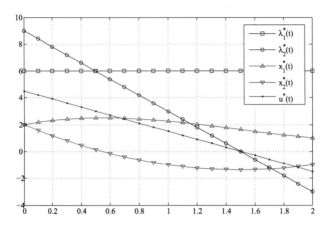

Fig. 3.1 The optimal control u^* and corresponding state x^* and costate λ^* trajectories for optimal control problems with fixed final time and fixed final state

The performance cost subject to the optimal control u^* is

$$J(u^*) = \int_{t_0}^{t_f} [u^*(t)]^2 dt = \int_0^2 \left[-3t + \frac{9}{2} \right]^2 dt = \frac{21}{2}. \tag{3.46}$$

□

3.2.2 Optimal Control with Fixed Final Time and Free Final State

In this section, it still supposes that t_f is fixed, say $\delta t_f = 0$. Thus for notational simplicity, here it considers $h(x(t_f)) \equiv h(x(t_f), t_f)$ since t_f is fixed.

3.2.2.1 Final State Arbitrarily Valued

Here it further considers that $x(t_f)$ is arbitrary, then δx_f could be any valued.
   Thus by (3.11), the boundary condition becomes

$$\frac{\partial h(x^*(t_f))}{\partial x} - \lambda^*(t_f) = 0 \tag{3.47}$$

must be satisfied.

Example 3.4 Consider the system and the performance cost function given in Example 3.1 with the t_0, t_f, and $x(t_0)$ specified as in Example 3.3, but the final state $x(t_f)$ being free.

Implement the optimal control and the corresponding state trajectory.

Solution. By Theorem 3.1, the necessary condition is the same as that for Example 3.1 except that the boundary condition is replaced by $x(0) = \begin{bmatrix} 2 \\ 2 \end{bmatrix}$ together with

$$
\begin{aligned}
& \left[\frac{\partial h(x^*(t_f))}{\partial x} - \lambda^*(t_f) \right]^\top \delta x_f \\
& + \left[\mathcal{H}(x^*(t_f), u^*(t_f), \lambda^*(t_f), t_f) + \frac{\partial h(x^*(t_f))}{\partial t} \right] \delta t_f \\
& = 2 \begin{bmatrix} x_1^*(t_f) - 1 \\ x_2^*(t_f) + 1 \end{bmatrix}^\top \begin{bmatrix} \delta x_{f1} \\ \delta x_{f2} \end{bmatrix}, \quad \text{since } \delta t_f = 0 \\
& = 2[x_1^*(t_f) - 1]\delta x_{f1} + 2[x_2^*(t_f) + 1]\delta x_{f2} = 0.
\end{aligned}
\tag{3.48}
$$

Hence

$$
x_1^*(t_f) = 1,
$$
$$
x_2^*(t_f) = -1.
$$

Consider that the final time $t_f = 2$, the optimal state should be given as

$$
x_1^*(2) = 1, \quad \text{and } x_2^*(2) = -1.
\tag{3.49}
$$

Thus the optimal solution is the same as that of Example 3.3.
Also by considering the final time $t_f = 4$, it has

$$
x_1^*(4) = 1, \quad \text{and } x_2^*(4) = -1.
$$

Thus the extrema strategy with $t_f = 4$ is given as

$$
x_1^*(t) = \frac{3}{32}t^3 - \frac{15}{16}t^2 + 2t + 2,
$$
$$
x_2^*(t) = \frac{9}{32}t^2 - \frac{15}{8}t + 2,
$$

which satisfies the boundary conditions of $x^*(0) = \begin{bmatrix} 2 \\ 2 \end{bmatrix}$ and $x^*(4) = \begin{bmatrix} 1 \\ -1 \end{bmatrix}$. □

3.2.2.2   Final State Lying on a Surface at Fixed Final Time t_f

Next, suppose that final state $x(t_f)$ satisfying the surface is given by $\phi(x(t_f)) = 0$, see an illustration for a two-dimensional state system in Fig. 3.2.

Theorem 3.3 *Suppose that u^* is an optimal control for Problem 3.1 with fixed final time t_f and the final state $x(t_f)$ at t_f satisfying the following situation:*

$$\phi(x(t_f)) \equiv \begin{bmatrix} \phi_1(x(t_f)) \\ \phi_2(x(t_f)) \\ \cdots \\ \phi_k(x(t_f)) \end{bmatrix} = 0, \tag{3.50}$$

for some k, with $1 \le k \le n - 1$, then the boundary condition is given as follows:

$$\frac{\partial h(x^*(t_f))}{\partial x} - \lambda^*(t_f) = d_1 \frac{\partial \phi_1(x^*(t_f))}{\partial x} + \cdots + d_k \frac{\partial \phi_k(x^*(t_f))}{\partial x}. \tag{3.51}$$

Proof Each $\phi_i(x(t_f))$ represents a hypersurface in n-dimensional state space. Thus, the final state $x(t_f)$ lies on the intersection of these k hypersurfaces, and then the variation of x^* at final time t_f, denoted by $\delta x(t_f)$, should be tangent to each of the hypersurfaces at the point $(x^*(t_f), t_f)$, see an illustration in Fig. 3.2 for a two-dimensional system.

Consequently, $\delta x(t_f)$ is orthogonal to each of the following (gradient) vectors:

$$\frac{\partial \phi_1(x^*(t_f))}{\partial x}, \ldots, \frac{\partial \phi_k(x^*(t_f))}{\partial x}, \tag{3.52}$$

Fig. 3.2 The optimal state trajectory x^* and an variation $x^* + \delta x$

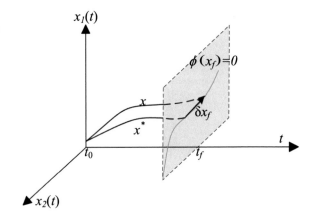

Fig. 3.3 A display of state trajectory terminating on a linear line at t_f

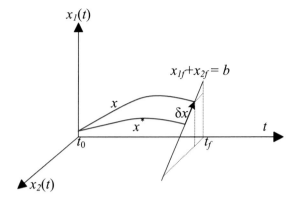

which implies that $\delta x(t_f)$ is orthogonal to the space spanned by $\frac{\partial \phi_i(x^*(t_f))}{\partial x}$, with $i = 1, \ldots, k$, that is to say $\delta x(t_f)$ is orthogonal to the linear combination of $\frac{\partial \phi_1(x^*(t_f))}{\partial x}$, $\ldots$, $\frac{\partial \phi_k(x^*(t_f))}{\partial x}$, i.e.,

$$\left[d_1 \frac{\partial \phi_1(x^*(t_f))}{\partial x} + \cdots + d_k \frac{\partial \phi_k(x^*(t_f))}{\partial x} \right]^\top \delta x(t_f) = 0, \qquad (3.53)$$

for any real valued d_i.

By (3.11) in Theorem 3.1, and due to $\delta t_f = 0$, the boundary condition becomes the following:

$$\left[\frac{\partial h(x^*(t_f))}{\partial x} - \lambda^*(t_f) \right]^\top \delta x(t_f) = 0. \qquad (3.54)$$

Hence by the above two equations, we can obtain the boundary condition at final time t_f that is specified as (3.51). $\qquad \square$

In an example below, we will study the situation with $\phi(\cdot)$ as a linear function like $x_{1f} + x_{2f} = b$ as illustrated in Fig. 3.3.

Example 3.5 Implement the optimal control solution to the problem defined in Example 3.1, except that the final state at t_f is required to satisfy the following:

$$\phi(x(t_f)) \equiv 2x_1(t_f) + 3x_2(t_f) - 3 = 0. \qquad (3.55)$$

Solution. By Theorem 3.3, we can obtain at the final time $t_f = 2$, the following holds:

$$\frac{\partial h(x^*(t_f))}{\partial x_1} - \lambda_1^*(t_f) = 2[x_1^*(t_f) - 1] - \lambda_1^*(t_f) = d_1 \frac{\partial \phi(x_1^*(t_f))}{\partial x_1} = 2d_1 \quad (3.56a)$$

$$\frac{\partial h(x^*(t_f))}{\partial x_2} - \lambda_2^*(t_f) = 2[x_2^*(t_f) + 1] - \lambda_2^*(t_f) = d_1 \frac{\partial \phi(x_2^*(t_f))}{\partial x_2} = 3d_1 \quad (3.56b)$$

$$2x_1^*(t_f) + 3x_2^*(t_f) - 3 = 0 \quad (3.56c)$$

with $t_f = 2$, then by the optimal state & costate trajectories given in Example 3.1, we obtain

$$\frac{1}{3}\xi_1 - 2\xi_2 + 10 = 2d_1, \quad (3.57a)$$

$$4\xi_1 - 3\xi_2 + 6 = 3d_1, \quad (3.57b)$$

$$\frac{13}{3}\xi_1 - 5\xi_2 + 15 = 0 \quad (3.57c)$$

with its solution as $\xi_1 = \frac{18}{7}$, $\xi_2 = \frac{183}{35}$ and $d_1 = \frac{1}{5}$; thus

$$\lambda_1^*(t) = \frac{18}{7}, \quad (3.58a)$$

$$\lambda_2^*(t) = -\frac{18}{7}t + \frac{183}{35}, \quad (3.58b)$$

$$x_1^*(t) = \frac{3}{14}t^3 - \frac{183}{140}t^2 + 2t + 2, \quad (3.58c)$$

$$x_2^*(t) = \frac{9}{14}t^2 - \frac{183}{70}t + 2, \quad (3.58d)$$

and thus the optimal control u^* is

$$u^*(t) = \frac{1}{2}\lambda_2^*(t) = -\frac{9}{7}t + \frac{183}{70}. \quad (3.59)$$

See Fig. 3.4 for an illustration of optimal control and its associated state and costate trajectories, respectively, for Example 3.5.

The performance cost subject to the optimal control u^* is

$$J(u^*) = [x_1^*(t_f) - 1]^2 + [x_2^*(t_f) + 1]^2 + \int_{t_0}^{t_f} [u^*(t)]^2 dt$$

$$= [x_1^*(2) - 1]^2 + [x_2^*(2) + 1]^2 + \int_0^2 \left[-\frac{9}{7}t + \frac{183}{70}\right]^2 dt,$$

which is approximately equal to 6.955.   □

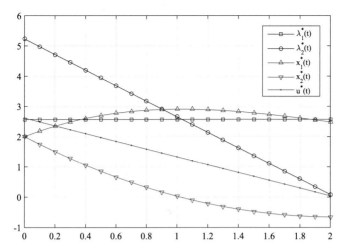

Fig. 3.4 The optimal control u^* and corresponding state x^* and costate λ^* trajectories for optimal control problems with fixed final time and unspecified final state on a linear curve

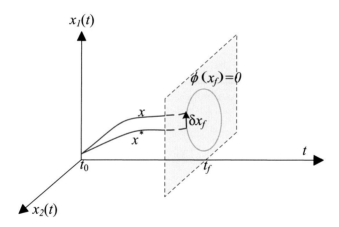

Fig. 3.5 A display of state trajectory which terminates on a circular at t_f

Notice that, different from the optimal control problem with fixed final time and state, as displayed in Fig. 3.4, for the above example, the final state value $x^*(t_f)$ subject to the optimal control u^* is $\begin{bmatrix} 2.485 \\ -0.657 \end{bmatrix}$.

Next, we will give another example such that the state at t_f has to be reached at a surface of the state space which is established from a nonlinear function of $\phi(x(t_f))$, as displayed in Fig. 3.5 such that

$$\phi(x(t_f)) = [x_1(t_f) - 1]^2 + x_2^2(t_f) - 1. \tag{3.60}$$

Example 3.6 Implement the optimal solution to the optimal control problems specified in Example 3.1 with the final state lies on a circle as defined in (3.60).

Solution. By (3.51), we obtain the boundary condition is given as follows:

$$\frac{\partial h(x^*(t_f))}{\partial x} - \lambda^*(t_f) = \begin{bmatrix} 2[x_1^*(t_f) - 1] \\ 2[x_2^*(t_f) + 1] \end{bmatrix} - \begin{bmatrix} \lambda_1^*(t_f) \\ \lambda_2^*(t_f) \end{bmatrix} = d_1 \begin{bmatrix} 2[x_1^*(t_f) - 1] \\ 2x_2^*(t_f) \end{bmatrix},$$
$$\tag{3.61a}$$

$$\phi(x^*(t_f)) = [x_1^*(t_f) - 1]^2 + [x_2^*(t_f)]^2 - 1 = 0. \tag{3.61b}$$

Hence by using the above boundary condition together with the optimal state and costate trajectories developed in (3.27), we can obtain the optimal solution. □

The previous sections focused upon the optimal control problems with the final time t_f fixed. Next, we will study the optimal solution to the problems with unspecified final time t_f.

3.2.3 Optimal Control with Free Final Time and Fixed Final State

In this section, suppose that the final time t_f is unspecified, but the final state $x(t_f)$ is fixed. Thus $\delta x_f = 0$, and δt_f is arbitrary. So by Theorem 3.1, the boundary condition becomes

$$\mathcal{H}(x^*(t_f), u^*(t_f), \lambda^*(t_f), t_f) + \frac{\partial h(x^*(t_f), t_f)}{\partial t} = 0. \tag{3.62}$$

Example 3.7 Implement the optimal solution to the optimal control problems specified in Example 3.1 with the final time is unspecified and the final state is fixed as $x(t_f) = \begin{bmatrix} 1 \\ -1 \end{bmatrix}$.

Solution. In this case, the boundary condition becomes

$$x_1^*(t_f) = \frac{1}{12}\xi_1 t_f^3 - \frac{1}{4}\xi_2 t_f^2 + 2t_f + 2 = 1, \tag{3.63a}$$

$$x_2^*(t_f) = \frac{1}{4}\xi_1 t_f^2 - \frac{1}{2}\xi_2 t_f + 2 = -1, \tag{3.63b}$$

together with

$$\mathcal{H}(x^*(t_f), u^*(t_f), \lambda^*(t_f), t_f) + \frac{\partial h(x^*(t_f), t_f)}{\partial t}$$
$$= [u^*(t_f)]^2 + \lambda_1^*(t_f)\dot{x}_2^*(t_f) - \lambda_2^*(t_f)u^*(t_f) = 0 \tag{3.64}$$

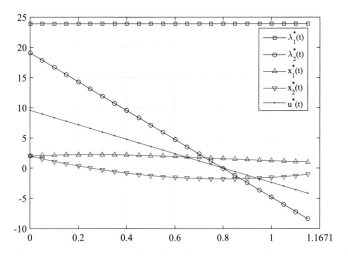

Fig. 3.6 A display of optimal control u^* and its associated state and costate trajectories which terminates at final time of 1.1671

with $\lambda_1^*(t_f) = \xi_1$, $\lambda_2^*(t_f) = -\xi_1 t_f + \xi_2$, $u^*(t_f) = \frac{1}{2}\lambda_2^*(t_f)$, and $\dot{x}_2^*(t_f) = -u^*(t_f)$. The boundary condition is equivalent to the following

$$\xi_1^2 t_f^2 + \xi_2^2 - 2\xi_1^2 = 0 \tag{3.65a}$$

$$\xi_1 t_f^3 - 3\xi_2 t_f^2 + 24 t_f + 12 = 0 \tag{3.65b}$$

$$\xi_1 t_f^2 - 2\xi_2 t_f + 12 = 0 \tag{3.65c}$$

with the solution of $\xi_1 = 23.904$, $\xi_2 = 19.091$ and $t_f = 1.1671$.

Thus, the optimal solution is given as

$$\lambda_1^*(t) = 23.904, \tag{3.66a}$$

$$\lambda_2^*(t) = -23.904t + 19.091, \tag{3.66b}$$

$$x_1^*(t) = 1.992t^3 - 4.773t^2 + 2t + 2, \tag{3.66c}$$

$$x_2^*(t) = 5.976t^2 - 9.545t + 2, \tag{3.66d}$$

with the final time of t_f equal to 1.1671, and the optimal control u^* as

$$u^*(t) = \frac{1}{2}\lambda_2^*(t) = -\frac{1}{2}\xi_1 t + \frac{1}{2}\xi_2 = -11.952t + 9.545. \tag{3.67}$$

The implemented optimal solution is displayed in Fig. 3.6. □

3.2.4 Optimal Control with Free Final Time and Free Final State

In this part, we will study the optimal control for the functional with free final time and free final state.

3.2.4.1 t_f and x_f Being Arbitrary and Independent with Each Other

Here consider that t_f and x_f are arbitrary and independent, then by Theorem 3.1, the coefficients of δt_f and δx_f in the boundary condition of (3.11) should be zero, respectively. Thus

$$\lambda^*(t_f) - \frac{\partial h(x^*(t_f), t_f)}{\partial x} = 0, \tag{3.68a}$$

$$\mathcal{H}(x^*(t_f), u^*(t_f), \lambda^*(t_f), t_f) + \frac{\partial h(x^*(t_f), t_f)}{\partial t} = 0. \tag{3.68b}$$

Notice that in case that it does not involve a terminal performance cost h, say $h(x(t_f), t_f) = 0$; then the boundary condition degenerates to the following:

$$\lambda^*(t_f) = 0, \tag{3.69a}$$

$$\mathcal{H}(x^*(t_f), u^*(t_f), \lambda^*(t_f), t_f) = 0. \tag{3.69b}$$

Example 3.8 Implement the optimal solution to the problem given in Example 3.1 except that the final time t_f and final state $x(t_f)$ are arbitrary and independent with each other.

Solution. By (3.68), the boundary condition for the underlying optimal control problem is

$$\lambda_1^*(t_f) - 2[x_1^*(t_f) - 1] = 0, \tag{3.70a}$$

$$\lambda_2^*(t_f) - 2[x_2^*(t_f) + 1] = 0, \tag{3.70b}$$

$$[u^*(t_f)]^2 + \lambda_1^*(t_f)x_2^*(t_f) - \lambda_2^*(t_f)u^*(t_f) = 0, \tag{3.70c}$$

which is equivalent to the following equations:

$$\xi_1 t_f^3 - 3\xi_2 t_f^2 + 24t_f - 6\xi_1 + 12 = 0, \tag{3.71a}$$

$$\xi_1 t_f^2 + 2[\xi_1 - \xi_2]t_f - 2\xi_2 + 12 = 0, \tag{3.71b}$$

$$2\xi_1^2 t_f^2 - 4\xi_1\xi_2 t_f + 8\xi_1 + \xi_2^2 = 0. \tag{3.71c}$$

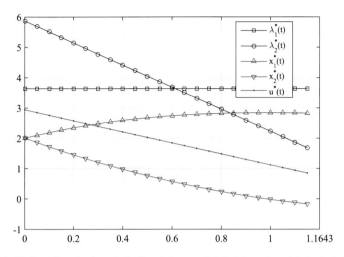

Fig. 3.7 A display of optimal control u^* and its associated state and costate trajectories which terminates at the final time of 1.1643

The solution to the above collection of equations is

$$\xi_1 = 3.637, \ \xi_2 = 5.868, \ \text{and } t_f = 1.1643. \tag{3.72}$$

Thus, the optimal solution is given as

$$\lambda_1^*(t) = 3.637, \tag{3.73a}$$
$$\lambda_2^*(t) = -3.637t + 5.868, \tag{3.73b}$$
$$x_1^*(t) = 0.303t^3 - 1.467t^2 + 2t + 2, \tag{3.73c}$$
$$x_2^*(t) = 0.909t^2 - 2.934t + 2, \tag{3.73d}$$

with the final time of t_f equal to 1.1643, and the optimal control u^* as

$$u^*(t) = \frac{1}{2}\lambda_2^*(t) = -\frac{1}{2}\xi_1 t + \frac{1}{2}\xi_2 = -1.8185t + 2.934. \tag{3.74}$$

The implemented optimal solution is displayed in Fig. 3.7.
The performance cost subject to the optimal control u^* is

$$J(u^*) = [x_1^*(t_f) - 1]^2 + [x_2^*(t_f) + 1]^2 + \int_{t_0}^{t_f} [u^*(t)]^2 dt$$

$$= [2.821 - 1]^2 + [-0.172 + 1]^2 + \int_0^2 [-1.8185t + 2.934]^2 \, dt = 8.695.$$

□

3.2.4.2 Final State x_f Evolving with Respect to Final Time t_f

Suppose that the final state $x(t_f)$ should lie on a trajectory $\theta(t_f)$, see an illustration in Fig. 2.15, say $x(t_f) = \phi(t_f)$ as stated in (2.132); thus following the analysis in Theorem 2.6, δx_f and δt_f are approximately related as (2.135) which is rewritten as follows:

$$\delta x_f \doteq \frac{d\phi(t_f)}{dt}\delta t_f.$$

Hence, by the boundary condition given in (3.11), we obtain the boundary condition for this case as

$$\mathcal{H}(x^*(t_f), u^*(t_f), \lambda^*(t_f), t_f) + \frac{\partial h(x^*(t_f), t_f)}{\partial t}$$
$$+ \left[\frac{\partial h(x^*(t_f), t_f)}{\partial x} - \lambda^*(t_f)\right]^T \frac{d\phi(t_f)}{dt} = 0 \qquad (3.75)$$

together with

$$x^*(t_0) = x_0 \text{ and } x^*(t_f) = \phi(t_f). \qquad (3.76)$$

Example 3.9 Implement the optimal solution to the problem given in Example 3.1 with a boundary condition such that the final time t_f is free and the final state $x(t_f)$ lies on $\begin{bmatrix} t_f - \frac{1}{2} \\ -\frac{t_f}{2} \end{bmatrix}$.

Solution. By (3.75), the boundary condition for the underlying optimal control problem is

$$\mathcal{H}(x^*(t_f), u^*(t_f), \lambda^*(t_f), t_f) + \frac{\partial h(x^*(t_f), t_f)}{\partial t}$$
$$+ \left[\frac{\partial h(x^*(t_f), t_f)}{\partial x} - \lambda^*(t_f)\right]^T \frac{d\phi(t_f)}{dt} = 0, \qquad (3.77a)$$
$$x^*(t_f) = \phi(t_f), \qquad (3.77b)$$

which implies that

$$[u^*(t_f)]^2 + \lambda_1^*(t_f)x_2^*(t_f) - \lambda_2^*(t_f)u^*(t_f)$$
$$+ \left[2[x_1^*(t_f) - 1] - \lambda_1^*(t_f)\right] - \frac{1}{2}\left[2[x_2^*(t_f) + 1] - \lambda_2^*(t_f)\right] = 0, \quad (3.78a)$$
$$x_1^*(t_f) = t_f - \frac{1}{2},$$
$$x_2^*(t_f) = -\frac{t_f}{2}; \qquad (3.78b)$$

then by (3.10) in Theorem 3.1, we get

$$\frac{\xi_1}{6}t_f^3 + at_f^2 + bt_f + c = 0, \tag{3.79a}$$

$$\frac{\xi_1}{12}t_f^3 - \frac{\xi_2}{4}t_f^2 + t_f + \frac{5}{2} = 0, \tag{3.79b}$$

$$\frac{\xi_1}{4}t_f^2 - \frac{1}{2}[\xi_2 - 1]t_f + 2 = 0 \tag{3.79c}$$

with $a \equiv -\frac{\xi_1}{4} - \frac{\xi_2}{2}, b \equiv -\frac{\xi_1}{2} + \frac{\xi_2}{2} + 4$ and $c \equiv -\frac{\xi_2^2}{4} + \xi_1 + \frac{\xi_2}{2} - 1$.
The solution to the above collection of equations is

$$\xi_1 = 1.1767, \xi_2 = 4.0683, \text{ and } t_f = 2.5773. \tag{3.80}$$

Thus, the optimal solution is given as

$$\lambda_1^*(t) = \xi_1 = 1.1767, \tag{3.81a}$$

$$\lambda_2^*(t) = -\xi_1 t + \xi_2 = -1.1767t + 4.0683, \tag{3.81b}$$

$$x_1^*(t) = \frac{\xi_1}{12}t^3 - \frac{\xi_2}{4}t^2 + 2t + 2 = 0.0981t^3 - 1.0171t^2 + 2t + 2, \tag{3.81c}$$

$$x_2^*(t) = \frac{\xi_1}{4}t^2 - \frac{\xi_2}{2}t + 2 = 0.2942t^2 - 2.0342t + 2, \tag{3.81d}$$

with the final time of t_f equal to 2.5773, and the optimal control u^* as

$$u^*(t) = \frac{1}{2}\lambda_2^*(t) = -\frac{\xi_1}{2}t + \frac{\xi_2}{2} = -0.5884t + 2.0342. \tag{3.82}$$

The implemented optimal solution is displayed in Fig. 3.8.
The performance cost subject to the optimal control u^* is

$$J(u^*) = [x_1^*(t_f) - 1]^2 + [x_2^*(t_f) + 1]^2 + \int_{t_0}^{t_f} [u^*(t)]^2 dt$$

$$= [2.0773 - 1]^2 + [-1.2887 + 1]^2 + \int_0^{2.5773} [-0.5884t + 2.0342]^2 dt$$

$$= 5.9335.$$

$\square$

3.2.4.3 Final State x_f Being on a Time-Invariant Surface

In Theorem 3.4 , we will specify a necessary condition of the extrema solution of those problems such that final state x_f is on a time-invariant surface.

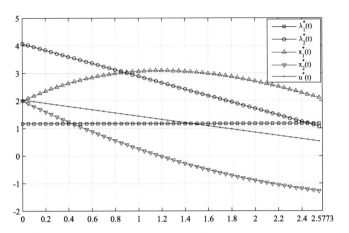

Fig. 3.8 A display of optimal control u^*, and its associated state and costate trajectories, which terminates at final time of 2.5773 in case the final state dependent upon the final time t_f

Theorem 3.4 *Suppose that u^* is an optimal control for Problem 3.1 with final time t_f and final state $x(t_f)$ unspecified, such that $x(t_f)$ satisfies the following situation:*

$$\phi(x(t_f)) = \begin{bmatrix} \phi_1(x(t_f)) \\ \phi_2(x(t_f)) \\ \cdots \\ \phi_k(x(t_f)) \end{bmatrix} = 0,$$

for some k, with $1 \le k \le n-1$, which is also considered in Theorem 3.3; then the boundary condition is given as

$$\frac{\partial h(x^*(t_f), t_f)}{\partial x} - \lambda^*(t_f) = \sum_{i=1}^{k} d_i \frac{\partial \phi_i(x^*(t_f))}{\partial x}, \tag{3.83a}$$

$$\mathcal{H}(x^*(t_f), u^*(t_f), \lambda^*(t_f), t_f) + \frac{\partial h(x^*(t_f), t_f)}{\partial t} = 0, \tag{3.83b}$$

together with the given boundary conditions of $x^*(t_0) = x_0$ and $\phi(x^*(t_f)) = 0$.

Proof Firstly, to the first order, the change in $x(t_f)$ must be in the plane tangent to the surface at the point $(x^*(t_f), t_f)$.

Each $\phi_i(x(t_f)) = 0$ represents a hypersurface in n-dimensional state space. Thus, the final state lies on the intersection of these k hypersurfaces, and then δx_f should be (to the first order) tangent to each of the hypersurfaces at the point $(x^*(t_f), t_f)$, see an illustration in Fig. 3.9 for a two-dimensional system.

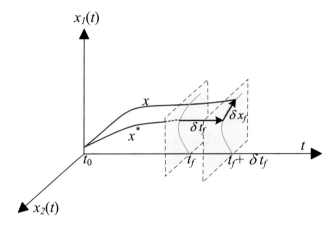

Fig. 3.9 A display of state trajectory which terminates on a surface with t_f unspecified

Consequently, δx_f is orthogonal to each of the following (gradient) vectors:

$$\frac{\partial \phi_1(x^*(t_f))}{\partial x}, \ldots, \frac{\partial \phi_k(x^*(t_f))}{\partial x}, \tag{3.84}$$

which implies that δx_f is orthogonal to the space spanned by $\frac{\partial \phi_i(x^*(t_f))}{\partial x}$, with $i = 1, \ldots, k$, that is to say δx_f is orthogonal to the linear combination of $\frac{\partial \phi_1(x^*(t_f))}{\partial x}, \ldots,$ $\frac{\partial \phi_k(x^*(t_f))}{\partial x}$, i.e.,

$$\left[d_1 \frac{\partial \phi_1(x^*(t_f))}{\partial x} + \cdots + d_k \frac{\partial \phi_k(x^*(t_f))}{\partial x} \right]^\top \delta x_f = 0, \tag{3.85}$$

for any real valued d_i.

Since δx_f is independent of δt_f, the coefficient of δt_f must be zero. Thus we can get the conclusion of (3.83b).

By (3.83b) together with (3.11) in Theorem 3.1, we can obtain

$$\left[\frac{\partial h(x^*(t_f), t_f)}{\partial x} - \lambda^*(t_f) \right]^\top \delta x_f = 0. \tag{3.86}$$

Then by (3.85) and (3.86), we can show the conclusion of (3.83a). □

Example 3.10 Suppose that the final time t_f is free and the final state x_f is required to lie on the following curve:

$$\phi(x(t)) = [x_1(t) - 2]^2 + 2[x_2(t) - 1]^2 - 4 = 0. \tag{3.87}$$

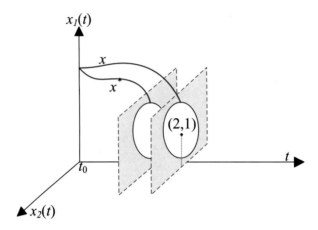

Fig. 3.10 A display of state trajectory which terminates on an oval with t_f unspecified

Solution. Since the final time t_f is free, the admissible points lie on the surface as shown in Fig. 3.10.

By (3.83a), we have

$$\frac{\partial h(x^*(t_f), t_f)}{\partial x} - \lambda^*(t_f) = d_1 \frac{\partial \phi(x^*(t_f))}{\partial x} = d_1 \begin{bmatrix} 2[x_1^*(t_f) - 2] \\ 4[x_2^*(t_f) - 1] \end{bmatrix}; \qquad (3.88)$$

then by Theorem 3.4, the boundary condition is composed of (3.88), together with (3.83b), the given initial condition $x^*(t_0) = x_0$ and

$$\phi(x^*(t_f)) = [x_1^*(t_f) - 2]^2 + 2[x_2^*(t_f) - 1]^2 - 4 = 0. \qquad (3.89)$$

$\square$

3.2.4.4 Final State $x(t_f)$ Lying on a Moving Surface Which Evolves with Time

In this part, consider a more general case that the final state, with t_f unspecified, shall lie on a moving surface $\phi(x(t), t) = 0$ which is related to time t.

Figure 3.11, displays a moving surface on a two-dimensional system with

$$\phi(x(t), t) \triangleq x_1(t) + x_2(t) - at = 0. \qquad (3.90)$$

Theorem 3.5 *Determine the optimal solution to the optimal control problems with unspecified final time and final state, such that the final state lying on a surface $\phi(x(t), t) = 0$, which evolves with respect to t, such that*

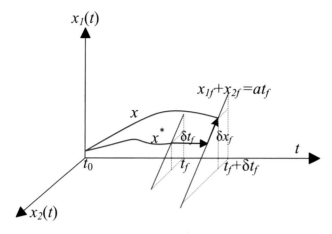

Fig. 3.11 A display of state trajectory which terminates on a surface with t_f unspecified

$$\phi(x(t), t) \equiv \begin{bmatrix} \phi_1(x(t), t) \\ \cdots \\ \phi_k(x(t), t) \end{bmatrix} = 0, \tag{3.91}$$

with $1 \leq k \leq n$; then the boundary condition is given as

$$\frac{\partial h(x^*(t_f), t_f)}{\partial x} - \lambda^*(t_f) = \sum_{i=1}^{k} d_i \frac{\partial \phi_i(x^*(t_f), t_f)}{\partial x}, \tag{3.92a}$$

$$\mathscr{H}(x^*(t_f), u^*(t_f), \lambda^*(t_f), t_f) + \frac{\partial h(x^*(t_f), t_f)}{\partial t} = \sum_{i=1}^{k} d_i \frac{\partial \phi_i(x^*(t_f), t_f)}{\partial t}, \tag{3.92b}$$

together with the given boundary conditions of $x^*(t_0) = x_0$ and $\phi(x^*(t_f), t_f) = 0$.

Proof Following the similar analysis for the case that $\phi(x(t))$ is independent on time, we can claim that $\begin{bmatrix} \delta x_f \\ \delta t_f \end{bmatrix}$ should be orthogonal to each of the following gradient vectors:

$$\begin{bmatrix} \frac{\partial \phi_1(x^*(t_f), t_f)}{\partial x} \\ \frac{\partial \phi_1(x^*(t_f), t_f)}{\partial t} \end{bmatrix}, \cdots, \begin{bmatrix} \frac{\partial \phi_k(x^*(t_f), t_f)}{\partial x} \\ \frac{\partial \phi_k(x^*(t_f), t_f)}{\partial t} \end{bmatrix}. \tag{3.93}$$

Thus $\begin{bmatrix} \delta x_f \\ \delta t_f \end{bmatrix}$ is orthogonal to any linear combination of the above vectors, say

Fig. 3.12 A display of time-variant surfaces which the final state lies in a circular specified in (3.96)

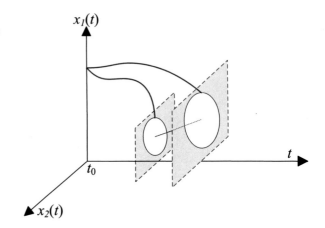

$$\left[\sum_{i=1}^{k} d_i \left[\begin{array}{c} \frac{\partial \phi_i(x^*(t_f),t_f)}{\partial x} \\ \frac{\partial \phi_i(x^*(t_f),t_f)}{\partial t} \end{array} \right] \right]^{\top} \left[\begin{array}{c} \delta x_f \\ \delta t_f \end{array} \right] = 0. \tag{3.94}$$

Also by (3.11) in Theorem 3.1, it has

$$\left[\begin{array}{c} \frac{\partial h(x^*(t_f),t_f)}{\partial x} - \lambda^*(t_f) \\ \mathscr{H}(x^*(t_f), u^*(t_f), \lambda^*(t_f), t_f) + \frac{\partial h(x^*(t_f),t_f)}{\partial t} \end{array} \right]^{\top} \left[\begin{array}{c} \delta x_f \\ \delta t_f \end{array} \right] = 0. \tag{3.95}$$

By the above results, we can obtain the conclusion.   □

Example 3.11 Suppose that the final state must lie on the following moving surface, as displayed in Fig. 3.12,

$$\phi(x(t_f), t_f) = [x_1(t_f) - 2]^2 + 2[x_2(t_f) - 1 - t_f]^2 - 4t_f^2 = 0. \tag{3.96}$$

Solution. Firstly, by the specification of $\phi(x(t), t)$, it has

$$\frac{\partial \phi(x(t_f), t_f)}{\partial x_1} = 2[x_1(t_f) - 2],$$

$$\frac{\partial \phi(x(t_f), t_f)}{\partial x_2} = 4[x_2(t_f) - 1 - t_f],$$

$$\frac{\partial \phi(x(t_f), t_f)}{\partial t} = -4[x_2(t_f) - 1 - t_f] - 8t_f = -4[x_2(t_f) - 1 + t_f];$$

then the boundary conditions are given as

$$\frac{\partial h(x^*(t_f), t_f)}{\partial x} - \lambda^*(t_f) = d_1 \begin{bmatrix} 2[x_1^*(t_f) - 2] \\ 4[x_2^*(t_f) - 1 - t_f] \end{bmatrix}, \tag{3.97a}$$

$$\mathcal{H}(x^*(t_f), u^*(t_f), \lambda^*(t_f), t_f) + \frac{\partial h(x^*(t_f), t_f)}{\partial t} = -4d_1[x_2^*(t_f) - 1 + t_f], \tag{3.97b}$$

together with the given boundary conditions of $x^*(t_0) = x_0$ and

$$\phi(x^*(t_f), t_f) = [x_1^*(t_f) - 2]^2 + 2[x_2^*(t_f) - 1 - t_f]^2 - 4t_f^2 = 0. \tag{3.98}$$

□

In the next two sections, Sects. 3.3 and 3.4, we will study the optimal control solutions of the linear-quadratic regulation and tracking problems, respectively, by applying the results developed in this part.

3.3 Linear-Quadratic Regulation Problems

We formulate a specific class of optimal control problems below.

Problem 3.2 Consider a state system denoted by $x(\cdot)$ subject to a control u such that linear state equations are

$$\dot{x}(t) = A(t)x(t) + B(t)u(t), \tag{3.99}$$

where $A(t)$ and $B(t)$ are $n \times n$ and $n \times m$ matrices, respectively.

The objective is to minimize the performance cost function as given below.

$$J(u) \triangleq \frac{1}{2}x^\top(t_f)Hx(t_f) + \frac{1}{2}\int_{t_0}^{t_f} \left[x^\top(t)Q(t)x(t) + u^\top(t)R(t)u(t) \right] dt. \tag{3.100}$$

□

Suppose that the final time t_f is fixed and the final state $x(t_f)$ is free. H and Q are real $n \times n$ matrices which are symmetric and positive semi-definite, respectively, and R is assumed to be real symmetric and positive definite. It also does not consider any bounds for the states and controls.

The optimal control problem specified above is called the linear-quadratic regulation (LQR for short) problem, which can be solved by directly applying the results developed in Sect. 3.2.

Theorem 3.6 *The optimal control for LQR problems is given as $u^*(t) = F(t)x(t)$, where $F(t)$ is a time-variant matrix, i.e., $u^*(t)$ is a linear, time variant combination of the system state x at t.*

Proof Apply the variational method, firstly define the Hamiltonian $\mathscr{H}$ for the underlying regulation problem in the following:

$$\mathscr{H}(x(t), u(t), \lambda(t), t)$$
$$= \frac{1}{2}x^{\top}(t)Q(t)x(t) + \frac{1}{2}u^{\top}(t)R(t)u(t) + [\lambda(t)]^{\top}[A(t)x(t) + B(t)u(t)]. \quad (3.101)$$

By Theorem 3.1, the necessary conditions for optimality are given as

$$\dot{x}^{*}(t) = A(t)x^{*}(t) + B(t)u^{*}(t), \quad (3.102a)$$

$$\dot{\lambda}^{*}(t) = -\frac{\partial \mathscr{H}^{*}}{\partial x} = -Q(t)x^{*}(t) - A^{\top}(t)\lambda^{*}(t), \quad (3.102b)$$

$$0 = \frac{\partial \mathscr{H}^{*}}{\partial u} = R(t)u^{*}(t) + B^{\top}(t)\lambda^{*}(t), \quad (3.102c)$$

with $\mathscr{H}^{*} \equiv \mathscr{H}(x^{*}(t), u^{*}(t), \lambda^{*}(t), t)$ and the boundary conditions given as

$$\frac{\partial h(x^{*}(t_f), t_f)}{\partial x} - \lambda^{*}(t_f) = Hx^{*}(t_f) - \lambda^{*}(t_f) = 0. \quad (3.103)$$

By (3.102c) and the positive definite of the matrix $A(t)$, we can obtain

$$u^{*}(t) = -R^{-1}(t)B^{\top}(t)\lambda^{*}(t), \quad (3.104)$$

by which we can get

$$\dot{x}^{*}(t) = A(t)x^{*}(t) - B(t)R^{-1}(t)B^{\top}(t)\lambda^{*}(t). \quad (3.105)$$

By organizing (3.102b) and (3.105), we can establish the following linear homogeneous differential equation:

$$\begin{bmatrix} \dot{x}^{*}(t) \\ \dot{\lambda}^{*}(t) \end{bmatrix} = \begin{bmatrix} A(t) & -B(t)R^{-1}(t)B^{\top}(t) \\ -Q(t) & -A^{\top}(t) \end{bmatrix} \begin{bmatrix} x^{*}(t) \\ \lambda^{*}(t) \end{bmatrix}. \quad (3.106)$$

The solution to the above linear differential equation has the form

$$\begin{bmatrix} x^{*}(t_f) \\ \lambda^{*}(t_f) \end{bmatrix} = \Phi(t_f, t)\begin{bmatrix} x^{*}(t) \\ \lambda^{*}(t) \end{bmatrix}, \quad (3.107)$$

where Φ represents the transition matrix for (3.106).

Partitioning the transition matrix, we have

$$\begin{bmatrix} x^{*}(t_f) \\ \lambda^{*}(t_f) \end{bmatrix} = \begin{bmatrix} \Phi_{11}(t_f, t) & \Phi_{12}(t_f, t) \\ \Phi_{21}(t_f, t) & \Phi_{22}(t_f, t) \end{bmatrix} \begin{bmatrix} x^{*}(t) \\ \lambda^{*}(t) \end{bmatrix}. \quad (3.108)$$

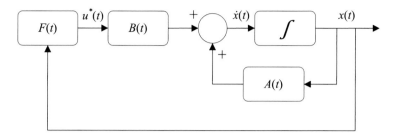

Fig. 3.13 A diagram of the optimal control for linear-quadratic regulation problems

By the boundary condition of (3.103),

$$\lambda^*(t_f) = Hx^*(t_f); \tag{3.109}$$

then by using $Hx^*(t_f)$ to replace $\lambda^*(t_f)$ in (3.108), we get that

$$x^*(t_f) = \Phi_{11}(t_f, t)x^*(t) + \Phi_{12}(t_f, t)\lambda^*(t), \tag{3.110a}$$
$$Hx^*(t_f) = \Phi_{21}(t_f, t)x^*(t) + \Phi_{22}(t_f, t)\lambda^*(t). \tag{3.110b}$$

Again by using the RHS of (3.110a) to replace the $x^*(t_f)$ in (3.110b), we obtain

$$H\Phi_{11}(t_f, t)x^*(t) + H\Phi_{12}(t_f, t)\lambda^*(t) = \Phi_{21}(t_f, t)x^*(t) + \Phi_{22}(t_f, t)\lambda^*(t). \tag{3.111}$$

When solved for $\lambda^*(t)$, it yields

$$\lambda^*(t) = \left[\Phi_{22}(t_f, t) - H\Phi_{12}(t_f, t)\right]^{-1} \left[H\Phi_{11}(t_f, t) - \Phi_{21}(t_f, t)\right]x^*(t)$$
$$= K(t)x^*(t), \tag{3.112}$$

with $K(t) \triangleq [\Phi_{22}(t_f, t) - H\Phi_{12}(t_f, t)]^{-1}[H\Phi_{11}(t_f, t) - \Phi_{21}(t_f, t)]$.
Thus by (3.112) together with (3.104), we obtain

$$u^*(t) = -R^{-1}(t)B^{\top}(t)K(t)x^*(t) \equiv F(t)x^*(t), \tag{3.113}$$

with $F(t) \triangleq -R^{-1}(t)B^{\top}(t)K(t)$, which is the conclusion. ☐

Figure 3.13 displays the diagram of the optimal control for the underlying linear optimal regulation problems.

To implement the feedback gain matrix $F(\cdot)$, we need to get the transition matrix $K(\cdot)$ defined in (3.112).

In case all of the involved matrices A, B, R, and Q are all time invariant, we may implement the transition matrix by evaluating the inversive Laplace transform of the matrix

$$\left[s\mathbb{I} - \begin{bmatrix} A & -BR^{-1}B^\top \\ -Q & -A^\top \end{bmatrix} \right]^{-1} \tag{3.114}$$

where $\mathbb{I}$ represents an identity matrix.

However, it is still challenging and time consuming to implement the solution to any high-dimensional state system.

If any of the matrices A, B, R, or Q is time variant, it may be required to apply a numerical procedure to determine $\Phi(t_f, t)$.

Nevertheless in the following, it gives an alternative method to solve the matrix K.

$K(t)$ satisfies the following matrix differential equations, which is called *the Riccati equation*,

$$\dot{K}(t) = -K(t)A(t) - A^\top(t)K(t) - Q(t) + K(t)B(t)R^{-1}(t)B^\top(t)K(t), \tag{3.115}$$

with the boundary condition of $K(t_f) = H$.

Even though all of the matrices A, B, R, and Q are time invariant, say

$$\dot{K}(t) = -K(t)A - A^\top K(t) - Q + K(t)BR^{-1}B^\top K(t). \tag{3.116}$$

K still evolves with respect to time t.

We will illustrate these results with the following examples.

Example 3.12 Implement the optimal control for the system

$$\dot{x}(t) = -ax(t) + u(t), \tag{3.117}$$

such that the following cost is minimized

$$J(u) = \frac{1}{2}Hx^2(t_f) + \frac{1}{4}\int_0^{t_f} u^2(t)dt, \tag{3.118}$$

with $H > 0$, and the final state $x(t_f)$ is free.

Solution. By the specification of the LQR problem, we have

$$A(t) = -a, B(t) = 1, Q(t) = 0, R(t) = \frac{1}{2}, H(t) = H. \tag{3.119}$$

Thus by (3.106),

$$\begin{bmatrix} \dot{x}^*(t) \\ \dot{\lambda}^*(t) \end{bmatrix} = \begin{bmatrix} -a & -2 \\ 0 & a \end{bmatrix} \begin{bmatrix} x^*(t) \\ \lambda^*(t) \end{bmatrix}, \qquad (3.120)$$

with the boundary condition of $x^*(0) = x_0$ and $\lambda^*(t_f) = Hx^*(t_f)$.
The optimal state and costate trajectories are given as

$$\begin{bmatrix} x^*(t) \\ \lambda^*(t) \end{bmatrix} = \begin{bmatrix} \exp(-at) & \frac{1}{a}\exp(-at) - \frac{1}{a}\exp(at) \\ 0 & \exp(at) \end{bmatrix} \begin{bmatrix} x^*(0) \\ \lambda^*(0) \end{bmatrix} \qquad (3.121)$$

with the boundary conditions of $x^*(0) = x_0$ and $\lambda^*(t_f) = Hx^*(t_f)$.
Thus we can have

$$\lambda^*(0) = \frac{aH}{[H+a]\exp(2at_f) - H} x_0. \qquad (3.122)$$

And we can get that the transition matrix as

$$\Phi(t_f, t) = \begin{bmatrix} \exp(-a[t_f - t]) & \frac{1}{a}\exp(-a[t_f - t]) - \frac{1}{a}\exp(a[t_f - t]) \\ 0 & \exp(a[t_f - t]) \end{bmatrix}. \qquad (3.123)$$

By (3.112), it has

$$K(t) \triangleq [\Phi_{22}(t_f, t) - H\Phi_{12}(t_f, t)]^{-1}[H\Phi_{11}(t_f, t) - \Phi_{21}(t_f, t)]$$

$$= \frac{aH}{[H+a]\exp(2a[t_f - t]) - H}; \qquad (3.124)$$

then by (3.113), the optimal control is

$$u^*(t) = -R^{-1}(t)B^\top(t)K(t)x^*(t) = -\frac{2aH}{[H+a]\exp(2a[t_f - t]) - H} x^*(t). \qquad (3.125)$$

We will study the optimal solutions to the regulation problems with respect to different values of the parameters of the system a and H.

CASE 1: Suppose that the initial state $x(0) = 3$, the parameter $a = 0.1$, the final time $t_f = 4$.
In this case, the optimal state and costate trajectories are given as

$$\begin{bmatrix} x^*(t) \\ \lambda^*(t) \end{bmatrix} = \begin{bmatrix} \exp(-0.1t) & 10\exp(-0.1t) - 10\exp(0.1t) \\ 0 & \exp(0.1t) \end{bmatrix} \begin{bmatrix} x^*(0) \\ \lambda^*(0) \end{bmatrix} \qquad (3.126)$$

with the boundary conditions of $x^*(0) = 3$ and $\lambda^*(t_f) = Hx^*(t_f)$. Thus

$$\lambda^*(0) = \frac{aH}{[H+a]\exp(2at_f) - H}x_0 = \frac{0.3H}{[H+0.1]\exp(0.8) - H}, \qquad (3.127)$$

since $t_f = 4$ and $x(0) = 3$.
And we can get that the transition matrix as

$$\Phi(t_f, t) = \begin{bmatrix} \exp(-0.1[4-t]) & 10\exp(-0.1[4-t]) - 10\exp(0.1[4-t]) \\ 0 & \exp(0.1[4-t]) \end{bmatrix}.$$
$$(3.128)$$

And the optimal control is given as

$$u^*(t) = -\frac{2aH}{[H+a]\exp(2a[t_f - t]) - H}x^*(t)$$

$$= -\frac{0.2H}{[H+0.1]\exp(0.2[4-t]) - H}x^*(t). \qquad (3.129)$$

Figure 3.14 displays the optimal solution x^*, λ^*, and u^* with $a = 0.1$ and $H = 1$.

CASE 2: Moreover, Fig. 3.15 displays the optimal solution x^* and u^* with $a = 0.1$ and different values of H, respectively.
As observed, with a and t_f fixed with a value of 0.1 and 4, respectively, larger the value of H is, closer the final state approaches to zero (Fig. 3.15).

CASE 3: Similarly, Fig. 3.16, displays the optimal solution x^* and u^* with respect to $H = 1$ and different values of a.
As observed, H and t_f are fixed with values of 1 and 4, respectively, larger the value of a, closer the final state approaches to zero and less efforts for the control is required.
Moreover, Fig. 3.17, displays the optimal solution x^*, λ^* and u^* with respect to $H = 1$ and a negative value of a. As observed, compared with those positive valued a, it takes more efforts to drive a state near to zero at the final time.

□

Notice that $K(t)$ can also be determined by solving the Riccati equation (3.115); then

$$\dot{K}(t) = aK(t) + aK(t) + 2K(t)K(t) = 2K^2(t) + 2aK(t).$$

And it can be verified that the solution to the above Riccati equation is equal to (3.125) as well.

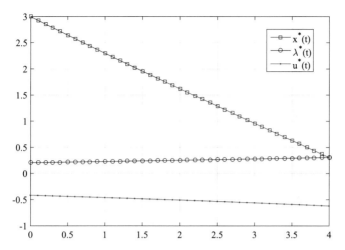

Fig. 3.14 A display of x^*, λ^* and u^* with $a = 0.1$ and $H = 1$

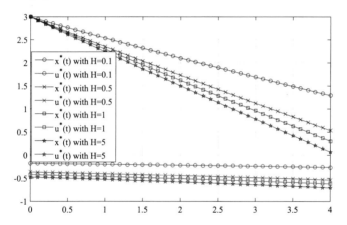

Fig. 3.15 A display of x^* and u^* with $a = 0.1$, and $H = 0.1, 0.5, 1$, and 5

3.3.1 Infinite-Interval Time-Invariant LQR Problems

In this part, we will study a specific class of LQR problems below.

Problem 3.3 Consider a state system denoted by $x(\cdot)$ subject to a control u such that linear state equations

$$\dot{x}(t) = Ax(t) + Bu(t), \tag{3.130}$$

where $A(t)$ and $B(t)$ are $n \times n$ and $n \times m$ matrices, respectively.

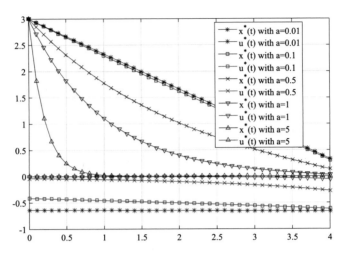

Fig. 3.16 A display of x^* and u^* with $a = 0.01, 0.1, 0.5, 1$, and 5, respectively

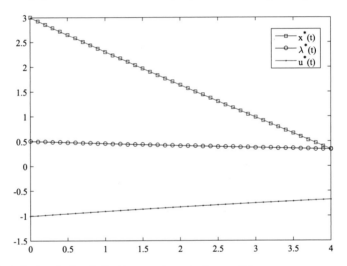

Fig. 3.17 A display of x^* and u^* with a negative valued $a = -0.1$

The objective is to minimize the performance cost function over an infinite interval as given below.

$$J(u) \triangleq \frac{1}{2} \int_0^\infty \left[x^\top(t) Q x(t) + u^\top(t) R u(t) \right] dt. \tag{3.131}$$

$\square$

Theorem 3.7 *The optimal control for the infinite-interval time-invariant LQR specified in Problem 3.3 is time invariant, such that*

$$u^*(t) = -R^{-1}B^{\top}Kx^*(t), \tag{3.132}$$

with K being the solution to the following algebra Riccati equation:

$$0 = -KA - A^{\top}K - Q + KBR^{-1}B^{\top}K. \tag{3.133}$$

$\square$

Example 3.13 Implement the optimal control solution to the infinite-interval time-invariant LQR problem for the following system:

$$\dot{x}_1(t) = x_2(t), \tag{3.134a}$$
$$\dot{x}_2(t) = x_1(t) - x_2(t) + u(t), \tag{3.134b}$$

which is to be controlled to minimize

$$J(u) = \int_0^{\infty} \left[\frac{1}{2}x_1^2(t) + \frac{1}{2}x_2^2(t) + u^2(t) \right] dt. \tag{3.135}$$

Solution. For the underlying problem, the associated matrices of A, B, Q, and R are specified below.

$$A = \begin{bmatrix} 0 & 1 \\ 1 & -1 \end{bmatrix}, B = \begin{bmatrix} 0 \\ 1 \end{bmatrix}, Q = \begin{bmatrix} 1 & 0 \\ 0 & 1 \end{bmatrix}, R = 2, H = \begin{bmatrix} 0 & 0 \\ 0 & 0 \end{bmatrix}. \tag{3.136}$$

Firstly the state system is completely controllable, since the rank of the matrix $\begin{bmatrix} B & AB \end{bmatrix} = \begin{bmatrix} 0 & 1 \\ 1 & -1 \end{bmatrix}$ is 2.

Thus the Riccati equation is given as

$$0 = -KA - A^{\top}K - Q + KBR^{-1}B^{\top}K$$
$$= -K\begin{bmatrix} 0 & 1 \\ 1 & -1 \end{bmatrix} - \begin{bmatrix} 0 & 1 \\ 1 & -1 \end{bmatrix}K - \begin{bmatrix} 1 & 0 \\ 0 & 1 \end{bmatrix} + \frac{1}{2}K\begin{bmatrix} 0 \\ 1 \end{bmatrix}\begin{bmatrix} 0 & 1 \end{bmatrix}K; \tag{3.137}$$

then we obtain

$$0 = \frac{1}{2}K_{12}^2 - 2K_{12} - 1, \tag{3.138a}$$

$$0 = \frac{1}{2}K_{12}K_{22} - K_{11} + K_{12} - K_{22}, \tag{3.138b}$$

$$0 = \frac{1}{2}K_{22}^2 - 2K_{12} + 2K_{22} - 1. \tag{3.138c}$$

By (3.113), the optimal control is given as

$$u^*(t) = -R^{-1}B^\top K x^*(t) = -\frac{1}{2}\begin{bmatrix} K_{12} & K_{22} \end{bmatrix} x^*(t). \qquad (3.139)$$

□

3.4   Linear-Quadratic Tracking Problems

In this part, we will study another specific class of optimal control problems.

Problem 3.4 Consider a state system denoted by $x(\cdot)$ subject to a control u such that linear state equation given as

$$\dot{x}(t) = A(t)x(t) + B(t)u(t), \qquad (3.140)$$

where $A(t)$ and $B(t)$ are $n \times n$ and $n \times m$ matrices, respectively.

Moreover, we consider a specific performance cost function as given below.

$$J = \frac{1}{2}[x(t_f) - r(t_f)]^\top H[x(t_f) - r(t_f)]$$
$$+ \frac{1}{2}\int_{t_0}^{t_f} \left[[x(t) - r(t)]^\top Q(t)[x(t) - r(t)] + u^\top(t)R(t)u(t) \right] dt. \quad (3.141)$$

□

Same as the assumption for the LQR problem studied in the last section, it still supposes that the final time t_f is fixed and the final state $x(t_f)$ is free. H and $Q(t)$, for all t, are real $n \times n$ matrices which are symmetric and positive semi-definite, respectively, and $R(t)$, for all t, is real symmetric and positive definite. It also does not consider any bounds for the states and controls.

Notice that by the comparison between (3.100) and (3.141), we can claim that the LQR problem formulated in Sect. 3.3 is a specific LQT problem in case the reference trajectory $r(t) = 0$ for all $t \in [t_0, t_f]$.

Due to the positive-definitive or semi-definite property of the matrices H, $Q(t)$, and $R(t)$ for all t, for notational simplicity, we may write J as the following compact form:

$$J = \frac{1}{2}\|x(t_f) - r(t_f)\|_H^2 + \frac{1}{2}\int_{t_0}^{t_f} \left[\|x(t) - r(t)\|_{Q(t)}^2 + \|u(t)\|_{R(t)}^2 \right] dt. \quad (3.142)$$

Theorem 3.8 *The optimal control for LQT problems is given as*

$$u^*(t) = F(t)x^*(t) + v(t),$$

where $F(t)$ is a time-variant matrix and $v(t)$ is a time-variant vector, respectively, i.e., $u^*(t)$ is the summation of a linear, time variant combination of the system state x at t and a time-variant vector $v(t)$.

Proof Apply the variational method, firstly define the Hamiltonian $\mathcal{H}$ for the underlying optimal tracking problem in the following:

$$
\begin{aligned}
&\mathcal{H}(x(t), u(t), \lambda(t), t) \\
&= \frac{1}{2}\|x(t) - r(t)\|_{Q(t)}^2 + \frac{1}{2}\|u(t)\|_{R(t)}^2 + [\lambda(t)]^\top [A(t)x(t) + B(t)u(t)].
\end{aligned} \tag{3.143}
$$

By Theorem 3.1, the necessary conditions for optimality are

$$
\dot{x}^*(t) = A(t)x^*(t) + B(t)u^*(t), \tag{3.144a}
$$

$$
\dot{\lambda}^*(t) = -\frac{\partial \mathcal{H}^*}{\partial x} = -Q(t)x^*(t) - A^\top(t)\lambda^*(t) + Q(t)r(t), \tag{3.144b}
$$

$$
0 = \frac{\partial \mathcal{H}^*}{\partial u} = R(t)u^*(t) + B^\top(t)\lambda^*(t), \tag{3.144c}
$$

with $\mathcal{H}^* \equiv \mathcal{H}(x^*(t), u^*(t), \lambda^*(t), t)$ and the boundary condition as

$$
\frac{\partial h(x^*(t_f), t_f)}{\partial x} - \lambda^*(t_f) = H[x^*(t_f) - r(t_f)] - \lambda^*(t_f) = 0. \tag{3.145}
$$

By (3.144c) and the definite positive of the matrix $A(t)$, we can obtain

$$
u^*(t) = -R^{-1}(t)B^\top(t)\lambda^*(t), \tag{3.146}
$$

by which we can get

$$
\dot{x}^*(t) = A(t)x^*(t) - B(t)R^{-1}(t)B^\top(t)\lambda^*(t). \tag{3.147}
$$

By organizing (3.144b) and (3.147), we can establish the following linear differential equation:

$$
\begin{bmatrix} \dot{x}^*(t) \\ \dot{\lambda}^*(t) \end{bmatrix} = \begin{bmatrix} A(t) & -B(t)R^{-1}(t)B^\top(t) \\ -Q(t) & -A^\top(t) \end{bmatrix} \begin{bmatrix} x^*(t) \\ \lambda^*(t) \end{bmatrix} + \begin{bmatrix} 0 \\ Q(t)r(t) \end{bmatrix}. \tag{3.148}
$$

The solution to the above linear differential equation has the form

$$
\begin{bmatrix} x^*(t_f) \\ \lambda^*(t_f) \end{bmatrix} = \Phi(t_f, t) \begin{bmatrix} x^*(t) \\ \lambda^*(t) \end{bmatrix} + \int_t^{t_f} \Phi(t_f, \tau) \begin{bmatrix} 0 \\ Q(\tau)r(\tau) \end{bmatrix} d\tau \tag{3.149}
$$

with Φ represents the transition matrix of (3.148).

If Φ is partitioned, these equations can be written as

$$\begin{bmatrix} x^*(t_f) \\ \lambda^*(t_f) \end{bmatrix} = \begin{bmatrix} \Phi_{11}(t_f,t) & \Phi_{12}(t_f,t) \\ \Phi_{21}(t_f,t) & \Phi_{22}(t_f,t) \end{bmatrix} \begin{bmatrix} x^*(t) \\ \lambda^*(t) \end{bmatrix} + \begin{bmatrix} \eta_1(t) \\ \eta_2(t) \end{bmatrix}. \tag{3.150}$$

By the boundary condition of (3.145),

$$\lambda^*(t_f) = Hx^*(t_f) - Hr(t_f); \tag{3.151}$$

then by using $Hx^*(t_f)$ to replace $\lambda^*(t_f)$ in (3.150), it gives

$$x^*(t_f) = \Phi_{11}(t_f,t)x^*(t) + \Phi_{12}(t_f,t)\lambda^*(t) + \eta_1(t), \tag{3.152a}$$
$$Hx^*(t_f) = \Phi_{21}(t_f,t)x^*(t) + \Phi_{22}(t_f,t)\lambda^*(t) + \eta_2(t); \tag{3.152b}$$

then it has

$$H[\Phi_{11}(t_f,t)x^*(t) + \Phi_{12}(t_f,t)\lambda^*(t) + \eta_1(t)] - Hr(t_f)$$
$$= \Phi_{21}(t_f,t)x^*(t) + \Phi_{22}(t_f,t)\lambda^*(t) + \eta_2(t). \tag{3.153}$$

Solving for $\lambda^*(t)$ yields

$$\lambda^*(t) = [\Phi_{22}(t_f,t) - H\Phi_{12}(t_f,t)]^{-1}[H\Phi_{11}(t_f,t) - \Phi_{21}(t_f,t)]x^*(t)$$
$$+ [\Phi_{22}(t_f,t) - H\Phi_{12}(t_f,t)]^{-1}[H\eta_1(t) - Hr(t_f) - \eta_2(t)]$$
$$\triangleq K(t)x^*(t) + s(t). \tag{3.154}$$

Thus by (3.154) together with (3.146), we obtain

$$u^*(t) = -R^{-1}(t)B^\top(t)K(t)x^*(t) - R^{-1}(t)B^\top(t)s(t)$$
$$\triangleq F(t)x^*(t) + v(t), \tag{3.155}$$

where $F(t) \triangleq -R^{-1}(t)B^\top(t)K(t)$ is the feedback gain matrix. $\qquad\qquad\square$

Figure 3.18 displays an illustration of the diagram of the optimal control for linear tracking problems.

In the following, we will determine the transition matrix Φ.

By differentiating both sides of (3.154) with respect to t, we can have

$$\dot{\lambda}^*(t) = \dot{K}^*(t)x^*(t) + K(t)\dot{x}^*(t) + \dot{s}(t), \tag{3.156}$$

by which together with (3.148), it has

$$[\dot{K}(t) + K(t)A(t) + Q(t)]\dot{x}^*(t) + \dot{s}(t) - Q(t)r(t)$$
$$+ [-K(t)B(t)R^{-1}(t)B^\top(t) + A^\top(t)]\lambda^*(t) = 0. \tag{3.157}$$

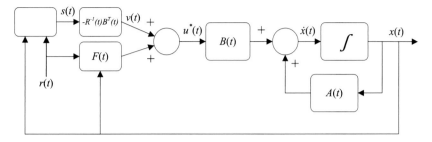

Fig. 3.18 A diagram of the optimal control for LQT problems

Then by (3.154), we obtains

$$0 = \left[\dot{K}(t) + Q(t) + K(t)A(t) + A^\top(t)K(t) - K(t)B(t)R^{-1}(t)B^\top(t)K(t) \right] x^*(t)$$
$$+ \left[\dot{s}(t) + A^\top(t)s(t) - K(t)B(t)R^{-1}(t)B^\top(t)s(t) - Q(t)r(t) \right]. \qquad (3.158)$$

Since the above equality holds for all $x^*(t)$ and $r(t)$, we can claim that

$$\dot{K}(t) = -K(t)A(t) - A^\top(t)K(t) - Q(t) + K(t)B(t)R^{-1}(t)B^\top(t)K(t), \qquad (3.159a)$$

$$\dot{s}(t) = -\left[A^\top(t) - K(t)B(t)R^{-1}(t)B^\top(t) \right] s(t) + Q(t)r(t). \qquad (3.159b)$$

By (3.151) and (3.154), we obtain

$$\lambda^*(t_f) = Hx^*(t_f) - Hr(t_f) = K(t_f)x^*(t_f) + s(t_f), \qquad (3.160)$$

which holds for all $x^*(t_f)$ and $r(t_f)$, hence the boundary conditions are

$$K(t_f) = H, \text{ and } s(t_f) = -Hr(t_f). \qquad (3.161)$$

Notice that it is straightforward to verify that in case the reference $r(t) = 0$ for all $t \in [t_0, t_f]$, the optimal solution to the LQT problems is the same as that to the LQR problems. It is consistent with the statement claimed earlier that the LQR problem formulated in Sect. 3.3 is a specific LQT problem in case $r(t) = 0$ for all $t \in [t_0, t_f]$.

The developed results will be demonstrated with some numerical examples below.

Example 3.14 Implement the optimal solution to the system defined in Example 3.12 such that the following performance cost function is minimized

$$J(u) = \frac{1}{2}H[x(t_f) - 1]^2 + \frac{1}{4}\int_0^{t_f} u^2(t)dt, \qquad (3.162)$$

with $H > 0$, and the final state $x(t_f)$ is free.

Solution. By the specification of the regulation problem, it has

$$A(t) = -a, B(t) = 1, r(t) = 1, Q(t) = 0, R(t) = \frac{1}{2}, H(t) = H. \qquad (3.163)$$

Thus by (3.148), we can obtain the following state and costate linear differential equation

$$\begin{bmatrix} \dot{x}^*(t) \\ \dot{\lambda}^*(t) \end{bmatrix} = \begin{bmatrix} A(t) & -B(t)R^{-1}(t)B^{\mathsf{T}}(t) \\ -Q(t) & -A^{\mathsf{T}}(t) \end{bmatrix} \begin{bmatrix} x^*(t) \\ \lambda^*(t) \end{bmatrix} + \begin{bmatrix} 0 \\ Q(t)r(t) \end{bmatrix}$$

$$= \begin{bmatrix} -a & -2 \\ 0 & a \end{bmatrix} \begin{bmatrix} x^*(t) \\ \lambda^*(t) \end{bmatrix}$$

with the boundary condition of $x^*(0) = x_0$ and

$$\lambda^*(t_f) = Hx^*(t_f) - Hr(t_f) = Hx^*(t_f) - H. \qquad (3.164)$$

Hence the optimal state and costate trajectories are given as

$$\begin{bmatrix} x^*(t) \\ \lambda^*(t) \end{bmatrix} = \begin{bmatrix} \exp(-at) & \frac{1}{a}\exp(-at) - \frac{1}{a}\exp(at) \\ 0 & \exp(at) \end{bmatrix} \begin{bmatrix} x^*(0) \\ \lambda^*(0) \end{bmatrix} \qquad (3.165)$$

with the boundary conditions of $x^*(0) = x_0$ and $\lambda^*(0) = \dfrac{aH[x_0 - \exp(at_f)]}{[H+a]\exp(2at_f) - H}$.

And we can get that the transition matrix as

$$\Phi(t_f, t) = \begin{bmatrix} \exp(-a[t_f - t]) & \frac{1}{a}\exp(-a[t_f - t]) - \frac{1}{a}\exp(a[t_f - t]) \\ 0 & \exp(a[t_f - t]) \end{bmatrix}. \qquad (3.166)$$

By (3.154), $\lambda^*(t) = K(t)x^*(t) + s(t)$, with $K(t)$ and $s(t)$ given below:

$$K(t) \triangleq [\Phi_{22}(t_f, t) - H\Phi_{12}(t_f, t)]^{-1}[H\Phi_{11}(t_f, t) - \Phi_{21}(t_f, t)]$$

$$= \frac{aH}{[H+a]\exp(2a[t_f - t]) - H}, \qquad (3.167a)$$

$$s(t) \triangleq [\Phi_{22}(t_f, t) - H\Phi_{12}(t_f, t)]^{-1}[H\eta_1(t) - Hr(t_f) - \eta_2(t)]$$

$$= -\frac{aH\exp(a[t_f - t])}{[H+a]\exp(2a[t_f - t]) - H}; \qquad (3.167b)$$

then by (3.155), the optimal control is

$$u^*(t) = -R^{-1}(t)B^{\mathsf{T}}(t)K(t)x^*(t) - R^{-1}(t)B^{\mathsf{T}}(t)s(t) = -2K(t)x^*(t) - 2s(t). \qquad (3.168)$$

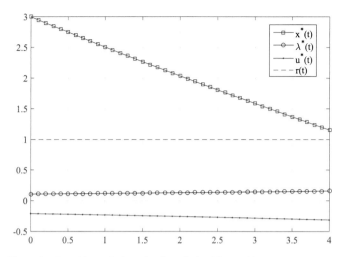

Fig. 3.19 The optimal tracking solution x^*, λ^*, and u^* with $a = 0.1$ and $H = 1$

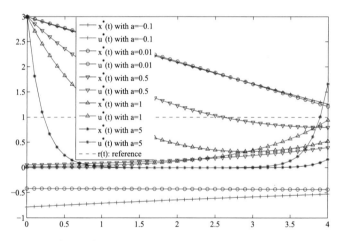

Fig. 3.20 Display of x^* and u^* with $a = -0.1, 0.01, 0.5, 1,$ and 5, respectively

We will study the optimal solutions to the LQT problems with respect to different values of the parameters of the system a and H.

Figure 3.19, displays the optimal tracking solution x^*, λ^* and u^* with respect to $a = 0.1$ and $H = 1$.

Moreover, Fig. 3.20, displays the optimal solution x^* and u^* with respect to $H = 1$ and different valued a respectively.   □

Example 3.15 Implement the optimal solution to the system defined in Example 3.12 such that the following performance cost function is minimized

$$J(u) = \frac{1}{2}H\left[x(t_f) - \frac{t_f}{4}\right]^2 + \int_0^{t_f}\left[\frac{1}{2}\left[x(t) - \frac{t}{4}\right]^2 + \frac{1}{4}u^2(t)\right]dt, \qquad (3.169)$$

with $H > 0$, and the final state $x(t_f)$ is free.

Solution. By the specification of the tracking problem, it has

$$A(t) = -a, B(t) = 1, r(t) = \frac{t}{4}, Q(t) = 1, R(t) = \frac{1}{2}, H(t) = H. \qquad (3.170)$$

By the Riccati equation for K and the differential equation for s specified in (3.159a) and (3.159b), respectively,

$$\dot{K}(t) = 2K^2(t) + 2aK(t) - 1, \qquad (3.171a)$$

$$\dot{s}(t) = [a + 2K(t)]s(t) + \frac{t}{4}. \qquad (3.171b)$$

By (3.161), it gets that the boundary conditions for K and s are given below respectively.

$$K(t_f) = H, s(t_f) = -\frac{H}{4}t_f. \qquad (3.172)$$

Thus, by (3.155), the optimal control is

$$u^*(t) = -R^{-1}(t)B^\top(t)K(t)x^*(t) - R^{-1}(t)B^\top(t)s(t) = -2K(t)x^*(t) - 2s(t). \qquad (3.173)$$

$\square$

Example 3.16 Implement the optimal solution to the problem with the system specified in Example 3.1, say

$$\dot{x}_1(t) = x_2(t)$$
$$\dot{x}_2(t) = -u(t)$$

with the initial state value as $x(t_0) = \begin{bmatrix} 2 \\ 2 \end{bmatrix}$, with $t_0 = 0$, and the final time as $t_f = 2$, such that the following performance cost function

$$J(u) = [x_1(t_f) - 1]^2 + [x_2(t_f) + 1]^2 + \int_{t_0}^{t_f} u^2(t)dt \qquad (3.174)$$

is minimized.

Solution. By the specification of the underlying control system, it has

$$A(t) = \begin{bmatrix} 0 & 1 \\ 0 & 0 \end{bmatrix}, \; B(t) = \begin{bmatrix} 0 \\ -1 \end{bmatrix}, \; Q(t) = \begin{bmatrix} 0 & 0 \\ 0 & 0 \end{bmatrix}, \; R(t) = 2, \; H = \begin{bmatrix} 2 & 0 \\ 0 & 2 \end{bmatrix}, \; r(t) = \begin{bmatrix} 1 \\ -1 \end{bmatrix}.$$

By the Riccati equation for $K(\cdot)$ specified in (3.159a)

$$\dot{K}_{11}(t) = \frac{1}{2} K_{12}^2(t), \tag{3.175a}$$

$$\dot{K}_{12}(t) = \frac{1}{2} K_{12}(t) K_{22}(t) - K_{11}(t), \tag{3.175b}$$

$$\dot{K}_{22}(t) = \frac{1}{2} K_{22}^2(t) - 2K_{12}(t), \tag{3.175c}$$

and, by the differential equation for $s(t)$ given in (3.159b), it has

$$\dot{s}_1(t) = \frac{1}{2} K_{12}(t) s_2(t), \tag{3.176a}$$

$$\dot{s}_2(t) = -s_1(t) + \frac{1}{2} K_{22}(t) s_2(t). \tag{3.176b}$$

By (3.161), it gets that the boundary conditions for K and s are given below, respectively.

$$K(t_f) = \begin{bmatrix} 2 & 0 \\ 0 & 2 \end{bmatrix}, \; s(t_f) = \begin{bmatrix} -2 \\ 2 \end{bmatrix}. \tag{3.177}$$

Thus, by (3.155), the optimal control is

$$u^*(t) = -R^{-1}(t) B^\top(t) K(t) x^*(t) - R^{-1}(t) B^\top(t) s(t)$$
$$= \frac{1}{2} \left[K_{12}(t) x_1^*(t) + K_{22}(t) x_2^*(t) + s_2(t) \right]. \tag{3.178}$$

□

3.5 Summary

In this chapter, optimal control problems by applying the results of the extremal of functional via the variational method developed in Chap. 2 are studied. It gives the necessary and sufficient conditions for the optimal solution to the optimal control problems concerning different boundary conditions on the final time and final state, respectively, and by applying these analyses the optimal control solutions for specific LQR and LQT problems are solved.

3.6   Exercises

Exercise 3.1 Implement the optimal control for a state system given by

$$\dot{x} = ax(t) + bu(t),$$

with the boundary condition of $x(0) = 2$ and $x(2) = 6$, such that the performance cost function

$$J(u) = \frac{1}{2} \int_0^2 \left[x^2(t) + u^2(t) \right] dt$$

is minimized.

Exercise 3.2 Given a system as

$$\ddot{x} = x(t) + u(t)$$

with the boundary conditions of

$$x(0) = 0, \ x(1) = 5, \ \dot{x}(0) = 1, \ \text{and} \ \dot{x}(1) = 3.$$

Determine the optimal control for this system subject to which the following performance cost function

$$J(u) = \frac{1}{4} \int_0^1 u^2(t) dt$$

is minimized.

Exercise 3.3 Consider a system given below

$$\dot{x}_1(t) = x_2(t),$$
$$\dot{x}_2(t) = -x_1(t) + 2x_2(t) + u(t)$$

with the boundary condition of $x(0) = \begin{bmatrix} 0 \\ 0 \end{bmatrix}$ and $x(1) = \begin{bmatrix} 2 \\ 2 \end{bmatrix}$ to be controlled to minimize the following performance cost function:

$$J(u) = \int_0^2 \left[x_1^2(t) + 2x_1(t)x_2(t) + x_2^2(t) + u^2(t) \right] dt.$$

Specify the costate equations for the system, and the control which minimizes the Hamiltonian $\mathcal{H}$.

Exercise 3.4 Consider the system given in Exercise 3.3 with $x(0) = \begin{bmatrix} 0 \\ 0 \end{bmatrix}$ and the final state $x(1)$ being free, and the following performance cost function

$$J(u) = \int_0^2 \left[x_1^2(t) + u^2(t) \right] dt$$

is to be minimized.

Specify the costate equations for the system, and the control which minimizes the Hamiltonian $\mathcal{H}$.

Exercise 3.5 (*Optimal Control Problems with Free Final Time and Fixed Final State*) Consider a state system $\dot{x}(t) = -2x(t) + u(t)$, with the initial state $x(0) = 3$ and final state $x(t_f) = 1$.

Determine the optimal control $u^*(t)$ and its associated final time t_f, such that the following performance cost function

$$J(u) = \int_{t_0}^{t_f} \left[2x^2(t) + u^2(t) \right] dt$$

is minimized.

Exercise 3.6 (*Optimal Control Problems with Free Final Time and Free Final State: Final State Lying on An Oval Surface*) Consider a two-dimensional system

$$\dot{x}_1(t) = x_2(t),$$
$$\dot{x}_2(t) = x_1(t) - u(t)$$

with $t_0 = 0$, $t_f = 4$, and the boundary condition as $x(t_0) = \begin{bmatrix} 1 \\ -1 \end{bmatrix}$ and $x(t_f)$ lying on the surface $\phi(x(t_f))$ as specified below:

$$\phi(x(t_f)) \triangleq [x_1(t_f) - 1]^2 + 2[x_2(t_f) - 2]^2 - 4 = 0.$$

Determine the necessary condition for the optimal control u^* and the associated state processes x^* such that the following cost function

$$J(u) = \int_{t_0}^{t_f} u^2(t) dt$$

is minimized.

Exercise 3.7 (*Optimal Control Problems with Free Final Time and Free Final State: Independence of Each Other*) Consider the state system defined in Exercise 3.6 with a given initial state $x(0) = \begin{bmatrix} 1 \\ -1 \end{bmatrix}$ and the final time t_f and final state $x(t_f)$ being unspecified.

Determine the optimal control u^* such that

$$J(u) = \frac{1}{2} \int_{t_0}^{t_f} \left[x_1^2(t) + u^2(t) \right] dt$$

is minimized.

Exercise 3.8 (*Optimal Control Problems with Free Final Time and Free Final State: δx_f and δt_f correlated with Each Other*) Find the optimal solution to the problem given in Example 3.1 with a boundary condition such that the final time t_f is free and the final state $x(t_f)$ lies on $\begin{bmatrix} t_f + 1 \\ \frac{1}{2} t_f^2 \end{bmatrix}$.

Exercise 3.9 (*Optimal Control Problems with Free Final Time and Free Final State: Final State $x(t_f)$ on A Time-invariant Surface*) Determine the optimal control u^* for the problems specified in Exercise 3.6 except that the final time t_f is unspecified.

Exercise 3.10 (*Optimal Control Problems with Free Final Time and Free Final State: Final State $x(t_f)$ on A Time-variant Surface*) Revisit Exercise 3.9 by considering a time-variant surface, denoted by $\phi(x(t_f), t_f)$, such that

$$\phi(x(t_f), t_f) \triangleq x_1(t_f) + 2[x_2(t_f) - 2t]^2 - 4 = 0.$$

Exercise 3.11 (*LQR Problems*) Find the optimal control solution to the following LQR problem for the following two-dimensional system:

$$\dot{x}_1(t) = x_2(t),$$
$$\dot{x}_2(t) = x_1(t) - x_2(t) + u(t),$$

which is to be controlled to minimize

$$J(u) = \frac{1}{2} \int_{t_0}^{t_f} \left[x_1^2(t) + x_2^2(t) + u^2(t) \right] dt.$$

Exercise 3.12 (*LQR Problems: Stationary Case*) Find the optimal control solution to the following infinite-interval LQR problem for the system specified in Exercise 3.11 such that the following infinite-interval performance cost function is minimized:

$$J(u) = \frac{1}{2} \int_0^\infty \left[x_1^2(t) + u^2(t) \right] dt.$$

Exercise 3.13 (*LQT Problems*) Consider a two-dimensional system as follows:

$$\dot{x}_1(t) = x_2(t),$$
$$\dot{x}_2(t) = x_1(t) - x_2(t) + u(t);$$

then find the optimal tracking control solution to the underlying system with the performance cost function specified in the following:

$$J(u) = [x_1(t_f) - 1]^2 + \int_{t_0}^{t_f} \left[[x_1(t) - 1]^2 + \frac{1}{8} u^2(t) \right] dt.$$

Exercise 3.14 (*LQT Problems*) Find the optimal solution to the system given in Exercise 3.13, such that the following performance cost function is minimized:

$$J(u) = \int_{t_0}^{t_f} \left[\left[x_1(t) - \frac{t}{4} \right]^2 + \frac{1}{8} u^2(t) \right] dt.$$

Chapter 4
Pontryagin's Minimum Principle

In Chap. 3, the optimal control problems with the assumption that the admissible controls and states are not constrained by any boundaries are studied. However, such constraints certainly occur in realistic systems. Controls generally have certain limitations, like energy consumed in a vehicle, the thrust of a rocket in a spaceship, and due to the concerns on the safety and restrictions on the structure of the system, the state may be constrained, e.g., the speed of a motor in a control system, the speed limit of a vehicle, the fuel stored in a rocket and the current limitation of an electric circuit.

In this chapter, we develop the Pontryagin's minimum principle for optimal control problems with constrained control and constrained system state in Sects. 4.1 and 4.2, respectively; then by applying the proposed Pontryagin's minimum principle, we study how to implement the optimal control solution for specific interesting optimal control problems with constraints, say minimum time, minimum fuel, and minimum energy problems in Sects. 4.3, 4.4, and 4.6, respectively. For the purpose of comparison, it further studies the optimal controls for minimum time-fuel and minimum time-energy problems in Sects. 4.5 and 4.7, respectively. In Sect. 4.8, a brief summary of the results developed in this chapter is given. Finally, in Sect. 4.9, some exercises for readers to assist them to grasp the stuff covered in this chapter are listed.

4.1 Pontryagin's Minimum Principle with Constrained Control

First, consider the effect of control constraints on the fundamental theorem derived in Theorem 2.1 in Chap. 2, and then show how the developed necessary conditions for the optimal control solutions without constraints on controls and states are modified.

As a result, the generalization of the fundamental theorem can lead to the so-called Pontryagin's minimum principle developed in Sects. 4.1 and 4.2.

By definition, the control u^* causes the functional J to have a relative minimum if

$$J(u) - J(u^*) = \Delta J \geq 0 \tag{4.1}$$

for all admissible controls sufficiently close to u^*. If $u = u^* + \delta u$, the increment in J can be expressed as

$$\Delta J(u^*, \delta u) = \delta(u^*, \delta u) + \text{higher-order terms.} \tag{4.2}$$

δJ is linear in δu and the higher-order terms approach zero as the norm of δu goes to zero. As discussed earlier, in case the control is unbounded, it could use the linearity of δJ with respect to δu, and the fact that δu can vary arbitrarily to show that a necessary condition for δu^* to be an extremal control is that the variation $\delta J(u^*, \delta u)$ must be zero for all admissible δu having a sufficiently small norm.

In the past, it does not consider the case that the admissible controls are not bounded, δu is arbitrary only if the extremal control is strictly within the boundary for all time in the interval $[t_0, t_f]$, as displayed in Figs. 4.1 and 4.2. In this situation, the boundary of the control set has not effect on the problem solution.

However, as displayed in Fig. 4.3, suppose that an extremal control lies on a boundary during a subinterval $[t_1, t_2]$ of the interval $[t_0, t_f]$; then there exists an admissible control variation denoted by $\widehat{\delta u}$, see Fig. 4.4 for an illustration. We can obtain that a necessary condition for the extremal solution u^* to minimize the performance cost function J is that $\delta J(u^*, \delta u) \geq 0$ if it only considers the admissible variations.

On the other hand, for variations $\delta\tilde{u}$, which are nonzero only for t not in the interval $[t_1, t_2]$ as, for example, in Fig. 4.4, it is necessary that $\delta J(u^*, \delta\tilde{u}) = 0$; the reasoning used in proving the fundamental theorem applies.

Considering all admissible variations with $\|\delta u\|$ small enough so that the sign of ΔJ is determined by δJ, thus a necessary condition for u^* to minimize J is

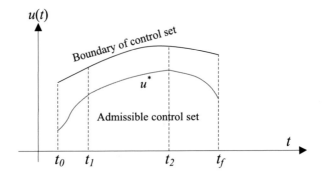

Fig. 4.1 An illustration of an optimal control inside a constrained control set

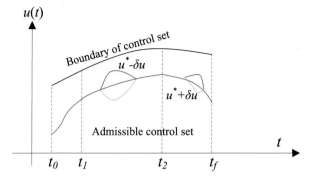

Fig. 4.2 An illustration of variations of an optimal control located inside a constrained control set

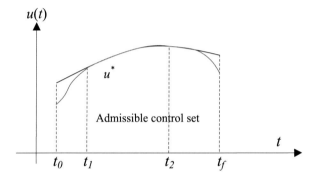

Fig. 4.3 A constrained control set and an illustration of admissible optimal control

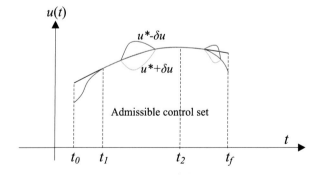

Fig. 4.4 A constrained control set and an illustration of inadmissible control

$$\delta J(u^*, \delta u) \geq 0. \tag{4.3}$$

Refer a function f defined on a closed interval $[t_0, t_f]$. The differential df is the linear part of the increment δf. Consider the end points t_0 and t_f of the interval, and admissible values of the time increment Δt, which are small enough so that the sign of Δf is determined by the sign of df. The necessary conditions for the function f to have relative minima at the end points of the interval are

$$df(t_0, \Delta t) \geq 0, \quad \text{admissible} \quad \Delta t \geq 0, \tag{4.4a}$$
$$df(t_f, \Delta t) \geq 0, \quad \text{admissible} \quad \Delta t \leq 0, \tag{4.4b}$$

and a necessary condition for f to have a relative minimum at an interior point t, $t_0 < t < t_f$, is

$$df(t, \Delta t) = 0. \tag{4.5}$$

For the control problem, the analogous necessary conditions are

$$\delta J(u^*, \delta u) \geq 0, \tag{4.6}$$

if u^* lies on the boundary during any portion of the time interval $[t_0, t_f]$, and

$$\delta J(u^*, \delta u) = 0, \tag{4.7}$$

if u^* lies within the boundary during the entire time interval $[t_0, t_f]$.

Theorem 4.1 *The necessary conditions for u^* to be an optimal control are*

$$\dot{x}^*(t) = \frac{\partial \mathcal{H}(x^*(t), u^*(t), \lambda^*(t), t)}{\partial \lambda}, \tag{4.8a}$$

$$\dot{\lambda}^*(t) = -\frac{\partial \mathcal{H}(x^*(t), u^*(t), \lambda^*(t), t)}{\partial x}, \tag{4.8b}$$

$$\mathcal{H}(x^*(t), u^*(t), \lambda^*(t), t) \leq \mathcal{H}(x^*(t), u(t), \lambda^*(t), t), \tag{4.8c}$$

and with the following boundary condition:

$$\left[\frac{\partial h(x^*(t_f), t_f)}{\partial x} - \lambda^*(t_f) \right]^\top \delta x_f$$
$$+ \left[\mathcal{H}(x^*(t_f), u^*(t_f), \lambda^*(t_f), t_f) + \frac{\partial h(x^*(t_f), t_f)}{\partial t} \right] \delta t_f = 0. \tag{4.9}$$

$u^*(t)$ *is a control that causes $\mathcal{H}(x^*(t), u(t), \lambda^*(t), t)$ minimum.*
Notice that the above conditions are necessary but not sufficient.

Proof Firstly, define the increment of the cost function J as follows:

$$\Delta J(u^*, \delta u) = \left[\frac{\partial h(x^*(t_f), t_f)}{\partial x} - \lambda^*(t_f)\right]^\top \delta x_f$$

$$+ \left[\mathcal{H}(x^*(t_f), u^*(t_f), \lambda^*(t_f), t_f) + \frac{\partial h(x^*(t_f), t_f)}{\partial t}\right] \delta t_f$$

$$+ \int_{t_0}^{t_f} \left[\left[\dot{\lambda}^*(t) + \frac{\partial \mathcal{H}(x^*(t), u^*(t), \lambda^*(t), t)}{\partial x}\right]^\top \delta x(t)\right.$$

$$+ \left[\frac{\partial \mathcal{H}(x^*(t), u^*(t), \lambda^*(t), t)}{\partial u}\right]^\top \delta u(t)$$

$$\left.+ \left[\frac{\partial \mathcal{H}(x^*(t), u^*(t), \lambda^*(t), t)}{\partial \lambda} - \dot{x}^*(t)\right]^\top \delta \lambda(t)\right] dt$$

$$+ \text{higher-order terms.}$$

If the state equations are satisfied, and $\lambda^*(t)$ is selected so that the coefficient of $\delta x(t)$ in the integral is identically zero, and the boundary condition equation is satisfied, it has

$$\Delta J(u^*, \delta u) = \int_{t_0}^{t_f} \left[\frac{\partial \mathcal{H}(x^*(t), u^*(t), \lambda^*(t), t)}{\partial u}\right]^\top \delta u(t) dt$$

$$+ \text{higher-order terms.}$$

The integrand is the first-order approximation to the change in $\mathcal{H}$ caused by a change in u alone; that is,

$$\left[\frac{\partial \mathcal{H}(x^*(t), u^*(t), \lambda^*(t), t)}{\partial u}\right]^\top \delta u(t)$$

$$\doteq \mathcal{H}(x^*(t), u^*(t) + \delta u(t), \lambda^*(t), t) - \mathcal{H}(x^*(t), u^*(t), \lambda^*(t), t). \qquad (4.10)$$

Hence we obtain that

$$\Delta J(u^*, \delta u) = \int_{t_0}^{t_f} [\mathcal{H}(x^*(t), u^*(t) + \delta u(t), \lambda^*(t), t)$$

$$+ \mathcal{H}(x^*(t), u^*(t), \lambda^*(t), t)] dt$$

$$+ \text{higher-order terms.}$$

In case that $u^* + \delta u$ is in a sufficiently small neighborhood of u^*, the higher-order terms are smaller compared with the lower-order one.

Hence for u^* to be a minimizing control, it is necessary

$$\int_{t_0}^{t_f} \left[\mathcal{H}(x^*(t), u^*(t) + \delta u(t), \lambda^*(t), t) - \mathcal{H}(x^*(t), u^*(t), \lambda^*(t), t) \right] dt \geq 0$$

$$(4.11)$$

for all admissible control u.

Next, we will verify below that a necessary condition for the optimality of u^* is that

$$\mathcal{H}(x^*(t), u^*(t) + \delta u(t), \lambda^*(t), t) \geq \mathcal{H}(x^*(t), u^*(t), \lambda^*(t), t) \qquad (4.12)$$

for all admissible $\delta u(t)$ and for all $t \in [t_0, t_f]$.

Define a control in a specific form of

$$u(t) = \begin{cases} u^*(t), & \text{with } t \notin [t_1, t_2] \\ u^*(t) + \delta u(t), & \text{with } t \in [t_1, t_2] \end{cases}, \qquad (4.13)$$

where $[t_1, t_2]$ is an arbitrarily small, but nonzero, time interval, and δu is an admissible control variation with $\|\delta u\| < \beta$.

Suppose that the inequality (4.12) is not satisfied for the control specified in (4.13). Thus for any $t \in [t_1, t_2]$, the following holds

$$\mathcal{H}(x^*(t), u(t), \lambda^*(t), t) < \mathcal{H}(x^*(t), u^*(t), \lambda^*(t), t), \qquad (4.14)$$

and hence we have the following analysis:

$$\int_{t_0}^{t_f} [\mathcal{H}(x^*(t), u(t), \lambda^*(t), t) - \mathcal{H}(x^*(t), u^*(t), \lambda^*(t), t)] dt$$

$$= \int_{t_1}^{t_2} [\mathcal{H}(x^*(t), u(t), \lambda^*(t), t) - \mathcal{H}(x^*(t), u^*(t), \lambda^*(t), t)] dt < 0. \qquad (4.15)$$

Since $[t_1, t_2]$ could be any subinterval in $[t_0, t_f]$, then in case that

$$\mathcal{H}(x^*(t), u(t), \lambda^*(t), t) < \mathcal{H}(x^*(t), u^*(t), \lambda^*(t), t), \qquad (4.16)$$

for any $t \in [t_0, t_f]$, it is always possible to design an admissible control subject to which $\Delta J < 0$. This is contradicted with the optimality of u^*.

Thus we can make a conclusion that a necessary condition for a control u^* to minimize the performance cost function J is the following:

$$\mathcal{H}(x^*(t), u^*(t), \lambda^*(t), t) \leq \mathcal{H}(x^*(t), u(t), \lambda^*(t), t) \qquad (4.17)$$

for all $t \in [t_0, t_f]$ and for all admissible controls u.

The optimal control minimizing the Hamiltonian is called *Pontryagin's minimum principle*.   □

As obtained above, the Pontryagin's minimum principle, it is developed for the optimal control problems with the control constrained in a closed and bounded set. Nevertheless, it is worth to notice that the Pontryagin's minimum principle can also be used for those problems without constraints on the controls, as studied in Chap. 3, by considering an arbitrarily large bound for controls, such that the optimal control will never reach the boundary of the control set.

Thus, in this situation, suppose that $u^*(t)$ minimizes the Hamiltonian, then the following must hold:

$$\frac{\partial \mathcal{H}(x^*(t), u^*(t), \lambda^*(t), t)}{\partial u} = 0.$$

Moreover, besides the above, suppose further the following

$$\frac{\partial^2 \mathcal{H}(x^*(t), u^*(t), \lambda^*(t), t)}{\partial u^2} > 0$$

holds as well; then it is sufficient to state that $u^*(t)$ minimizes the Hamiltonian.

Example 4.1 Consider a system such that the state equation is given as

$$\dot{x}_1(t) = x_2(t),$$
$$\dot{x}_2(t) = -x_2(t) + u(t), \tag{4.18a}$$

with the initial condition of $x(t_0) = x_0$, the final time t_f is specified, and the final state $x(t_f)$ is free.

The performance cost function is specified as

$$J(u) = \int_{t_0}^{t_f} \frac{1}{2} \left[x_1^2(t) + u^2(t) \right] dt. \tag{4.19}$$

(I) Give a necessary condition for an unconstrained control to minimize J.
(II) Give a necessary condition for optimal control in case

$$-1 \le u(t) \le 1, \tag{4.20}$$

for all $t \in [t_0, t_f]$.

Solution of (I). The Hamiltonian of the underlying problem is specified as

$$\mathcal{H}(x(t), u(t), \lambda(t))$$
$$= \frac{1}{2} x_1^2(t) + \frac{1}{2} u^2(t) + \lambda_1(t) x_2(t) - \lambda_2(t) x_2(t) + \lambda_2(t) u(t), \tag{4.21}$$

from which the costate equations are

$$\dot{\lambda}_1^*(t) = -\frac{\partial \mathscr{H}^*}{\partial x_1} = -x_1^*(t), \tag{4.22}$$

$$\dot{\lambda}_2^*(t) = -\frac{\partial \mathscr{H}^*}{\partial x_2} = -\lambda_1^*(t) + \lambda_2^*(t) \tag{4.23}$$

with $\mathscr{H}^* \equiv \mathscr{H}(x^*(t), u^*(t), \lambda^*(t))$.

Since the control values are unconstrained, by applying Theorem 3.1 in Chap. 3, it is necessary that

$$\frac{\partial \mathscr{H}^*}{\partial u} = u^*(t) + \lambda_2^*(t) = 0, \tag{4.24}$$

which implies that $u^*(t) = -\lambda_2^*(t)$ for all t.

Besides, we have

$$\frac{\partial^2 \mathscr{H}^*}{\partial u^2} = 1 > 0; \tag{4.25}$$

then we can obtain the control given below

$$u^*(t) = -\lambda_2^*(t), \tag{4.26}$$

Which minimizes the Hamiltonian.

Also, we can obtain the boundary condition $\lambda^*(t_f) = 0$.

Solution of (II). It is obvious that the state and costate equations and the boundary condition for $\lambda^*(t_f)$ remain unchanged as for case (I).

However due to the constraints on the control in this situation, the optimal control to minimize the following Hamiltonian

$$\mathscr{H}(x^*(t), u(t), \lambda^*(t))$$
$$= \frac{1}{2}[x_1^*(t)]^2 + \frac{1}{2}u^2(t) + \lambda_1^*(t)x_2^*(t) - \lambda_2^*(t)x_2^*(t) + \lambda_2^*(t)u(t) \tag{4.27}$$

is different from that specified in (I).

To specify the control that minimizes $\mathscr{H}$, it leaves all of the terms containing $u(t)$ in the $\mathscr{H}$ as shown below

$$\frac{1}{2}u^2(t) + \lambda_2^*(t)u(t). \tag{4.28}$$

For those instants at which the optimal control is unsaturated, by (4.28), it can be obtained that

$$u^*(t) = -\lambda_2^*(t), \tag{4.29}$$

as displayed in Fig. 4.5 when $\lambda_2^*(t)$ satisfies the inequality of $|\lambda_2^*(t)| \leq 1$.

Suppose that there exist instances at which $|\lambda_2^*(t)| > 1$, then the control which minimizes $\mathscr{H}$ is given as

$$u^*(t) = \begin{cases} -1, & \text{in case } \lambda_2^*(t) > 1 \\ 1. & \text{in case } \lambda_2^*(t) < -1 \end{cases}.$$

In summary, it is given the optimal control $u^*(t)$ is a saturation function of $\lambda_2^*(t)$, such that

$$u^*(t) = \begin{cases} -1, & \text{in case } \lambda_2^*(t) > 1 \\ -\lambda_2^*(t), & \text{in case } -1 \leq \lambda_2^*(t) \geq 1, \\ 1, & \text{in case } \lambda_2^*(t) < -1 \end{cases}$$

as displayed in Fig. 4.5. $\qquad\qquad\qquad\qquad\qquad\qquad\qquad\qquad\qquad\qquad\qquad\square$

By the analysis above, we get that, in order to implement the optimal control u^* explicitly, the state and costate equations should be solved explicitly. The state-costate trajectories in these two cases, say constrained and unconstrained on the control, will be exactly identical with each other only in case that the state values are those such that the bounded control does not saturate.

Here it is given some necessary conditions for the optimality of the Hamiltonian $\mathscr{H}$ for some specific cases as stated below.

- Suppose the final time t_f is fixed and the Hamiltonian $\mathscr{H}$ does not depend explicitly upon time t; then

$$\mathscr{H}\left(x^*(t), u^*(t), \lambda^*(t)\right) = \xi_1, \tag{4.30}$$

for all $t \in [t_0, t_f]$, i.e., the Hamiltonian $\mathscr{H}$ is constant valued on an extremal trajectory.

Fig. 4.5 Constrained and unconstrained optimal controls for Example 4.1

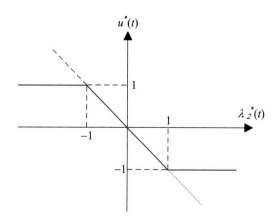

- Suppose the final time t_f is free and the Hamiltonian $\mathcal{H}$ does not explicitly depend upon time t; then

$$\mathcal{H}\left(x^*(t), u^*(t), \lambda^*(t)\right) = 0, \tag{4.31}$$

for all $t \in [t_0, t_f]$, i.e., the Hamiltonian $\mathcal{H}$ is equal to zero on an extremal trajectory.

4.2 Pontryagin's Minimum Principle with Constrained State Variable

Now consider problems in which there may be inequality constraints that involve the state variables as well as the controls. It will be assumed that the state constraints are of the form

$$f(x(t), t) \equiv \begin{bmatrix} f_1(x(t), t) \\ \cdots \\ f_l(x(t), t) \end{bmatrix} \geq 0, \tag{4.32}$$

where f is a vector function of x and t, which has continuous first and second partial derivatives with respect to $x(t)$.

In the following, the collection of inequalities will be transformed into a single equality constraint, and then the performance cost function is augmented with this equality constraint, as it was considered earlier with the state equations.

Define a new variable $\dot{x}_{n+1}(t)$ as below

$$\dot{x}_{n+1}(t) \triangleq \sum_{i=1}^{l} [f_i(x(t), t)]^2 I_{f_i < 0}, \tag{4.33}$$

where $I_{f_i < 0}$ represents an indicator function, say

$$I_{f_i < 0} = \begin{cases} 1, & \text{in case } f_i < 0 \\ 0. & \text{otherwise} \end{cases}. \tag{4.34}$$

Notice that $\dot{x}_{n+1}(t) \geq 0$ for all t, and that $\dot{x}_{n+1}(t) = 0$ only when all of the constraints are satisfied.

Also suppose that $x_{n+1}(t)$ is given as below

$$x_{n+1}(t) = \int_{t_0}^{t} \dot{x}_{n+1}(t) dt + x_{n+1}(t_0), \tag{4.35}$$

and satisfies the boundary conditions of $x_{n+1}(t_0) = 0$ and $x_{n+1}(t_f) = 0$.

Since, by the definition, $\dot{x}_{n+1}(t) \geq 0$ for all t, the above boundary conditions implies that $\dot{x}_{n+1}(t) = 0$, for all $t \in [t_0, t_f]$. However this occurs only in case that the constraints are satisfied for all t.

The objective is to minimize the functional

$$J(u) = h(x(t_f), t_f) + \int_{t_0}^{t_f} g(x(t), u(t), t)dt \qquad (4.36)$$

subject to the constraints due to the following state equation:

$$\dot{x}(t) = a(x(t), u(t), t), \qquad (4.37)$$

the constraints on the control variables, and the state inequality constraints of $f(x(t), t) \geq 0$, for all t.

Here it first forms the Hamiltonian as below

$$\mathcal{H}\big(x(t), u(t), \lambda(t), t\big)$$
$$= g(x(t), u(t), t) + \sum_{i=1}^{n} \lambda_i(t)a_i(t)(x(t), u(t), t) + \lambda_{n+1}(t) \sum_{i=1}^{l} [f_i(x(t), t)]^2 I_{f_i < 0}$$
$$\triangleq g(x(t), u(t), t) + \lambda^\top(t)a(x(t), u(t), t), \qquad (4.38)$$

where it uses $a(\cdot)$ as an extended vector function such that

$$a_{n+1}(x(t), t) \triangleq \sum_{i=1}^{l} [f_i(x(t), t)]^2 I_{f_i < 0}. \qquad (4.39)$$

Notice that as observed from (4.38), the Hamiltonian does not contain $x_{n+1}(t)$ explicitly.

Hence by the proposed results concerning the necessary condition of optimal control solution, we have the state and costate equations given as

$$\dot{x}_i^*(t) = a_i(x^*(t), u^*(t), t), \qquad (4.40a)$$
$$\dot{\lambda}_i^*(t) = -\frac{\partial \mathcal{H}(x^*(t), u^*(t), \lambda^*(t), t)}{\partial x_i}, \qquad (4.40b)$$

for all $i = 1, \ldots, n+1$, such that $\dot{\lambda}_{n+1}^* = 0$, since $x_{n+1}(t)$ does not appear explicitly in $\mathcal{H}$.

The boundary conditions $x^*(t_0)$ are specified with $x_{n+1}^*(t_0) = 0$ and $x_{n+1}^*(t_f) = 0$, and the remaining boundary conditions at $t = t_f$ can be determined by using the results obtained in earlier parts.

Thus the necessary conditions for optimality is (4.40) together with the following optimality of the Hamiltonian $\mathcal{H}$:

$$\mathcal{H}(x^*(t), u^*(t), \lambda^*(t), t) \le \mathcal{H}(x^*(t), u(t), \lambda^*(t), t). \tag{4.41}$$

Now revisit the optimal control problem studied earlier in Example 4.2 in the following by considering certain constraints for states this time.

Example 4.2 Consider a system with the state equation given below

$$\dot{x}_1(t) = x_2(t), \tag{4.42a}$$
$$\dot{x}_2(t) = -x_2(t) + u(t), \tag{4.42b}$$

which is to be controlled to minimize the following performance cost function

$$J(u) = \int_{t_0}^{t_f} \frac{1}{2} \left[x_1^2(t) + u^2(t) \right] dt, \tag{4.43}$$

such that $x(t_0)$ is specified, and t_f is given but the final state $x(t_f)$ is free.
It also supposes that the admissible control values are constrained by

$$-1 \le u(t) \le 1, \ \text{ for all } t \in [t_0, t_f]. \tag{4.44}$$

Moreover, it is assumed that the state satisfies the following constraint

$$-2 \le x_2(t) \le 2 \tag{4.45}$$

for all $t \in [t_0, t_f]$.
Solution. By the constraints on the state x_2, we obtain

$$x_2(t) + 2 \ge 0, \tag{4.46a}$$
$$2 - x_2(t) \ge 0; \tag{4.46b}$$

then we have

$$f_1(x(t)) \triangleq x_2(t) + 2 \ge 0, \tag{4.47a}$$
$$f_2(x(t)) \triangleq 2 - x_2(t) \ge 0. \tag{4.47b}$$

Thus the Hamiltonian $\mathscr{H}$ is given by

$$\mathscr{H}(x(t), u(t), \lambda(t)) \tag{4.48}$$

$$\triangleq \frac{1}{2}x_1^2(t) + \frac{1}{2}u^2(t) + \lambda_1(t)x_2(t) - \lambda_2(t)x_2(t) + \lambda_2(t)u(t) \tag{4.49}$$

$$+ \lambda_3(t)\left[[x_2(t)+2]^2 I_{x_2(t)+2<0} + [2 - x_2(t)]^2 I_{2-x_2(t)<0}\right]. \tag{4.50}$$

The necessary conditions for optimality are

$$\dot{x}_1^*(t) = x_2^*(t), \tag{4.51a}$$

$$\dot{x}_2^*(t) = -x_2^*(t) + u^*(t), \tag{4.51b}$$

$$\dot{x}_3^*(t) = [x_2^*(t)+2]^2 I_{x_2^*(t)+2<0} + [2 - x_2^*(t)]^2 I_{2-x_2^*(t)<0}, \tag{4.51c}$$

with $x_1^*(t_0) = x_1$, $x_2^*(t_0) = x_2$, $x_3^*(t_0) = 0$, and

$$\dot{\lambda}_1^*(t) = -\frac{\partial \mathscr{H}^*}{\partial x_1} = -x_1^*(t), \tag{4.52a}$$

$$\dot{\lambda}_2^*(t) = -\frac{\partial \mathscr{H}^*}{\partial x_2} = -\lambda_1^*(t) + \lambda_2^*(t)$$

$$- 2\lambda_3^*(t)[x_2^*(t)+2]I_{x_2^*(t)+2<0} + 2\lambda_3^*(t)[2 - x_2^*(t)]I_{2-x_2^*(t)<0} \tag{4.52b}$$

$$\dot{\lambda}_3^*(t) = -\frac{\partial \mathscr{H}^*}{\partial x_3} = 0, \text{ which implies that } \lambda_3^*(t) \text{ is a constant for all } t \tag{4.52c}$$

with $\mathscr{H}^* \equiv \mathscr{H}^*(x^*(t), u^*(t), \lambda^*(t))$.

Also we have

$$u^*(t) = \begin{cases} -1, & \text{in case } 1 < \lambda_2^*(t) \\ -\lambda_2^*(t), & \text{in case } -1 \le \lambda_2^*(t) \le 1 \text{ .} \\ 1. & \text{in case } \lambda_2^*(t) < -1 \end{cases}$$

$\square$

Comparing these necessary conditions with the results obtained in Example 4.1, we can observe that the expressions for the optimal controls in terms of the extremal costates are the same; however the equations for $\dot{\lambda}_2^*(t)$ are different because of the presence of the state inequality constraints, so the trajectories and control histories will generally not be the same.

In the discussion of state and control inequality constraints, it was not considered constraints that include both the states and controls, that is,

$$f(x(t), u(t), t) \ge 0. \tag{4.53}$$

In the remainder of this chapter, we shall consider several specific applications of the Pontryagin's minimum principle.

More specifically, in Sects. 4.3, 4.4, and 4.6, we study the optimal control solutions for the minimum time, minimum fuel, and minimum energy problems, respectively. And in Sect. 4.5 (and Sect. 4.7, respectively), we study the optimal controls for performance costs composed of elapsed time and consumed fuels (consumed energies, respectively).

4.3   Minimum Time Problems

In this section, we study a class of optimal control problems such that the objective is to drive a system state from an initial state to zero in minimum time denoted by t^*.

Consider a system

$$\dot{x}(t) = f(x(t), u(t), t) \tag{4.54}$$

with $f(\cdot)$ representing a function of corresponding arguments.

The objective is to find out an optimal control to minimize

$$J(u) = \int_{t_0}^{t_f} dt = t_f - t_0, \tag{4.55}$$

from the initial state x_0 to zero in minimum time t^*.

Here suppose that the control variables satisfy the following constraints

$$|u_i(t)| \leq M_i, \text{ with } i = 1, 2, \ldots, m, \text{ for all } t \in [t_0, t^*] \tag{4.56}$$

for certain positive valued M_i.

In this section, we will study how to use Pontryagin's minimum principle to determine the optimal control.

4.3.1   Optimal Control Solution for Minimum Time Problems

Here it focuses on the study of the minimum time problems for the state system with the following specific form.

The state equation of the system is of the form

$$\dot{x}(t) = f(x(t), t) + B(x(t), t)u(t), \tag{4.57}$$

where B is an $n \times m$ matrix which may be dependent on the state x and time t.

As mentioned earlier, we still consider the inequality constraints for control as given below:

$$M_{i-} \le u_i(t) \le M_{i+}, \quad i = 1, 2, \ldots, m, \quad t \in [t_0, t^*]. \tag{4.58}$$

For the underlying problem, the Hamiltonian is specified as

$$\mathscr{H}(x(t), u(t), \lambda(t), t) = 1 + [\lambda(t)]^\top [f(x(t), t) + B(x(t), t)u(t)]; \tag{4.59}$$

then by the Pontryagin's minimum principle, we have

$$\begin{aligned} &1 + [\lambda^*(t)]^\top [f(x^*(t), t) + B(x^*(t), t)u^*(t)] \\ &\le 1 + [\lambda^*(t)]^\top [f(x^*(t), t) + B(x^*(t), t)u(t)], \end{aligned} \tag{4.60}$$

which implies that

$$[\lambda^*(t)]^\top B(x^*(t), t)u^*(t) \le [\lambda^*(t)]^\top B(x^*(t), t)u(t). \tag{4.61}$$

If we write the matrix B in the following column form

$$B(x^*(t), t) = \begin{bmatrix} B_1(x^*(t), t) & B_2(x^*(t), t) & \cdots & B_m(x^*(t), t) \end{bmatrix}; \tag{4.62}$$

then we have

$$[\lambda^*(t)]^\top B(x^*(t), t)u(t) = \sum_{i=1}^m [\lambda^*(t)]^\top B_i(x^*(t), t)u_i(t). \tag{4.63}$$

Hence, due to the mutual independence of all the control components with each other, we can obtain each optimal control u_i^* satisfies the following

$$[\lambda^*(t)]^\top B_i(x^*(t), t)u_i^*(t) \le [\lambda^*(t)]^\top B_i(x^*(t), t)u_i(t) \tag{4.64}$$

for all $i = 1, 2, \ldots, m$.

Consequently, the form of the optimal control is given as

$$u_i^*(t) = \begin{cases} M_{i+}, & \text{in case } [\lambda^*(t)]^\top B_i(x^*(t), t) < 0 \\ M_{i-}, & \text{in case } [\lambda^*(t)]^\top B_i(x^*(t), t) > 0 \, . \\ \text{Undetermined}, & \text{in case } [\lambda^*(t)]^\top B_i(x^*(t), t) = 0 \end{cases}$$

The coefficient of $u_i(t)$ is as shown in the upper figure in Fig. 4.6. The trajectory of $u_i^*(t)$ will be as shown in the lower figure in Fig. 4.6.

In case $[\lambda^*(t)]^\top B_i(x^*(t), t)$ passes through zero, a switching of the control $u_i^*(t)$ occurs.

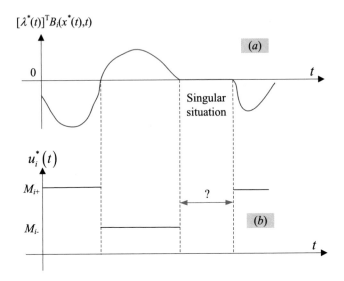

Fig. 4.6 The value of optimal control with respect to its coefficient in the Hamiltonian

In case $[\lambda^*(t)]^\top B_i(x^*(t), t)$ is zero for some time, then the coefficient of $u_i(t)$ in the Hamiltonian is zero, so the necessary condition that $u_i^*(t)$ minimizes the Hamiltonian $\mathscr{H}$ provides no information about how to implement the value of $u_i^*(t)$. It is called a singular condition. In this textbook, we only consider the normal condition such that no singular conditions occur.

The specific form of optimal control given above is the well-known *bang-bang control*, say the optimal control to obtain a minimum period of time is to make maximum effort over the whole interval.

4.3.2 Minimum Time Problems for Linear Time-Invariant Systems

In this section, we consider an n-dimensional linear time-invariant (LTI) system driven by an m-dimensional control with the state equation given as

$$\dot{x}(t) = Ax(t) + Bu(t), \tag{4.65}$$

where the control satisfies the following constraints

$$|u_i(t)| \leq 1 \tag{4.66}$$

for all $i = 1, 2, \ldots, m$.

It is well known that the optimal control, if it exists, is bang-bang.

In the following, some results for the optimal control problems defined above without any verifications are given .

Theorem 4.2 *Considering the minimum time problem for the LTI system defined in this section, suppose that all of the eigenvalues of the matrix A have non-positive real parts; then there exists an optimal control which drives any initial state x_0 to the origin.* □

Theorem 4.3 *Considering the minimum time problem for the LTI system defined in this section, suppose that it exists an optimal control, then it is unique.* □

Theorem 4.4 *Considering the minimum time problem for the LTI system defined in this section, suppose that an optimal control exists, then the Pontryagin's minimum principle is both necessary and sufficient condition for the optimal control of the underlying system.* □

Theorem 4.5 *Considering the minimum time problem for the LTI system defined in this section, suppose that the eigenvalues of matrix A are all real, and there exits an optimal control for the underlying problems, then each of the control components can switch at most $n - 1$ times.* □

As a consequence, by the results stated above, we can claim that the minimum time problem, for an n-dimensional time-invariant system where the matrix A has all real, non-positive eigenvalues, has a unique optimal control with components that each switches at most $n - 1$ times.

Example 4.3 Implement the optimal control satisfying the constraints of $|u(t)| \leq 1$ such that the following LTI state system

$$\dot{x}_1(t) = x_2(t) \tag{4.67}$$
$$\dot{x}_2(t) = u(t) \tag{4.68}$$

can be driven from any initial state x_0 to the origin in a minimum time.

Solution. By the specification of the given LTI system, we have $A = \begin{bmatrix} 0 & 1 \\ 0 & 0 \end{bmatrix}$, and $B = \begin{bmatrix} 0 \\ 1 \end{bmatrix}$.

Since the eigenvalues of A are both equal to zero, by the earlier statement the optimal control exists, which is unique, and has at most one switching.

The Hamiltonian is given as

$$\mathcal{H}(x(t), u(t), \lambda(t)) = 1 + \lambda_1(t)x_2(t) + \lambda_2(t)u(t). \tag{4.69}$$

Thus, the Pontryagin's minimum principle indicates that the optimal control $u^*(t)$ must satisfy

$$\lambda_2^*(t)u^*(t) \leq \lambda_2^*(t)u(t). \tag{4.70}$$

The optimal control u^* should satisfy the following:

$$u^*(t) = \begin{cases} -1, & \text{in case } \lambda_2^*(t) > 0 \\ 1, & \text{in case } \lambda_2^*(t) < 0 \end{cases} \triangleq -\text{sgn}(y)(\lambda_2^*(t)) \tag{4.71}$$

where $\text{sgn}(y)$ represents a signum function on y.

Also, by the specification of the Hamiltonian, the costate equations are

$$\dot{\lambda}_1^*(t) = 0, \tag{4.72}$$

$$\dot{\lambda}_2^*(t) = -\lambda_1^*(t); \tag{4.73}$$

then it is straightforward to obtain that the costate solution is in the following form

$$\dot{\lambda}_1^*(t) = \xi_1 \tag{4.74}$$

$$\dot{\lambda}_2^*(t) = -\xi_1 t + \xi_2 \tag{4.75}$$

for some constant valued ξ_1 and ξ_2.

Since there can be at most one switching for the two-dimensional system, the optimal control for a specific initial state must be one of the following forms:

$$u^*(t) = \begin{cases} 1, & \text{for all } t \in [t_0, t^*], \text{ or} \\ -1, & \text{for all } t \in [t_0, t^*], \text{ or} \\ 1, & \text{for all } t \in [t_0, t_1), \text{ and } -1, \text{ for all } t \in [t_1, t^*], \text{ or} \\ -1, & \text{for all } t \in [t_0, t_1), \text{ and } 1, \text{ for all } t \in [t_1, t^*]. \end{cases} \tag{4.76}$$

The corresponding optimal state trajectories can be found by integrating the state equations subject to the control $u = \pm 1$, respectively,

$$x_2(t) = \pm t + \xi_3, \tag{4.77a}$$

$$x_1(t) = \pm \frac{1}{2} t^2 + \xi_3 t + \xi_4, \tag{4.77b}$$

based upon which we can obtain the evolutions of x_1 with respect to x_2 subject to $u = 1$ and $u = -1$ as below

$$x_1(t) = \begin{cases} \frac{1}{2} x_2^2(t) + \xi_5, & \text{with } u = 1, \\ -\frac{1}{2} x_2^2(t) + \xi_6, & \text{with } u = -1, \end{cases} \tag{4.78}$$

as displayed in Fig. 4.7 with $u = 1$ and Fig. 4.8 with $u = -1$, respectively.

As illustrated in Fig. 4.7 and Fig. 4.8 of the relationships between x_1 and x_2, we can claim that the following statements subject to different optimal controls:

- Suppose that the optimal control $u^*(t) = 1$ for $t \in [t_0, t^*]$; then the initial state x_0 must lie on the curve A-0 in Fig. 4.7.
- Suppose that the optimal control $u^*(t) = -1$ for $t \in [t_0, t^*]$; then the initial state x_0 must lie on the curve B-0 in Fig. 4.8.
- Suppose that the optimal control $u^*(t) = 1$ for $t \in [t_0, t_1)$, and $u^*(t) = -1$ for $t \in [t_1, t^*]$. In this case, the system state must lie on the curve B-0 for $t \in [t_1, t^*]$. This has been done by a control of $u^* = 1$ during $[t_0, t_1)$; thus, the optimal trajectory consists of an initial segment like a trajectory in Fig. 4.7 which is followed by a switching of the control to -1 at t_1 when the trajectory intersects the curve B-0, and then evolves along B-0 with $u^* = -1$ until reaching the origin. Hence the curve B-0 is considered as a switching curve. As a consequence, by observing Figs. 4.7 and 4.8, we get that for all the states initially lying below the curve A-0-B, the corresponding optimal control will be $u^* = 1$ followed by $u^* = -1$.
- Following the similar analysis above, it can be claimed that for all those states initially lying above the curve A-0-B, the corresponding optimal control will be $u^* = -1$ followed by $u^* = 1$, and the switching occurs at time t_1 when the trajectory intersects the curve A-0. Hence, in this case, $u^*(t) = -1$ for $t \in [t_0, t_1)$, and $u^*(t) = 1$ for $t \in [t_1, t^*]$.

Thus, it can be obtained that the optimal state trajectories always terminate at the curve A-0 or B-0. And the curve A-0-B, as shown in Fig. 4.9, is the switching curve in case the initial state does not lie on the curve A-0 or B-0 .

Hence by (4.78), the switching curve is given as

$$x_1(t) = -\frac{1}{2}x_2(t)|x_2(t)|, \tag{4.79}$$

which is a specific form of (4.78) with $\xi_5 = \xi_6 = 0$.

Figure 4.10, displays the evolutions of optimal state with the initial state not lying on the switching curve A-0-B.

Before further analysis, we define a so-called switching function denoted by $\phi(x(t))$ such that

$$\phi(x(t)) \triangleq x_1(t) + \frac{1}{2}x_2(t)|x_2(t)|. \tag{4.80}$$

Notice that

- $\phi(x(t)) > 0$ implies that $x(t)$ lies above the switching curve A-0-B.
- $\phi(x(t)) < 0$ implies that $x(t)$ lies below the switching curve A-0-B.
- $\phi(x(t)) = 0$ implies that $x(t)$ lies on the switching curve A-0-B.

Thus the optimal control u^* with respect to the switching function $\phi(\cdot)$ is given as below:

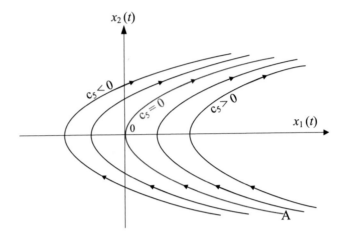

Fig. 4.7 An illustration of evolution of x_1 with respect to x_2 subject to $u = 1$

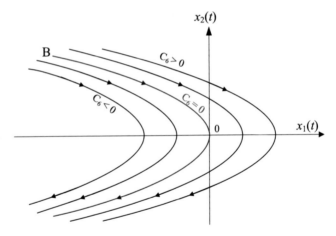

Fig. 4.8 An illustration of evolution of x_1 with respect to x_2 subject to $u = -1$

$$u^*(t) = \begin{cases} -1, & \text{in case } \phi(x(t)) > 0 \\ 1, & \text{in case } \phi(x(t)) < 0 \\ -1, & \text{in case } \phi(x(t)) = 0 \text{ and } x_2(t) > 0 \\ 1, & \text{in case } \phi(x(t)) = 0 \text{ and } x_2(t) < 0 \\ 0. & \text{in case } x(t) = 0 \end{cases} \qquad (4.81)$$

$\square$

In the following, it may be generalized the procedure developed in Example 4.3 to solve n-dimensional linear time-invariant systems driven by a control with a single component. Suppose that all of the eigenvalues of matrix A are real and non-positive

Fig. 4.9 An illustration of switching curve

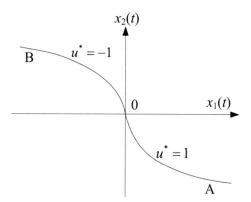

Fig. 4.10 An illustration of evolution of optimal state with initial state not lying on A-0-B

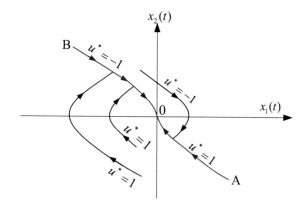

valued; then, by Theorems 4.2–4.5 given earlier, there exists a unique optimal control for minimum time problems with any initial state and it switches at most $n - 1$ times.

More specifically, to obtain the optimal control u^* as below:

- First specify the set of state values, from which the origin can be reached subject to control $u = 1$, which is denoted by $\mathscr{R}_+$, similarly, denote by $\mathscr{R}_-$ the set of state values from which the origin can be reached subject to $u = -1$ alone.
  Also define

$$\mathscr{R}_0 \triangleq \mathscr{R}_+ \cup \mathscr{R}_-, \tag{4.82}$$

  that is to say, $\mathscr{R}_0$ represents the set of state values from which the origin can be reached without any switchings.
- It further specifies the set of state values, denoted by $\mathscr{R}_{-+}$, from which the set $\mathscr{R}_+$ can be reached by applying $u = -1$. Hence, the origin can be reached from any state in $\mathscr{R}_{-+}$ by firstly applying $u = -1$, until reaching the set $\mathscr{R}_+$ at some time; then $u = 1$ is applied until the origin is reached.

Similarly, it specifies the set of states, denoted by $\mathscr{R}_{+-}$, from which the set $\mathscr{R}_-$ can be reached by applying $u = 1$. Hence, the origin can be reached from any state in $\mathscr{R}_{+-}$ by firstly applying $u = 1$, until reaching the set $\mathscr{R}_-$ at some time; then $u = -1$ is applied until the origin is reached.

Hence the set of states from which the origin can be reached with at most one switching, denoted by $\mathscr{R}_1$, is given as

$$\mathscr{R}_1 = \mathscr{R}_0 \cup \mathscr{R}_{+-} \cup \mathscr{R}_{-+}. \tag{4.83}$$

- Following the above analysis continuously, we can define the set of states, denoted by $\mathscr{R}_{n-2}$, from which the origin can be reached with at most $n - 2$ switchings. Hence all states which are not located in the set $\mathscr{R}_{n-2}$ require exactly $n - 1$ switchings to reach the origin. We can express $\mathscr{R}_{n-2}$ in the following form:

$$\phi(x(t)) = 0, \tag{4.84}$$

which is not explicitly dependent upon the time t.   □

Based upon the above analysis, it can be obtained that the optimal control can be applied to any state value. More specifically, the switching function $\phi(x(t))$ defines a switching hypersurface that divides the state space into two subspaces.

Suppose that from the first subspace the control $u^* = 1$ is applied to drive the system to $\mathscr{R}_{n-2}$ where the control switches to -1, until the system reaches $\mathscr{R}_{n-3}$, where the control again switches to 1, etc., until the origin is reached.

Similarly, we can have that from another subspace the control sequence is reversed, say $u^* = -1$ is firstly applied to drive the state system to $\mathscr{R}_{n-2}$ where the control switches to 1, etc., until the origin is reached as well.

In the below, we study another example to demonstrate the procedure given above.

Example 4.4 Implement the optimal control for driving the state system

$$\dot{x}_1(t) = x_2(t), \tag{4.85}$$
$$\dot{x}_2(t) = -ax_2(t) + u(t), \tag{4.86}$$

from an arbitrary initial sate x_0 to zero in a minimum time, such that the admissible controls are constrained by $|u(t)| \leq 1$ and the parameter a is positive real valued.

Solution. By the specification of the system, it can be obtained that the eigenvalues of this state system are 0 and $-a$ which are real and non-positive; then by applying Theorems 4.2–4.5, the optimal control exists is unique and has at most $n - 1 = 1$ switching. And the forms of the optimal control are the same as those as given for Example 4.3.

Next specify the sets $\mathscr{R}_+$ and $\mathscr{R}_-$ by solving the differential equations with $u = \pm 1$. The solutions are

$$x_2(t) = \xi_1 \exp(-at) \pm \frac{1}{a}[1 - \exp(-at)], \tag{4.87a}$$

$$x_1(t) = -\frac{\xi_1}{a} \exp(-at) \pm \frac{1}{a}t \pm \frac{1}{a^2} \exp(-at) + \xi_2, \tag{4.87b}$$

subject to the control $u = \pm 1$.

These equations define two collections of curves; set $x_1(t) = x_2(t) = 0$ and $t = 0$ solve for ξ_1 and ξ_2 to obtain

$$x_2(t) = \pm \frac{1}{a}[1 - \exp(-at)], \tag{4.88a}$$

$$x_1(t) = \pm \frac{1}{a} \pm \frac{1}{a^2} \exp(-at) \mp \frac{1}{a^2}. \tag{4.88b}$$

To specify $\mathcal{R}_+$ by using the upper sign with $u = 1$, the first equation for time t in the above is solved, and applied it in the second equation to obtain the relationship

$$x_1(t) = -\frac{1}{a^2} \ln\left(-a\left[x_2(t) - \frac{1}{a}\right]\right) - \frac{1}{a}x_2(t). \tag{4.89}$$

Hence $\mathcal{R}_+$ is specified as the set of states satisfying the above relationship.

Similarly, we can have that $\mathcal{R}_-$ is composed of all of the states satisfying the following relationship:

$$x_1(t) = \frac{1}{a^2} \ln\left(a\left[x_2(t) + \frac{1}{a}\right]\right) - \frac{1}{a}x_2(t). \tag{4.90}$$

Thus, by (4.89) and (4.90), the set $\mathcal{R}_0$ is given as

$$\mathcal{R}_0 = \left\{ x_1(t), x_2(t) : x_1(t) = \frac{x_2(t)}{|x_2(t)|} \frac{1}{a^2} \ln\left(a\left[|x_2(t)| + \frac{1}{a}\right]\right) - \frac{1}{a}x_2(t) \right\}. \tag{4.91}$$

The switching function is then obtained as

$$\phi(x(t)) = x_1(t) - \frac{x_2(t)}{|x_2(t)|} \frac{1}{a^2} \ln\left(a\left[|x_2(t)| + \frac{1}{a}\right]\right) + \frac{1}{a}x_2(t). \tag{4.92}$$

The switching curves with different parameters of a equal to 0.5, 1, and 2 are displayed in Fig. 4.11, respectively.

Some optimal state trajectories with parameter $a = 0.5$ with respect to different initial state values are displayed in Fig. 4.12, respectively. □

In this section, it is clarified that the optimal control for minimum time problems of state systems in the form of (4.57) is the so-called bang-bang control, i.e., the optimal control switches between the upper bound and lower bound of the control set. A procedure for implementing this type of optimal controls for the linear time-invariant state system is discussed in detail and demonstrated with two 2-dimensional state systems.

Though the procedure is clear, it may be challenging to implement, since

- For high-dimensional state systems, say with $n > 3$, it is generally challenging to obtain an analytical expression for the switching hypersurface.
- The specified procedure is impracticable for nonlinear state systems in general, since it is difficult to solve the differential equations in an analytical way.
- Even in those cases such that an exact expression for the switching hypersurface can be specified, the actual implementation of the optimal control may be complicated. Alternatively, certain suboptimal solutions may be acceptable.

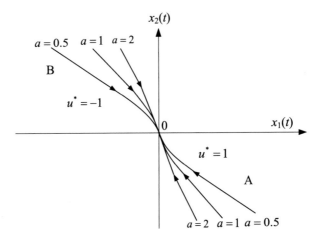

Fig. 4.11 The switching curves with respect to different parameters a for minimum time problems with two-dimensional state

Fig. 4.12 The switching curves with respect to with a specific parameter $a = 0.5$ for minimum time problems with two-dimensional state

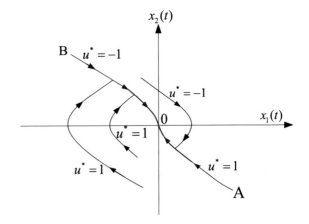

4.4 Minimum Fuel Problems

In the last section, it was studied a class of optimal control problems such that the objective is to drive a system state from an initial value to the origin in a minimum time.

Here, we specify another class of problems such that, rather than the elapsed time, the control efforts are considered as the performance cost. This type of problems arises in many fields like aerospace, etc., where the resources are limited for achieving preferred objectives.

Assume that the state equations of a system are of the form

$$\dot{x}(t) = f(x(t), t) + B(x(t), t)u(t), \tag{4.93}$$

where $f(\cdot)$ is a function of corresponding arguments, and $B(\cdot)$ is an $n \times m$ matrix that may be dependent on the state and/or time.

The performance cost to be minimized is defined as

$$J(u) \triangleq \int_{t_0}^{t_f} \left[\sum_{i=1}^{m} \gamma_i |u_i(t)| \right] dt, \tag{4.94}$$

where $\gamma_i > 0$, for all i, represents the weighting factors of control components.

Also consider that the controls satisfy the following inequality constraints:

$$M_{i-} \leq u_i(t) \leq M_{i+}, \quad i = 1, 2, \ldots, m, \quad t \in [t_0, t_f]. \tag{4.95}$$

Notice that, for the minimum fuel problems, the final time t_f may be specified or unspecified. Suppose that t_f is specified, it makes sense only in the case that t_f is not less than the minimum time such that the system can reach the origin from the given initial state.

The Hamiltonian $\mathcal{H}$ for the above minimum fuel problem is given as

$$\mathcal{H}(x(t), u(t), \lambda(t), t)$$
$$\triangleq \sum_{i=1}^{m} |u_i(t)| + \lambda^\top(t) f(x(t), t) + \lambda^\top(t) B(x(t), t) u(t). \tag{4.96}$$

By the Pontryagin's minimum principle, it can be obtained that

$$\sum_{i=1}^{m} |u_i^*(t)| + [\lambda^*(t)]^\top f(x^*(t), t) + [\lambda^*(t)]^\top B(x^*(t), t) u^*(t)$$
$$\leq \sum_{i=1}^{m} |u_i(t)| + [\lambda^*(t)]^\top f(x^*(t), t) + [\lambda^*(t)]^\top B(x^*(t), t) u(t), \tag{4.97}$$

for all admissible control u.

It deals with the matrix $B(x^*(t), t)$ into the following form of column vectors:

$$B(x^*(t), t) = \begin{bmatrix} B_1(x^*(t), t) & B_2(x^*(t), t) & \cdots & B_m(x^*(t), t) \end{bmatrix}. \qquad (4.98)$$

As always considered in this book, We still suppose that all the components of u are independent of each other; then by (4.97), it can be obtained that

$$|u_i^*(t)| + [\lambda^*(t)]^\top B_i(x^*(t), t)u_i^*(t) \leq |u_i(t)| + [\lambda^*(t)]^\top B_i(x^*(t), t)u_i(t), \quad (4.99)$$

for all $i = 1, \ldots, m$.

Also, we have

$$|u_i(t)| + [\lambda^*(t)]^\top B_i(x^*(t), t)u_i(t)$$
$$= \begin{cases} \Gamma^+ \equiv \left[1 + [\lambda^*(t)]^\top B_i(x^*(t), t)\right] u_i(t), & \text{if } u_i(t) \geq 0 \\ \Gamma^- \equiv \left[-1 + [\lambda^*(t)]^\top B_i(x^*(t), t)\right] u_i(t), & \text{if } u_i(t) \leq 0 \end{cases}, \qquad (4.100)$$

for all $i = 1, \ldots, m$, based upon which we have the following analysis:

- In case $[\lambda^*(t)]^\top B_i(x^*(t), t) > 1$, the minimum value of Γ^+ is zero with $u_i(t) = 0$ since $u_i(t) \geq 0$, and the minimum value of Γ^- is $-\left[-1 + [\lambda^*(t)]^\top B_i(x^*(t), t)\right] < 0$ with $u_i(t) = -1$ since $u_i(t) \leq 0$; thus $|u_i(t)| + [\lambda^*(t)]^\top B_i(x^*(t), t)u_i(t)$ is minimized with $u_i^*(t) = -1$.
- In case $[\lambda^*(t)]^\top B_i(x^*(t), t) = 1$, the minimum value of Γ^+ is zero with $u_i(t) = 0$ since $u_i(t) \geq 0$, and Γ^- is zero for all $u_i(t) \leq 0$; thus $|u_i(t)| + [\lambda^*(t)]^\top B_i(x^*(t), t)u_i(t)$ is minimized with any control $u_i(t) \leq 0$.
- In case $[\lambda^*(t)]^\top B_i(x^*(t), t) \in [0, 1)$, the minimum value of Γ^+ is zero with $u_i(t) = 0$ since $u_i(t) \geq 0$, and the minimum value of Γ^- is zero with $u_i(t) = 0$ since $u_i(t) \leq 0$; thus $|u_i(t)| + [\lambda^*(t)]^\top B_i(x^*(t), t)u_i(t)$ is minimized with $u_i^*(t) = 0$.
- Following the similar discussions in case $[\lambda^*(t)]^\top B_i(x^*(t), t) \geq 0$ as given above, we can get the optimal control solution in case $[\lambda^*(t)]^\top B_i(x^*(t), t) < 0$ as well.

Hence, in summary, we can get the optimal control as below:

$$u_i^*(t) \begin{cases} = -1, & \text{in case } [\lambda^*(t)]^\top B_i(x^*(t), t) > 1 \\ = 0, & \text{in case } [\lambda^*(t)]^\top B_i(x^*(t), t) \in (-1, 1) \\ = 1, & \text{in case } [\lambda^*(t)]^\top B_i(x^*(t), t) < -1 \\ \leq 0, \text{ undertermined}, & \text{in case } [\lambda^*(t)]^\top B_i(x^*(t), t) = 1 \\ \geq 0, \text{ undertermined}, & \text{in case } [\lambda^*(t)]^\top B_i(x^*(t), t) = -1 \end{cases} \qquad (4.101)$$

Figure 4.13, displays the evolution of the optimal control u^* with respect to $[\lambda^*(t)]^\top B_i(x^*(t), t)$.

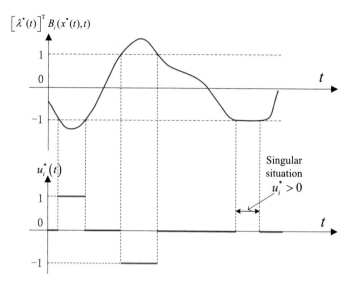

Fig. 4.13 An illustration of evolution of the optimal control u^* with respect to $[\lambda^*(t)]^\top B_i(x^*(t), t)$

Notice that different from the minimum time problems where the optimal control is the so-called bang-bang, the optimal control for the minimum fuel problems given in (4.101) may be called *bang-off-bang* if no singular conditions are included.

In the rest of this section, we will study some minimum fuel problems with linear state systems.

In the example below, we study the implementation of minimum fuel problems for a two-dimensional system.

Example 4.5 Consider a two-dimensional system given below

$$\dot{x}_1(t) = x_2(t), \tag{4.102a}$$
$$\dot{x}_2(t) = u(t) \tag{4.102b}$$

with an initial state x_0 and $|u(t)| \leq 1$.

Suppose that the objective is to drive the system state from the initial value x_0 to the origin, such that

$$J(u) = \int_{t_0}^{t_f} |u(t)| dt \tag{4.103}$$

is minimized.

Solution. Firstly as usual specifies the Hamiltonian $\mathscr{H}$ is specified as below

$$\mathscr{H}(x(t), u(t), \lambda(t), t) \triangleq |u(t)| + \lambda_1(t)x_2(t) + \lambda_2(t)u(t); \tag{4.104}$$

then it can be claimed that the minimization of the Hamiltonian is equivalent with the following

$$\min_{|u(t)| \leq 1} \{|u^*(t)| + \lambda_2(t)u(t)\}. \tag{4.105}$$

Thus, by applying (4.101), it can be obtained that the optimal control can be specified as below:

$$u^*(t) \begin{cases} = -1, & \text{in case } \lambda_2^*(t) > 1 \\ = 0, & \text{in case } \lambda_2^*(t) \in (-1, 1) \\ = 1, & \text{in case } \lambda_2^*(t) < -1 \\ \in [-1, 0], & \text{in case } \lambda_2^*(t) = 1 \\ \in [0, 1], & \text{in case } \lambda_2^*(t) = -1 \end{cases}, \tag{4.106}$$

which is displayed in Fig. 4.14.

Also (4.106) can be rewritten as a cleaner form:

$$u^*(t) = \begin{cases} -\text{sgn}(\lambda_2^*(t)), & \text{in case } |\lambda_2^*(t)| > 1 \\ 0, & \text{in case } |\lambda_2^*(t)| < 1 \\ \text{undetermined}, & \text{in case } |\lambda_2^*(t)| = 1 \end{cases}, \tag{4.107}$$

where $\text{sgn}(y)$ represents a signum function on y.

Furthermore, subject to $u^*(t)$ as specified above,

$$\min_{|u(t)| \leq 1} \{|u(t)| + \lambda_2^*(t)u(t)\} = \begin{cases} 1 - \lambda_2^*(t), & \text{in case } \lambda_2^*(t) > 1 \\ 0, & \text{in case } \lambda_2^*(t) < -1 \\ 1 + \lambda_2^*(t), & \text{otherwise} \end{cases}, \tag{4.108}$$

which is illustrated in Fig. 4.15.

Also, from the Hamiltonian, the costate equations are given as

$$\dot{\lambda}_1^*(t) = 0 \tag{4.109}$$

$$\dot{\lambda}_2^*(t) = -\lambda_1^*(t); \tag{4.110}$$

then it is straightforward to obtain that the costate solution is of the form

$$\lambda_1^*(t) = \xi_1, \tag{4.111}$$

$$\lambda_2^*(t) = -\xi_1 t + \xi_2, \tag{4.112}$$

for some constant valued ξ_1 and ξ_2.

Suppose that $\xi_1 \neq 0$, then there is no singular situation for the underlying minimum fuel problems. By the above results, it can be obtained that the optimal control could be one of the following sequences:

$$\{0\}, \{1\}, \{-1\}, \{0, 1\}, \{0, -1\}, \{1, 0\}, \{-1, 0\}, \{1, 0, -1\}, \{-1, 0, 1\}. \quad (4.113)$$

Notice that, since $\lambda_2^*(t)$ is continuous with respect to time t, the optimal control can not be switched between 1 and -1.

The corresponding optimal state trajectories can be found by integrating the state equations with $u = \pm 1$, respectively,

$$x_2(t) = \pm t + \xi_3, \quad (4.114a)$$

$$x_1(t) = \pm \frac{1}{2} t^2 + \xi_3 t + \xi_4, \quad (4.114b)$$

based upon which, it can be obtained the evolutions of x_1 with respect to x_2 subject to $u = 1$ and -1 as below

$$x_1(t) = \begin{cases} \dfrac{1}{2} x_2^2(t) + \xi_5, & \text{with } u = 1 \\ -\dfrac{1}{2} x_2^2(t) + \xi_6, & \text{with } u = -1 \end{cases} \quad (4.115)$$

with $\xi_5 = \xi_3 - \frac{1}{2}\xi_4^2$ and $\xi_6 = \xi_3 + \frac{1}{2}\xi_4^2$.

See the evolutions of x_1 with respect to x_2 subject to $u(t) = 1$ and $u(t) = -1$ in Figs. 4.7 and 4.8, respectively.

Figure 4.16 displays a switching curve for the minimum fuel problems of two-dimensional state systems.

By the state equations given in (4.102), with the control $u(t) = 0$,

$$x_1(t) = \xi_2 t + \xi_1, \quad (4.116)$$

$$x_2(t) = \xi_2, \quad (4.117)$$

which are displayed in Fig. 4.17. □

Lemma 4.1, gives an interesting result related to Example 4.5.

Lemma 4.1 *Denote by J^* the minimum fuel consumed such that the two-dimensional state system specified in (4.102) is driven from its initial state x_0 to the origin. Then we can get that*

$$J^* = |x_{20}| \quad (4.118)$$

with $|x_{20}|$ as the absolute value of the initial value of $x_2(t)$.

Proof By the state equation (4.102),

Fig. 4.14 The value of optimal control $u^*(t)$ with respect to $\lambda_2^*(t)$

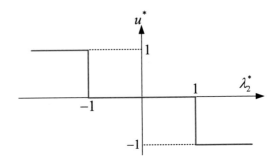

Fig. 4.15 The evolution of $|u^*(t)| + \lambda_2(t)u^*(t)$ with respect to $\lambda_2^*(t)$

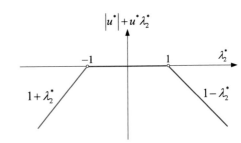

Fig. 4.16 A switching curve for two-state minimum fuel problems with $u = \pm 1$

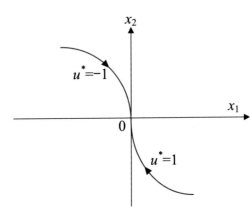

Fig. 4.17 A switching curve for two-state minimum fuel problems with $u = 0$

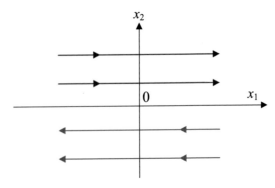

Fig. 4.18 An illustration of subspaces for the state systems

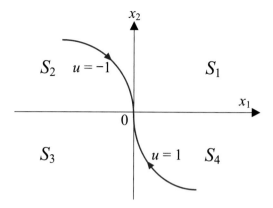

$$x_2(t) = x_{20} + \int_{t_0}^{t} u(s)ds; \tag{4.119}$$

for any t; then we have

$$x_{20} = -\int_{t_0}^{t_f} u(t)dt, \tag{4.120}$$

since $x_2(t_f) = 0$.

Hence we have

$$|x_{20}| = \left| \int_{t_0}^{t_f} u(t)dt \right| \leq \int_{t_0}^{t_f} |u(t)|dt \equiv J(x_0; u), \tag{4.121}$$

which implies the conclusion. □

As illustrated, it can be claimed that the system cannot be driven to the origin from any nonzero state subject to zero control.

It will give the optimal control sequence as below, before that we define a collection of subspaces for the states in the following:

- S_1 as the region to the right of $\mathscr{R}_0$ curve and with x_2 being positive valued.
- S_3 as the region to the left of $\mathscr{R}_0$ curve and with x_2 being negative valued.
- S_2 as the region to the left of $\mathscr{R}_0$ curve and with x_2 being positive valued.
- S_4 as the region to the right of $\mathscr{R}_0$ curve and with x_2 being negative valued.

See an illustration in Fig. 4.18.

The optimal control sequence will be specified below with respect to the initial state values.

- In case the initial states x_0 are located on the curves of $\mathscr{R}_+$ and $\mathscr{R}_-$, the optimal control $u^*(t) = 1$ and $u^*(t) = -1$, respectively.

Fig. 4.19 An illustration of optimal control for two-state minimum fuel problems

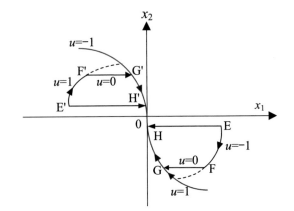

Fig. 4.20 An illustration of an ε optimal control for two-state minimum fuel problems

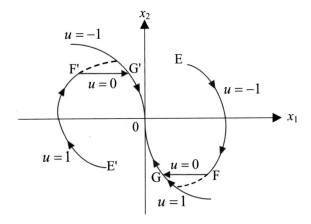

- In case the initial states x_0 are located in S_2, as illustrated in Fig. 4.19, the system can be driven to the origin with the control sequences of $\{0, -1\}$ and $\{1, 0, -1\}$. However, it is straightforward to claim that the optimal control is $\{0, -1\}$ and $\{1, 0, -1\}$ is not optimal, say the optimal state curve is E'-H'-0, not E'-F'-G'-0. Similarly in case the initial states x_0 are located in S_4, the system can be driven to the origin with the control sequences of $\{0, 1\}$ and $\{-1, 0, 1\}$. And only $\{0, 1\}$ is optimal, say the optimal state curve is E-H-0, not E-F-G-0.
- Furthermore, suppose that x_0 is in the subspace of S_1. The system can reach the origin with the control sequence of $\{-1, 0, 1\}$, like a state curve of E-F-G-0 displayed in Fig. 4.20. However we can get that for each admissible control $\widehat{u}$ with sequence $\{-1, 0, 1\}$, we can always define another control such that its switching from -1 to 0 is implemented later than that of $\widehat{u}$.

  Consequently in case x_0 is located in S_1, there does not exist the optimal control. Nevertheless, for any $\varepsilon > 0$, it can determine a suboptimal control with sequence $\{-1, 0, 1\}$ subject to which the performance fuel cost is not more than $J^* + \epsilon$.

Notice that as claimed earlier, $\{-1, 1\}$ is not an admissible control sequence for the underlying control problem. Also, it is worth to state that it costs more fuel resources with the control sequence $\{-1, 1\}$ than any other controls with sequence $\{-1, 0, 1\}$.

- Parallel with the above analysis in case of x_0 in S_1, we can get the similar conclusion that, in case of x_0 in S_3, there does not exist any optimal control and however for any $\varepsilon > 0$, it can specify a control with sequence $\{1, 0, -1\}$ subject to which the performance fuel cost is not more than $J^* + \epsilon$. See a state curve of E'-F'-G'-0 displayed in Fig. 4.20.

In the following, we will study how to implement the optimal control of the minimum fuel problems with some numerical examples.

Example 4.6 Consider a state system specified below

$$\dot{x}(t) = -ax(t) + u(t) \tag{4.122}$$

with $a > 0$, the initial state $x(t_0) = x_0$, the final state $x(t_f) = 0$, the final time t_f being free and

$$|u(t)| \leq 1. \tag{4.123}$$

Implement the optimal control such that the following performance cost is minimized:

$$J(u) = \int_0^{t_f} |u(t)| dt. \tag{4.124}$$

Solution. By the state equation and the performance cost function, the Hamiltonian is specified as

$$\mathscr{H}(x(t), u(t), \lambda(t)) = |u(t)| - \lambda(t)ax(t) + \lambda(t)u(t). \tag{4.125}$$

Thus, the costate equation is

$$\dot{\lambda}^*(t) = -\frac{\partial \mathscr{H}^*}{\partial x} = a\lambda^*(t), \tag{4.126}$$

which implies that

$$\lambda^*(t) = \xi_1 \exp(\alpha t), \tag{4.127}$$

where ξ_1 is a constant of integration.

From (4.101), the form of the optimal control with $b_i = B = 1$ is given as

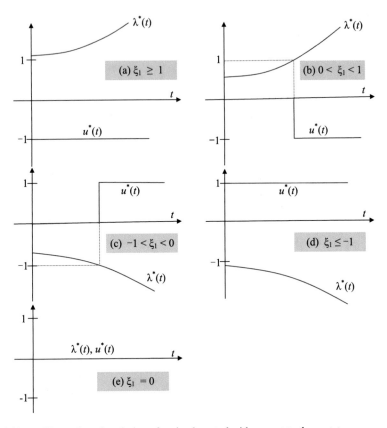

Fig. 4.21   An illustration of evolution of optimal control with respect to the costate

$$u^*(t) = \begin{cases} 1, & \text{in case } \lambda^*(t) < -1 \\ 0, & \text{in case } \quad -1 < \lambda^*(t) < 1 \\ -1, & \text{otherwise} \end{cases} . \qquad (4.128)$$

As observed from (4.128), in case $\lambda^*(t)$ passes through the value of 1 or -1, the control is switched. Besides, it may be possible that $\lambda^*(t)$ remains at the constant value of 1 or -1 during a finite interval of time. However, due to (4.127) and $a > 0$, it is straightforward to state that this cannot occur.

As demonstrated in Fig. 4.21, there are five distinct candidate forms for the costate $\lambda^*(t)$ with respect to the value of ξ_1.

It may be denoted the forms of the optimal control u^* by

$$\{-1\}, \{0, -1\}, \{0, 1\}, \{1\}, \text{ and } \{0\},$$

respectively.

$$u^*(t) = \begin{cases} -1, \text{ with } t \in [t_0, t_f], & \text{in case } 1 \leq \xi_1, \\ \begin{cases} 0, & t \in [t_0, t_1) \\ -1, & t \in [t_1, t_f], \end{cases} \text{ for some time } t_1 & \text{in case } 0 < \xi_1 < 1, \\ \begin{cases} 0, & t \in [t_0, \widehat{t_1}) \\ 1, & t \in [\widehat{t_1}, t_f], \end{cases}, \text{ for some time } \widehat{t_1} & \text{in case } -1 < \xi_1 < 0, \\ 1, \text{ with } t \in [t_0, t_f], & \text{in case } \xi_1 \leq -1, \\ 0, \text{ with } t \in [t_0, t_f], & \text{in case } \xi_1 = 0. \end{cases}$$

$$(4.129)$$

The solution of the state equation is given as

$$x(t) = \exp(-a[t - t_0])x_0 + \exp(-a[t - t_0]) \int_{t_0}^{t} \exp(a[\tau - t_0])u(\tau)d\tau. \quad (4.130)$$

Suppose that the optimal control is identically equal to zero during the whole interval of $[t_0, t_f]$; then, at the final time t_f

$$x(t_f) = \exp(-a[t_f - t_0])x_0, \quad (4.131)$$

which approaches to zero if time goes to infinity, since the system is stable with $\alpha > 0$.

Thus we have that the system cannot reach zero at any finite final time t_f with zero-valued control. However, as stated, the underlying optimal control problem is to set the final state at zero, say $x(t_f) = 0$.

Moreover, suppose that the initial value of $x_0 > 0$, then it is obvious to claim that the optimal control has to be in the form of $\{-1\}$ or $\{0, -1\}$.

- First consider that the optimal control is in the form of $\{-1\}$, say $u(t) = -1$ for all $t \in [t_0, t_f]$; then it can be shown by (4.130) that $x(t_f) = 0$ implies

$$t_f = \frac{1}{a} \ln(ax_0 + 1) + t_0. \quad (4.132)$$

Thus, the consumed fuel subject to this form of control is equal to

$$J = t_f - t_0 = \frac{1}{a} \ln(ax_0 + 1).$$

- Alternatively, consider that the control is in the form of $\{0, -1\}$, say $u(t) = 0$ for any time $t \in [t_0, t_1)$ for certain valued time t_1, and $u(t) = -1$ for all $t \in [t_1, t_f]$; then by (4.130) the state at the final time is given as

$$x(t_f) = \exp(-a[t_f - t_0])x_0 + \exp(-a[t_f - t_0]) \int_{t_1}^{t_f} [-\exp(a[\tau - t_0])]d\tau,$$

$$(4.133)$$

by which and due to the final state $x(t_f) = 0$, it can be obtained that

$$0 = x_0 - \int_{t_1}^{t_f} \exp(a[\tau - t_0])d\tau, \qquad (4.134)$$

by which, we get that

$$t_f = \frac{1}{a} \ln \left[ax_0 + \exp(a[t_1 - t_0]) \right] + t_0, \qquad (4.135)$$

Hence we get that the consumed fuel subject to this form of control is given as

$$J = t_f - t_1 = \frac{1}{a} \ln \left[ax_0 + \exp(a[t_1 - t_0]) \right] - [t_1 - t_0]. \qquad (4.136)$$

Thus in case that the final time t_f is free, by (4.136), it can be obtained that the consumed fuel J monotonically decreases with respect to the switching time t_1, and furthermore, J approaches to zero, as t_1 goes to infinity, see Fig. 4.22.

Consequently, in case that the final time t_f is free, there does not exist any optimal control, since we always could apply a control to drive the system state to zero with less consumed fuel.

Following the same analysis, we can analyze the properties of the optimal controls in case the initial state $x_0 < 0$.   □

In the example studied above, it was verified that if the final time t_f is not specified and there is no penalty cost related to the elapsed time as well, the system might approach the zero value without consuming any level of fuel.

Thus it is natural to study in Example 4.7 the optimal control for the minimum fuel problem with a certain fixed final time. Before that we firstly give some discussions below.

The possible forms for optimal controls and the corresponding state solution are given in (4.129) and (4.130), respectively.

As analyzed in the minimum time problems in the last section, the fixed final time t_f is equal to the minimum time denoted by t^* for the system to reach the value of zero from an initial state x_0, and the optimal control is a bang-bang control.

For the one-dimensional system considered in Example 4.7, the optimal control $u^*(t)$ is -1 over the whole interval $[t_0, t_f]$, and the final time specified in (4.132) with $x_0 > 0$.

And following the similar analysis, we can have, considering $x_0 < 0$, the final time t_f is the minimum time t^* by applying the control equal to 1 over $[t_0, t_f]$, such that

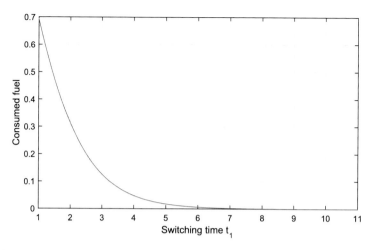

Fig. 4.22 The evolution of consumed fuel with respect to the switching time t_1 with $a = 1, x_0 = 1$ and $t_0 = 1$

$$t_f = \frac{1}{a} \ln(-ax_0 + 1) + t_0. \tag{4.137}$$

Thus (4.132) and (4.137) specify the minimum time t^* for the system defined in Example 4.7 with the initial state $x_0 > 0$ and $x_0 < 0$, respectively.

Consequently, for the minimum fuel problem the final time t_f has to be equal to or larger than the minimum time t^* which is determined by the initial value x_0 in (4.132) and (4.137), respectively.

Vice versa, by the above analysis we can also specify that

$$\pm \frac{1}{a} [\exp(a[t_f - t_0]) - 1] \tag{4.138}$$

are the largest and smallest values of x_0 from which the value zero can be reached at any specified final time t_f, that is to say, any initial states x_0 which satisfies the inequality below

$$|x_0| > \frac{1}{a} [\exp(a[t_f - t_0]) - 1] \tag{4.139}$$

cannot be driven to the zero from the initial value at the final time t_f.

Thus given any final time t_f, for the minimum fuel problems, it needs to be considered that the initial value x_0 satisfies the following constraint:

$$|x_0| \leq \frac{1}{a} [\exp(a[t_f - t_0]) - 1]. \tag{4.140}$$

If (4.140) is an equality, this means that $t_f = t^*$; otherwise, $t_f > t^*$, and the form of the optimal control must be as shown in Part (b) or Part (c) in Fig. 4.21.

Notice that, as stated earlier, the optimal control must be nonzero during some part of the time interval, because it was shown that the system will not reach zero with just zero-valued control.

For analytical simplicity, consider that the initial time $t_0 = 0$ for Example 4.7 studied below.

Example 4.7 Implement the optimal control for the minimum fuel problem with a fixed final time t_f such that the state system and the performance cost function are given in (4.122) and (4.124), respectively.

Solution. In Example 4.6, it was specified the optimal control u^* with $x_0 > 0$, and following the same procedure, we can figure out the optimal control with $x_0 < 0$ as well.

As shown in Part (b) in Fig. 4.21 and stated earlier, with $x_0 > 0$, the optimal control is in the form of $u^* = \{0, -1\}$.

Thus we define a control u, such that $u(t) = 0$ for all $t \in [0, t_1)$, where t_1 represents a switching time, and $u(t) = -1$ for all $t \in [t_1, t_f]$, and by applying this control u in (4.130) and considering $t_0 = 0$, it can be obtained that

$$x(t_f) = 0 = \exp(-at_f)x_0 - \frac{1}{a}\exp(-at_f)[\exp(at_f) - \exp(at_1)]. \tag{4.141}$$

Hence we can get that the switching time t_1 in case $x_0 > 0$ is given as

$$t_1 = \frac{1}{a}\ln(\exp(at_f) - ax_0). \tag{4.142}$$

Similarly, in case $x_0 < 0$, the optimal control u^* is in the form of $u^* = \{0, 1\}$, see an illustration in Part (c) in Fig. 4.21; then in this case, we get that

$$x(t_f) = 0 = \exp(-at_f)x_0 + \frac{1}{a}\exp(-at_f)[\exp(at_f) - \exp(a\widehat{t_1})]. \tag{4.143}$$

Denote by $\widehat{t_1}$ the switching time for the optimal control in case $x_0 < 0$; then it yields that

$$\widehat{t_1} = \frac{1}{a}\ln(\exp(at_f) + ax_0). \tag{4.144}$$

Consequently, by (4.142) and (4.144), the optimal control is given as

$$u^*(t) = \begin{cases} 0, & \text{in case } x_0 > 0 \text{ and } t < \dfrac{1}{a}\ln(\exp(at_f) - ax_0) \\[2mm] -1, & \text{in case } x_0 > 0 \text{ and } \dfrac{1}{a}\ln(\exp(at_f) - ax_0) \le t \le t_f \\[2mm] 0, & \text{in case } x_0 < 0 \text{ and } t < \dfrac{1}{a}\ln(\exp(at_f) + ax_0) \\[2mm] 1, & \text{in case } x_0 < 0 \text{ and } \dfrac{1}{a}\ln(\exp(at_f) + ax_0) \le t \le t_f \end{cases}. \tag{4.145}$$

It is worth to state that, since $u^*(t)$ is determined with respect to x_0 and t, the optimal control specified in (4.145) is in an *open-loop* form, that is the optimal control at time t, $u^*(t)$, is an expression on x_0 and t

$$u^*(t) = e(x_0, t). \tag{4.146}$$

Nevertheless, specify the control at any time as a feedback of the system state, that is, u^* can be expressed as the following form:

$$u^*(t) = \widehat{e}(x(t), t). \tag{4.147}$$

In the following, we will study how to re-organize the optimal control in the form of (4.147).

Firstly we have

$$x(t_f) = \exp(-a[t_f - t])x(t) + \exp(-at_f) \int_t^{t_f} \exp(ax)u(\tau)d\tau, \tag{4.148}$$

for all $t \in [0, t_f]$.

As it was obtained, during the latter part of the time interval, the optimal control is set to be 1 in case $x(t) < 0$ or -1 in case $x(t) > 0$.

Thus, suppose that $x(t) > 0$; then it is

$$x(t_f) = 0 = \exp(-a[t_f - t])x(t) - \exp(-at_f) \int_t^{t_f} \exp(ax)d\tau, \tag{4.149}$$

for all $t \ge t_1$, by which we can get that

$$x(t) = \frac{1}{a}[\exp(a[t_f - t]) - 1], \tag{4.150}$$

for all $t \ge t_1$.

Moreover, due to the specification of the optimal control, during the time interval before the switching time t_1, the optimal control is zero; then by (4.122) and with $t_0 = 0$, we have

$$x(t) = \exp(-at)x_0, \tag{4.151}$$

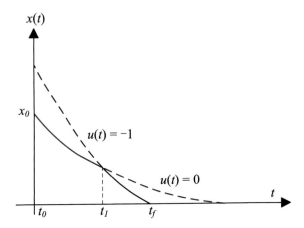

Fig. 4.23 The state trajectory subject to a control in the form of $\{0, -1\}$ with $x_0 > 0$

for all $t \in [0, t_1)$.

By (4.150) and (4.151), we can get that the optimal control is switched from 0 to -1 at the time t_1, such that

$$\exp(-at_1)x_0 = \frac{1}{a}[\exp(a[t_f - t_1]) - 1], \tag{4.152}$$

that is to say, the control is switched whenever the state system with $x(0) = x_0$ subject to zero control intersects the state with $x(t_f) = 0$ subject to the control $u = -1$ during the whole time interval $[0, t_f]$.

Figure 4.23 shows the state trajectory subject to a control in the form of $\{0, -1\}$ with initial state $x_0 > 0$.

Consequently, by the above analysis and the state dynamics displayed in Fig. 4.23, with $x_0 > 0$, the optimal control of the system switches from 0 to -1, in case that the state system $x(t)$ is equal to

$$\theta(t) \equiv \frac{1}{a}[\exp(a[t_f - t]) - 1] \tag{4.153}$$

at some t which is denoted by the switching time t_1.

Notice that the above conclusion holds only in case the initial state satisfies the following constraint:

$$x_0 \in \left(0, \frac{1}{a}[\exp(at_f) - 1]\right), \tag{4.154}$$

since as stated in earlier parts, the minimum fuel problem degenerates to the minimum time problem with $x_0 = \frac{1}{a}[\exp(at_f) - 1]$, and there is no solution with $x_0 > \frac{1}{a}[\exp(at_f) - 1]$.

Fig. 4.24 The state
trajectory subject to a control
in the form of {0, 1} with
$x_0 < 0$

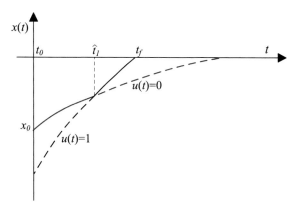

For comparison, Fig. 4.24 displays the state trajectory subject to a control in the
form of {0, 1} with initial state $x_0 < 0$.

As illustrated in Fig. 4.24, it can be verified that with the initial state $x_0 \in \left(-\frac{1}{a}[\exp(at_f) - 1], 0\right)$, the optimal control of the system switches from 0 to 1 in
case that the state system satisfies the following

$$x(t) = -\theta(t) \tag{4.155}$$

at some time t which is the switching time $\widehat{t}_1$.

In summary, the optimal control for the underlying problem is given as

$$u^*(t) = \begin{cases} -1, & \text{in case } x(t) \geq \theta(t) \\ 0, & \text{in case } |x(t)| < \theta(t) \ , \\ 1, & \text{in case } x(t) \leq -\theta(t) \end{cases} \tag{4.156}$$

which can be written in the following compact form

$$u^*(t) = \begin{cases} 0, & \text{in case } |x(t)| < \theta(t) \\ -\text{sgn}(x(t)), & \text{otherwise} \end{cases} \tag{4.157}$$

with $\text{sgn}(x) = \begin{cases} 1, & \text{in case } x > 0 \\ -1, & \text{otherwise} \end{cases}$.

Figure 4.25 displays a diagram of the system subject to the optimal control given
in (4.156) or (4.157). □

In Example 4.7, it was studied the minimum fuel problems with a fixed final time.
It may be interesting to study how to specify the final time and how the minimum
consumed fuel depends on the final time t_f.

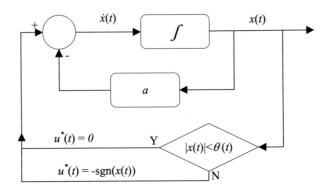

Fig. 4.25 An implementation of a fuel-optimal control

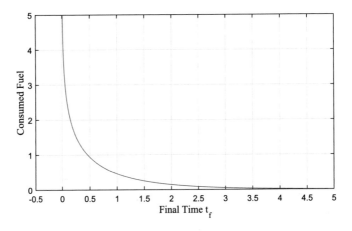

Fig. 4.26 The dependence of consumed fuel on specified final time t_f

By (4.142) and (4.144), we get that the optimal control switches from 0 to -1 or 1 at the switching time t_1 such that

$$t_1 = \frac{1}{a} \ln(\exp(at_f) - a|x_0|), \tag{4.158}$$

and then remains at 1 or -1 until the final time t_f.

Thus, the minimum consumed fuel is specified as

$$J(u^*) = t_f - t_1 = t_f - \frac{1}{a} \ln(\exp(at_f) - a|x_0|), \tag{4.159}$$

which can be directly verified to be monotonically decreasing with respect to the final time t_f, see an illustration in Fig. 4.26 with $|x_0| = 1$ and $a = 1$.

Actually, this phenomenon has been briefly discussed in Examples 4.5–4.7, say there exists the tradeoff between the consumed fuel and the elapsed time. It is due to that since in these examples to drive the system state to the origin and with no control applied, the state of these systems could approach the origin.

However, in case that the state moves further away from the target set with no control applied, it may be faced the different forms of the optimal control solutions.

4.5 Performance Cost Composed of Elapsed Time and Consumed Fuel

In this part, we further study the optimal problems considering the tradeoff between the elapsed time and the consumed fuel such that

$$J(u) \triangleq \int_{t_0}^{t_f} [\beta + |u(t)|]dt \tag{4.160}$$

with the final time t_f given free, and the weighting parameter $\beta \in [0, \infty)$.

Thus the value of β represents the relative importance of elapsed time and consumed fuel. More especially, as β goes to zero, this problem becomes a minimum fuel problem which has been studied in Sect. 4.4, while as β goes to infinity, this problem becomes a minimum time problem which has been studied in Sect. 4.3.

We will study the underlying optimal control problem via a simple example.

Example 4.8 Implement the optimal control for the system with the state process given as

$$\dot{x}(t) = -ax(t) + u(t), \tag{4.161}$$

which is also considered in Example 4.6 as well, with the initial condition of $x(t_0) = x_0$, the parameter $a > 0$, and the control satisfying the constraint of $|u(t)| \le 1$, for all t, such that the performance cost function (4.160) is minimized.

Notice that for analytical simplicity, as considered in Example 4.7, we still suppose that the initial time $t_0 = 0$ for this problem and Example 4.9.

Solution. Firstly the Hamiltonian is specified as

$$\mathscr{H}(x(t), u(t), \lambda(t)) = \beta + |u(t)| - a\lambda(t)x(t) + \lambda(t)u(t), \tag{4.162}$$

and by Theorem 4.1, the costate equation is given as

$$\dot{\lambda}^*(t) = -\frac{\partial \mathscr{H}(x^*(t), u^*(t), \lambda^*(t))}{\partial x} = a\lambda^*(t). \tag{4.163}$$

thus we have

$$\lambda^*(t) = \xi \exp(at), \tag{4.164}$$

with some constant value ξ.

By the necessary condition on the optimal control u^*, we have

$$\mathscr{H}(x^*(t), u^*(t), \lambda^*(t)) = \min_{u(t) \in [-1,1]} \mathscr{H}(x^*(t), u(t), \lambda^*(t)); \tag{4.165}$$

then it can be obtained that u^* is specified as

$$u^*(t) = \begin{cases} 1, & \text{in case } \lambda^*(t) < -1 \\ 0, & \text{in case } -1 < \lambda^*(t) < 1 \\ -1, & \text{in case } \lambda^*(t) > 1 \\ \text{undetermined, non-negative,} & \text{in case } \lambda^*(t) = -1 \\ \text{undetermined, non-positive,} & \text{in case } \lambda^*(t) = 1 \end{cases} \tag{4.166}$$

In case that $a > 0$, by the dynamics of $\lambda^*(t)$, we have that $\lambda^*(t)$ can not stay at 1 or -1 for any nonzero time interval. Hence there does not exist any singular case.

By (4.164) and (4.166), the optimal control is in one of the cases as displayed in Fig. 4.21. In the following, it will analyze these cases in the following.

Here it will first verify by contradiction that $u(t) = 0$, for all $t \in [0, t_f]$, can not be an optimal control.

Instead suppose that $u(t) = 0$, for all $t \in [0, t_f]$, is an optimal control; then it implies that

$$\mathscr{H}(x^*(t), u(t), \lambda^*(t)) = \beta - a\lambda^*(t)x^*(t), \tag{4.167}$$

with $u(t) = 0$, for all $t \in [0, t_f]$.

Here, it is the case that the final time t_f is free and t does not appear explicitly in the Hamiltonian; then by (4.31), we have

$$\mathscr{H}(x^*(t), u^*(t), \lambda^*(t)) = 0, \tag{4.168}$$

for all $t \in [0, t_f]$; then we have

$$x^*(t) = \frac{\beta}{a\lambda^*(t)} = \frac{\beta}{a\xi \exp(at)}, \tag{4.169}$$

for all $t \in [0, t_f]$.

By the boundary condition $x^*(t_f) = 0$, t_f converges to the infinity by (4.169) for all $\alpha < 1$.

Consequently, it can be obtained that $u(t) = 0$, for all $t \in [0, t_f]$, can not be an optimal control.

Suppose that $u^* = \{0, -1\}$ is the form of the optimal control, $\lambda^*(t)$ will pass through 1 at some time t_1 at which the control switches.

Moreover, since $\mathscr{H}(x^*(t), u^*(t), \lambda^*(t)) = 0$ for all t; then

$$\mathscr{H}(x^*(t_1), u^*(t_1), \lambda^*(t_1)) = \beta + |u^*(t_1)| - a\lambda(t)x^*(t_1) + \lambda(t_1)u^*(t_1) = 0, \tag{4.170}$$

which implies that

$$x^*(t_1) = \frac{\beta}{a}. \tag{4.171}$$

We get that when $x^*(t)$ passes through the value β/a at some time t; then the control will switch from 0 to -1 at that time.

By (4.130) and (4.171), it can be obtained the state processes subject to the optimal control such that

$$x(t) = x_0 \exp(-at), \text{ with } x(t) > \frac{\beta}{a}, \tag{4.172a}$$

$$x(t) = \frac{\beta}{a}\exp(-a[t-t_1]) - \frac{1}{a}[1 - \exp(-a[t-t_1])], \text{ with } 0 < x(t) \le \frac{\beta}{a}. \tag{4.172b}$$

As analyzed in previous examples, a control in the form $\{0, -1\}$ cannot drive the state system $\dot{x}(t) = -ax(t) + u(t)$ from a negative initial state to the origin; hence, (4.172) only holds with $x_0 > 0$.

The state processes subject to the optimal control with different initial values of x_0 are displayed in Fig. 4.27.

Notice that in case $0 < x_0 \le \beta/a$, the optimal control is to apply $u^*(t) = -1$ until the state system reaches the origin at the final time.

The above statement may hold, since, as the weighting factor in the performance cost function β goes to infinity, all the state processes start from the initial state $x_0 \le \beta/a$ and then the optimal control will be the bang-bang control for the minimum time problems.

Vice verse, as β goes to zero, the value of β/a approaches close to the value zero, and the optimal strategy approaches to that indicated by Example 4.6 with free final time; let the system state to as near the origin as possible before applying control.

The readers can show that for $x_0 < -\beta/a$, the optimal strategy is to allow the system to coast, with $u^*(t) = 0$, until it reaches $x(t) = -\beta/a$, where the optimal control switches to $u^*(t) = 1$.

As a consequence, the optimal control is summarized as

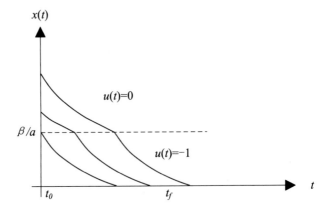

Fig. 4.27 Several optimal trajectories for a time-fuel performance cost

$$u^*(t) = \eta(x(t)) \triangleq \begin{cases} 0, & \text{in case } \frac{\beta}{a} < x(t) \\ -1, & \text{in case } 0 < x(t) \le \frac{\beta}{a} \\ 1, & \text{in case } -\frac{\beta}{a} \le x(t) < 0, \\ 0, & \text{in case } x(t) < -\frac{\beta}{a} \\ 0, & \text{in case } x(t) = 0 \end{cases} \qquad (4.173)$$

for all $t \in [0, t_f]$, which is displayed in Figs. 4.28 and 4.29.

In solving this example the reader should note that it was able to determine the optimal control using only the form of the costate solution—there was no need to solve for the constant of integration ξ_1.

It also exploited the necessary condition that

$$\mathcal{H}(x^*(t), u^*(t), \lambda^*(t)) = 0, \forall t \in [0, t_f], \qquad (4.174)$$

in case that the final time t_f is free and the Hamiltonian $\mathcal{H}$ is not explicitly dependent on the time t, to determine the optimal control and to show that the singular condition could not arise. □

Besides the example just studied above, in the following, study another example for the optimal control problems with a two-dimensional state system.

Example 4.9 Implement the optimal control to drive the state system

$$\dot{x}_1(t) = x_2(t), \qquad (4.175a)$$
$$\dot{x}_2(t) = u(t), \qquad (4.175b)$$

from an initial state $x(0) = x_0 \ne 0$ to the final state $x(t_f) = 0$, such that the performance cost function defined below

Fig. 4.28 The optimal control for Example 4.8

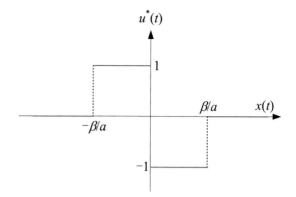

Fig. 4.29 An implementation of the weighted-time-fuel-optimal control of Example 4.8

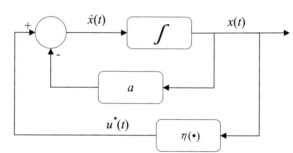

$$J(u) = \int_0^{t_f} [\beta + |u(t)|]dt \qquad (4.176)$$

is minimized, where the controls satisfy the constraints

$$|u(t)| \le 1, \qquad (4.177)$$

the final time t_f is free, and the weighting factor $\beta > 0$.

Solution. Firstly, it is straightforward to verify that the presence of weighting factor β in the Hamiltonian

$$\mathscr{H}(x(t), u(t), \lambda(t)) = \beta + |u(t)| + \lambda_1(t)x_2(t) + \lambda_2(t)u(t) \qquad (4.178)$$

does not effect the form of the optimal control which has been specified in (4.101). Thus, we get that

$$u^*(t) = \begin{cases} 1, & \text{in case } \lambda_2^*(t) < -1 \\ 0, & \text{in case } -1 < \lambda_2^*(t) < 1 \\ -1, & \text{in case } 1 < \lambda_2^*(t) \\ \text{undetermined, } \geq 0, & \text{in case } \lambda_2^*(t) = -1 \\ \text{undetermined, } \leq 0, & \text{in case } \lambda_2^*(t) = 1 \end{cases} . \tag{4.179}$$

The costate equations

$$\dot{\lambda}_1^*(t) = -\frac{\partial \mathcal{H}(x^*(t), u^*(t), \lambda^*(t))}{\partial x_1} = 0 \tag{4.180a}$$

$$\dot{\lambda}_2^*(t) = -\frac{\partial \mathcal{H}(x^*(t), u^*(t), \lambda^*(t))}{\partial x_2} = -\lambda_1^*(t) \tag{4.180b}$$

have solutions of the form

$$\lambda_1^*(t) = \xi_1, \tag{4.181a}$$

$$\lambda_2^*(t) = -\xi_1 t + \xi_2. \tag{4.181b}$$

By (4.181), it is obvious to claim that λ_2^* can change sign at most once, hence, excluding the singular form, the optimal control is in one of the following forms

$$u^* = \{0\}, \{1\}, \{-1\}, \{0, 1\}, \{0, -1\}, \{1, 0\}, \{-1, 0\}, \{1, 0, -1\}, \{-1, 0, 1\}. \tag{4.182}$$

In the following firstly verify by contradiction that the singular condition could not occur for the underlying problem.

Suppose that there exists a finite time interval during which $\lambda_2^*(t)$ remain at the value of 1 or -1; then by (4.181) we can get that $\xi_1 = 0$ and ξ_2 is equal to 1 or -1.

Substituting $\lambda_2^*(t) = \pm 1$ in (4.178), and using (4.178) and the definition of the absolute value function, it can be obtained that

$$\mathcal{H}(x^*(t), u^*(t), \lambda^*(t)) = \beta > 0. \tag{4.183}$$

However since the Hamiltonian $\mathcal{H}$ is explicitly independent of time t and t_f is free as well, by (4.31), the Hamiltonian must be zero subject to the optimal control and state.

Thus, get a contradiction by assuming there exists a singular condition.

In the next, it will analyze the control in the forms specified in (4.182).

Firstly, we have that the system ends with a time interval during which $u = 0$ can not reach the value zero.

Thus next, consider the candidates of the optimal control in the following forms

$$\{-1\}, \{0, -1\}, \{1, 0, -1\}. \tag{4.184}$$

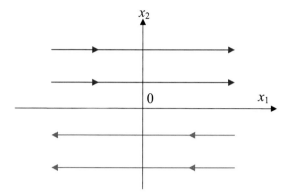

Fig. 4.30 Trajectories for $u = 0$

Moreover, the state equations of this problem are identical with those considered in the minimum time problem in Example 4.3; then the state trajectories should terminate in the curve B-0 as displayed in Fig. 4.8.

For any time interval during which $u(t) = 0$, the state equations are given as

$$\dot{x}_1(t) = x_2(t), \tag{4.185a}$$
$$\dot{x}_2(t) = 0, \tag{4.185b}$$

which implies that

$$x_2(t) = \xi_3, \tag{4.186a}$$
$$x_1(t) = \xi_3 t + \xi_4, \tag{4.186b}$$

with some constant valued ξ_3 and ξ_4.

Thus, the state $x_1(t)$ increases or decreases on time t, in case $x_2(t)$ is larger or smaller than zero over the time interval during which the control $u(t) = 0$, see an illustration in Fig. 4.30.

Notice that in case $x_2(t) = 0$ when the control switches to zero, x_1 remains as a constant until the control switches to nonzero. And the state trajectory segments generated by the control of $u = 1$ should be same as those displayed in Fig. 4.7.

The state trajectories x_0-C_1-D_1-0 and x_0-C_2-D_2-0 illustrated in Fig. 4.31 are typical candidates for the optimal trajectory with a given initial state x_0.

Hence it is to specify the point on the curve x_0-C_3, where the optimal control switches from 1 to 0. In case this switching state point is specified, the whole optimal state trajectory is specified as well.

Denote by t_1 the time when the optimal control switches from 1 to 0, and by t_2 the time when the optimal control switches from 0 to -1 respectively.

As displayed in Fig. 4.31, t_1 occurs on the curve x_0-C_3 and t_2 occurs on the curve C_3-0 respectively.

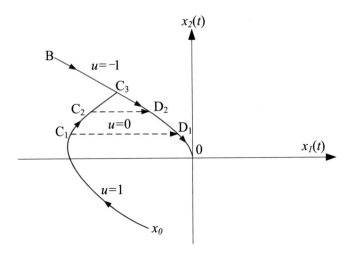

Fig. 4.31 Some typical candidates for the optimal state trajectory with a given initial state x_0

By (4.78), it can be obtained that, on the curve C_3-0, the following holds

$$x_1^*(t) = -\frac{1}{2}[x_2^*(t)]^2, \tag{4.187}$$

which implies that

$$x_1^*(t_2) = -\frac{1}{2}[x_2^*(t_2)]^2. \tag{4.188}$$

Furthermore, by the state equation (4.185) subject to a zero-valued control, we have

$$x_1^*(t_2) = x_1^*(t_1) + x_2^*(t_1)[t_2 - t_1], \tag{4.189}$$

and by (4.181) and (4.179), it can be obtained that

$$\lambda_2^*(t_1) = -\xi_1 t_1 + \xi_2 = -1, \tag{4.190a}$$
$$\lambda_2^*(t_2) = -\xi_1 t_2 + \xi_2 = 1, \tag{4.190b}$$

which implies that

$$t_2 - t_1 = -\frac{2}{\xi_1}. \tag{4.191}$$

With $\lambda_2^*(t_1) = -1$ and $\lambda_2^*(t_2) = 1$, and by the necessary condition that the Hamiltonian $\mathscr{H}$ should be equal to zero for the underlying problem, it can be obtained that

$$\beta + \xi_1 x_2^*(t_1) = 0, \tag{4.192a}$$
$$\beta + \xi_1 x_2^*(t_2) = 0. \tag{4.192b}$$

In the following, based upon the above analysis, it will verify that $x_1^*(t_1)$ and $x_2^*(t_1)$ satisfy the following relationship:

$$x_1^*(t_1) = -\frac{\beta + 4}{2\beta}[x_2^*(t_1)]^2. \tag{4.193}$$

- Verification of (4.193):
  Firstly by (4.192), it can be obtained that

$$x_2^*(t_1) = x_2^*(t_2), \tag{4.194a}$$
$$\xi_1 = \frac{-\beta}{x_2^*(t_1)}, \tag{4.194b}$$

by which together with (4.191), it can get

$$t_2 - t_1 = -\frac{2}{\xi_1} = \frac{2}{\beta}x_2^*(t_1). \tag{4.195}$$

Then by (4.189) and (4.195), it can be obtained that

$$x_1^*(t_2) = x_1^*(t_1) + x_2^*(t_1)[t_2 - t_1] = x_1^*(t_1) + \frac{2}{\beta}[x_2^*(t_1)]^2. \tag{4.196}$$

By (4.188) and (4.194a), we get that

$$-\frac{1}{2}[x_2^*(t_1)]^2 = x_1^*(t_1) + \frac{2}{\beta}[x_2^*(t_1)]^2, \tag{4.197}$$

which implies the conclusion of (4.193). □

The values of the optimal state x_1^* and x_2^* satisfying the relationship given in (4.193) are those states at which the control should switch from 1 to 0.

Besides, suppose that the optimal control u^* is in the form of $u^* = \{-1, 0, 1\}$ and denote by $\widehat{t}_1$ the switching time when the control switches from -1 to 0, and by $\widehat{t}_2$ the time when the optimal control switches from 0 to 1 respectively.

Thus following the similar analysis, it can be verified as well that at the switching time $\widehat{t}_1$, the values of the optimal state x_1^* and x_2^* satisfy the following relationship

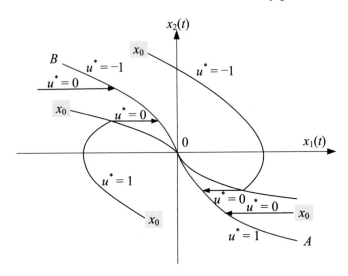

Fig. 4.32 Typical optimal state trajectories for time-fuel-optimal problems with different initial states

$$x_1^*(\widehat{t_1}) = \frac{\beta+4}{2\beta}[x_2^*(\widehat{t_1})]^2. \tag{4.198}$$

Notice that the optimal control u^* can be specified by the developed results of (4.193), (4.198), (4.188) and its counterpart for $u^*(t) = 1$ given in (4.199) below

$$x_1^*(\widehat{t_2}) = \frac{1}{2}[x_2^*(\widehat{t_2})]^2. \tag{4.199}$$

The optimal control for the underlying problem is time-invariant. And in Fig. 4.32, it displays optimal trajectories with different initial state values respectively.

Moreover, for comparison, it is also interesting to discuss how the optimal state trajectories evolve with respect to the weighting factor β. In Fig. 4.33, it illustrates the switching curves of the optimal state trajectories with distinct values of weighting factor β, with $\beta_1 > \beta_2$.

Moreover, as observed that as β increases, the switching curves approaches to the optimal state trajectory for the minimum time problems studied in Example 4.3. And as β goes to infinity, these two-state trajectories merge with each other, and the time interval of $u^* = 0$ approaches zero as well.

Besides, as β goes to zero, the time interval of $u^* = 0$ approaches infinity, and hence the state trajectories approach the optimal trajectories for the minimum fuel problems.

In Fig. 4.34, it illustrates the evolutions of the elapsed time $t_f - t_0$ and the consumed fuel on the weighting parameter β. As observed, as β increases, the elapsed time decreases and while the consumed fuel increases.   □

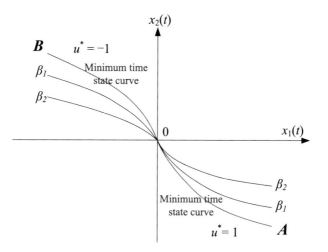

Fig. 4.33 Switching curves for minimum time-fuel problems

Fig. 4.34 The evolutions of the elapsed time and the consumed fuel on the weighting parameter β

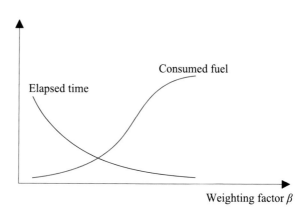

4.6 Minimum Energy Problems

In this section, it will study the minimum energy problems for a linear time-variant state system

$$\dot{x}(t) = A(t)x(t) + B(t)u(t), \tag{4.200}$$

where $x(t)$ is an n-dimension state and $u(t)$ is m-dimension control with

$$|u_i(t)| \le 1, \ \text{for all } i = 1, \ldots, m. \tag{4.201}$$

The objective is to drive the state system $x(t)$ specified above from an initial value x_0 to the origin such that the following consumed energy is minimized

$$J(x_0; u) \triangleq \frac{1}{2} \int_{t_0}^{t_f} [u(t)]^\top R(t) u(t) dt, \tag{4.202}$$

where $R(t)$ is assumed to be a positive definite $m \times m$ diagonal matrix.

As usual, firstly define the Hamiltonian for the underlying optimal control problems

$$\mathscr{H}(x(t), u(t), \lambda(t))$$
$$\triangleq \frac{1}{2} [u(t)]^\top R(t) u(t) + [\lambda(t)]^\top A(t) x(t) + [\lambda(t)]^\top B(t) u(t) \tag{4.203}$$

The state and costate equations $x^*(t)$ and $\lambda^*(t)$ subject to the optimal control $u^*(t)$ are given as

$$\dot{x}^*(t) = A(t) x^*(t) + B(t) u^*(t), \tag{4.204a}$$
$$\dot{\lambda}^*(t) = -[A(t)]^\top \lambda^*(t), \tag{4.204b}$$

with the boundary conditions

$$x^*(t_0) = x_0, \text{ and } x^*(t_f) = 0. \tag{4.205}$$

Concerning the discussion on the final time t_f for the minimum energy problems here, please refer the statement on the specification of the final time t_f for the minimum fuel problems given in Sect. 4.3.

By applying the Pontryagin's minimum principle, the necessary condition for the optimal control is specified as below:

$$\frac{1}{2} [u^*(t)]^\top R(t) u^*(t) + [\lambda^*(t)]^\top A(t) x^*(t) + [\lambda^*(t)]^\top B(t) u^*(t)$$
$$\leq \frac{1}{2} [u(t)]^\top R(t) u(t) + [\lambda^*(t)]^\top A(t) x^*(t) + [\lambda^*(t)]^\top B(t) u(t), \tag{4.206}$$

for all instants $t \in [t_0, t_f]$, which implies that

$$\frac{1}{2} [u^*(t)]^\top R(t) u^*(t) + [\lambda^*(t)]^\top B(t) u^*(t)$$
$$\leq \frac{1}{2} [u(t)]^\top R(t) u(t) + [\lambda^*(t)]^\top B(t) u(t). \tag{4.207}$$

Define a matrix $Q^*(t)$ such that

$$Q^*(t) \triangleq R^{-1}(t) [B(t)]^\top \lambda^*(t), \tag{4.208}$$

for all $t \in [t_0, t_f]$, where $R^{-1}(t)$ represents the inverse of the matrix $R(t)$ since $R(t)$ is assumed to be invertible; then

$$[\lambda^*(t)]^\top B(t)u^*(t) = [u^*(t)]^\top [B(t)]^\top \lambda^*(t) = [u^*(t)]^\top R(t)Q^*(t). \qquad (4.209)$$

Thus by the above, we can get that

$$\frac{1}{2}[u^*(t)]^\top R(t)u^*(t) + [u^*(t)]^\top R(t)Q^*(t)$$
$$\leq \frac{1}{2}[u(t)]^\top R(t)u(t) + [u(t)]^\top R(t)Q^*(t). \qquad (4.210)$$

Furthermore, by the specification of $Q^*(t)$, we have

$$\frac{1}{2}[Q^*(t)]^\top R(t)Q^*(t) = \frac{1}{2}[\lambda^*(t)]^\top B(t)R^{-1}(t)[B(t)]^\top \lambda^*(t); \qquad (4.211)$$

then by adding $[Q^*(t)]^\top R(t)Q^*(t)$ to both sides of (4.210), it can be obtained that

$$[u^*(t) + Q^*(t)]^\top R(t)[u^*(t) + Q^*(t)] \leq [u(t) + Q^*(t)]^\top R(t)[u(t) + Q^*(t)]. \qquad (4.212)$$

Thus, we have

$$[z^*(t)]^\top R(t)z^*(t) \leq [z(t)]^\top R(t)z(t), \qquad (4.213)$$

with $z^*(t) \equiv u^*(t) + Q^*(t)$ and $z(t) \equiv u(t) + Q^*(t)$.

We have supposed that $R(t)$ is a positive definite diagonal matrix, for all $t \in [t_0, t_f]$; then $R(t)$ is the diagonal matrix of its positive eigenvalues denoted by $r_j(t)$, with $j = 1, \ldots, m$. Then

$$\left\{ [z(t)]^\top R(t)z(t) \right\} = \sum_{i=1}^m r_i(t)\widehat{z}_i^2(t), \qquad (4.214)$$

which implies that

$$\min_{|u(t) \leq 1|} \left\{ [z(t)]^\top R(t)z(t) \right\}$$
$$= \min_{|u(t) \leq 1|} \left\{ \sum_{i=1}^m r_i(t)\widehat{z}_i^2(t) \right\}$$
$$= \sum_{i=1}^m \left\{ r_i(t) \min_{|u_i(t) \leq 1|} \widehat{z}_i^2(t) \right\}. \qquad (4.215)$$

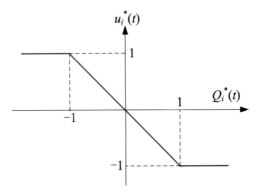

Fig. 4.35 An illustration of optimal control for minimum energy problems

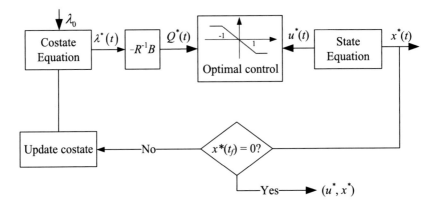

Fig. 4.36 An implementation of optimal control for minimum energy problems

As a result, by (4.212), (4.213) and (4.215), and we can get that the optimal control $u^*(t)$ is specified as

$$
u_i^*(t) = \begin{cases}
-Q_i^*(t), & \text{in case } |Q_i^*(t)| \leq 1 \\
1, & \text{in case } Q_i^*(t) < -1 , \\
-1, & \text{in case } Q_i^*(t) > 1
\end{cases}
\tag{4.216}
$$

see an illustration in Fig. 4.35.

The implementation of the optimal control specified in (4.216) could be performed following the diagram as displayed in Fig. 4.36.

Suppose that there are no constraints on the control, then the optimal control specified for minimum energy problems degenerates to the following unconstrained form

$$u^*(t) = -Q^*(t),$$ (4.217)

which is consistent with the result developed earlier.

In the following, it simply demonstrates the proposed result for the minimum energy problems of a one-dimensional state system.

Example 4.10 Consider a one-dimensional state system as below

$$x(t) = ax(t) + u(t),$$ (4.218)

with $a < 0$ and $|u(t)| \leq 1$, for all t.

The objective is to drive the state system $x(t)$ specified above from an initial value x_0 to the origin such that

$$J(x_0; u) \triangleq \frac{1}{2} \int_{t_0}^{t_f} ru^2(t)dt,$$ (4.219)

with $r > 0$.

Solution. By applying the proposed result, for the underlying minimum energy problems, we have

$$A(t) = a, \; B(t) = 1, \; R(t) = r,$$

for all $t \in [t_0, t_f]$.

The Hamiltonian for the underlying optimal control problems is given as

$$\mathcal{H}(x(t), u(t), \lambda(t)) = \frac{r}{2}u^2(t) + a\lambda(t)x(t) + \lambda(t)u(t).$$ (4.220)

The state and costate equations $x^*(t)$ and $\lambda^*(t)$ with the optimal control $u^*(t)$ are given as

$$\dot{x}^*(t) = ax^*(t) + u^*(t),$$ (4.221a)
$$\dot{\lambda}^*(t) = -a\lambda^*(t)$$ (4.221b)

with the boundary conditions

$$x^*(t_0) = x_0, \; \text{and} \; x^*(t_f) = 0.$$ (4.222)

It is straightforward to get that the solution of the costate $\lambda^*(t)$ is

Fig. 4.37 The evolution of optimal control with respect to $\lambda^*(t)$ for minimum energy problems

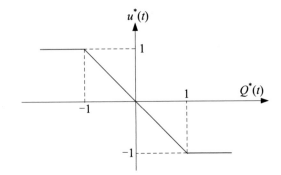

$$\lambda^*(t) = \lambda_0 \exp(-at) \tag{4.223}$$

with λ_0 being an initial value of the costate.

By the specification of $Q(t)$ given in (4.208)

$$Q^*(t) \triangleq R^{-1}(t)[B(t)]^\top \lambda^*(t) = \frac{\lambda^*(t)}{r}, \quad \text{for all } t \in [t_0, t_f]; \tag{4.224}$$

then by (4.216), we can get the optimal control $u^*(t)$ specified as

$$u^*(t) = \begin{cases} -Q^*(t) \equiv -\frac{\lambda^*(t)}{r}, & \text{in case } |\lambda^*(t)| \leq r \\ 1, & \text{in case } \lambda^*(t) < -r , \\ -1, & \text{in case } \lambda^*(t) > r \end{cases} \tag{4.225}$$

which is illustrated in Fig. 4.37.

As displayed, during any interval where $|\lambda^*(t)| \leq r$, the control is unsaturated; then $u^*(t) = -\frac{\lambda^*(t)}{r}$; while during any interval where $|\lambda^*(t)| > r$, the control reaches the bounds of the control set, say $u^*(t) = -1$ or 1 in this situation.

Notice that without considering the constraints on the control $|u(t)| \leq 1$, by (3.104) in Theorem 3.6, we can get necessary condition for the underlying problems given as

$$u^*(t) = -R^{-1}(t)B^\top(t)\lambda^*(t) = -\frac{\lambda^*(t)}{r}, \tag{4.226}$$

which is consistent with (4.225) in case that the boundary on control is not considered.

In case $\lambda_0 = 0$, we have $\lambda^*(t) = 0$ for all t, by (4.225), $u^*(t) = 0$; then the associated state $x^*(t) = x_0 \exp(at)$ which cannot reach the origin with $x_0 \neq 0$. As a result, λ_0 cannot be equal to zero.

The possible trajectories of costate $\lambda^*(t)$ are displayed in Fig. 4.38.

In the following, we will specify the specification of optimal control $u^*(t)$ with respect to the value of costate $\lambda^*(t)$ in different cases.

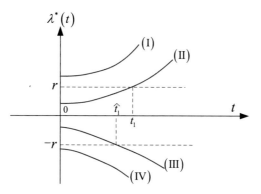

Fig. 4.38 Trajectories of $\lambda^*(t)$ with different initial values

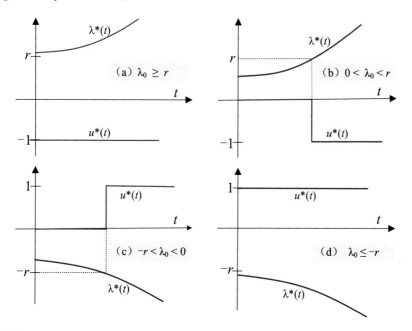

Fig. 4.39 The evolution of optimal control on costate $\lambda^*(t)$ with different initial values

- (I). In case $\lambda_0 > r$, by (4.223), $\lambda^*(t) > r$ for all t; then by (4.225), we get the optimal control $u^*(t) = -1$ for all t, see (a) in Fig. 4.39.
- (II). In case $0 < \lambda_0 < r$, by (4.223), there exists some time t_1, such that

$$\lambda^*(t) \begin{cases} \in (0, r], & \text{with } t \leq t_1 \\ > r, & \text{with } t > t_1 \end{cases};$$

then by (4.225), we get that the optimal control $u^*(t) = \begin{cases} -\frac{\lambda^*(t)}{r}, & \text{with } t \leq t_1 \\ -1, & \text{with } t > t_1 \end{cases}$,

see (b) in Fig. 4.39.

- (III). Similarly, in case $-r < \lambda_0 < 0$, by (4.223), there exists time $\widehat{t_1}$, such that

$$\lambda^*(t) \begin{cases} \in [-r, 0), & \text{with } t \leq \widehat{t_1} \\ < -r, & \text{with } t > \widehat{t_1} \end{cases};$$

then by (4.225), we get the optimal control $u^*(t) = \begin{cases} -\frac{\lambda^*(t)}{r}, & \text{with } t \leq \widehat{t_1} \\ 1, & \text{with } t > \widehat{t_1} \end{cases}$, see

(c) in Fig. 4.39.

- (IV). In case $\lambda_0 < -r$, by (4.223), $\lambda^*(t) < -r$ for all t; then by (4.225), we get the optimal control $u^*(t) = 1$ for all t, see (d) in Fig. 4.39.

Up to now, we have specified the optimal control $u^*(t)$ with respect to the costate $\lambda^*(t)$; next, it gives the state-dependent optimal control solution, say to figure out how to implement $u^*(t)$ on the state $x^*(t)$,

By the state equation for the underlying one-dimensional state system, we have for any admissible control

$$x(t_f) = 0 = \exp(a[t_f - t])x(t) + \exp(at_f) \int_t^{t_f} \exp(-as)u(s)ds. \tag{4.227}$$

Thus it implies that

$$x(t) = -\exp(at) \int_t^{t_f} \exp(-as)u(s)ds, \tag{4.228}$$

which is negative (positive respectively) in case that $u(s)$ is non-negative (non-positive, respectively) during the interval of $[t, t_f]$.

Also the controls given in (I) and (II) above are non-positive valued for all $t \in [t_0, t_f]$; thus these controls correspond to positive states.

And the controls given in (III) and (IV) above are non-negative valued for all $t \in [t_0, t_f]$; thus these controls correspond to negative states.

Moreover, since the final time t_f is unspecified and the Hamiltonian as given in (4.220) does not depend on time t explicitly; then by (4.31), we get that

$$\mathscr{H}(x^*(t), \lambda^*(t), u^*(t)) = \frac{r}{2}[u^*(t)]^2 + a\lambda^*(t)x^*(t) + \lambda^*(t)u^*(t) = 0, \tag{4.229}$$

for all $t \in [t_0, t_f]$; thus

$$x^*(t) = -\frac{u^*(t)}{a}\left[\frac{r}{2}\frac{u^*(t)}{\lambda^*(t)} + 1\right]. \tag{4.230}$$

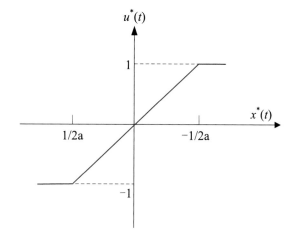

Fig. 4.40 The optimal control $u^*(t)$ with respect to state $x^*(t)$

Suppose that the control saturates with -1 at time $t = t_1$; then by (4.225) and (4.230),

$$x^*(t_1) = \frac{1}{2a}. \tag{4.231}$$

Also from (II), $u^*(t) = -1$ for all $t \in [t_1, t_f]$, and hence by the state equation

$$x^*(t) < x^*(t_1), \quad \text{for all } t > t_1. \tag{4.232}$$

As a result, it can be obtained that the optimal control

$$u^*(t) = \begin{cases} -2ax^*(t), & \text{with } |x^*(t)| \le -\frac{1}{2a} \\ 1, & \text{with } x^*(t) > -\frac{1}{2a} \\ -1, & \text{with } x^*(t) < \frac{1}{2a} \\ 0, & \text{with } x^*(t) = 0 \end{cases}, \tag{4.233}$$

which is displayed in Fig. 4.40. □

4.7 Performance Cost Composed of Elapsed Time and Consumed Energy

So far we have studied the optimal control for the minimum time problems, the minimum fuel problems and also the minimum time-fuel problems which consider the tradeoff between the elapsed time and the consumed fuel during the time interval.

In the following, we will study the characteristics of the minimum time-energy problems via a numerical example below.

Example 4.11 Implement the optimal control for a system with the state equation as given as

$$\dot{x}(t) = -ax(t) + u(t), \tag{4.234}$$

which has been applied in Examples 4.6 and 4.8 as well, with $a > 0$, which is to be driven from a given initial state, $x(0) = x_0$, to the origin at a final time t_f by a control such that the following performance cost

$$J(u) = \int_0^{t_f} \left[\beta + u^2(t)\right] dt \tag{4.235}$$

is minimized, with the weighting factor $\beta > 0$ and where t_f is free and the control $u(t)$ satisfies the following constraint

$$|u(t)| \le 1. \tag{4.236}$$

Solution. As usual, we first specify the Hamiltonian $\mathcal{H}$ for the underlying problem in the following:

$$\mathcal{H}(x(t), u(t), \lambda(t)) \triangleq \beta + u^2(t) - \lambda(t)ax(t) + \lambda(t)u(t). \tag{4.237}$$

The costate equation is given as

$$\dot{\lambda}^*(t) = a\lambda^*(t); \tag{4.238}$$

then its solution is specified as

$$\lambda^*(t) = \xi_1 \exp(at), \tag{4.239}$$

for all $t \in [0, t_f]$, with some constant valued ξ_1.

In the following, we will specify the optimal control u^* subject to which the Hamiltonian $\mathcal{H}$ defined in (4.237) is globally minimized.

- For $|u(t)| < 1$, the control that minimizes the Hamiltonian $\mathcal{H}$ is the solution of the equation

$$\frac{\partial \mathcal{H}(x^*(t), u^*(t), \lambda^*(t))}{\partial u} = 2u^*(t) + \lambda^*(t) = 0. \tag{4.240}$$

Notice that $\mathcal{H}$ is quadratic in $u(t)$ and

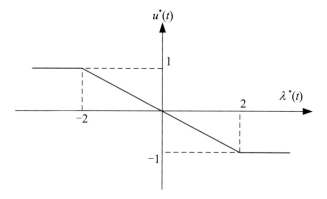

Fig. 4.41 The relationship between an extremal control and costate

$$\frac{\partial^2 \mathcal{H}(x^*(t), u^*(t), \lambda^*(t))}{\partial u^2} = 2 > 0. \tag{4.241}$$

So suppose that $|u^*(t)| < 1$; then we get that the optimal control u^* which globally minimizes the Hamiltonian that is given as

$$u^*(t) = -\frac{1}{2}\lambda^*(t), \tag{4.242}$$

which is equivalent to the following:

$$|\lambda^*(t)| < 2. \tag{4.243}$$

- If $|\lambda^*(t)| \geq 2$, then the control that minimizes $\mathcal{H}$ is

$$u^*(t) = \begin{cases} 1, & \text{in case } \lambda^*(t) \leq -2 \\ -1, & \text{in case } 2 \leq \lambda^*(t) \end{cases}. \tag{4.244}$$

Consequently, by (4.242) and (4.244), it can be obtained that

$$u^*(t) = \begin{cases} 1, & \text{in case } \lambda^*(t) \leq -2 \\ -\frac{1}{2}\lambda^*(t), & \text{in case } -2 < \lambda^*(t) < 2 \\ -1, & \text{in case } 2 \leq \lambda^*(t) \end{cases}. \tag{4.245}$$

The relationship between an optimal control u^* and an optimal costate λ^* specified in (4.245) is illustrated in Fig. 4.41.

There does not exist any singular conditions for the underlying problem, since the control $u(t)$ at any time t is uniquely determined by the value of the costate $\lambda^*(t)$ at that time.

By (4.245) together with the dynamics of costate $\lambda^*(t)$ specified in (4.239), in (I)–(IV) given below, we will study the specific form of the optimal control with respect to the dynamics of $\lambda^*(t)$.

Before that, firstly we state that $\lambda^*(t) = 0$ for $t \in [0, t_f]$, in case $\xi_1 = 0$, cannot be a candidate for the optimal costate dynamics, since $u^*(t) = 0$ for $t \in [0, t_f]$ in this case, and as we have verified earlier the system cannot reach the origin subject to a control ending with zero.

All the other forms for the costate $\lambda^*(t)$ are illustrated in Fig. 4.42.

(I) In case $\xi_1 \geq 2$, by (4.239), $\lambda^*(t) > 2$, for all $t \in [0, t_f]$; then by applying (4.245), we get that

$$u^*(t) = -1, \tag{4.246}$$

for all $t \in [0, t_f]$, or equivalently to say u^* is in the form of $\{-1\}$, see an illustration in Fig. 4.43.

(II) In case $\xi_1 \in (0, 2)$, by (4.239), the costate $\lambda^*(t)$ will increase from ξ_1 at initial time 0, and may be larger than 2 after some time t_1 before the system reaches the value zero, see the curve (II) in Fig. 4.42; then by applying (4.245), we get the optimal control u^* may be in one of the following forms

$$\left\{ -\frac{1}{2}\lambda^*(t) \right\}, \text{ or } \left\{ -\frac{1}{2}\lambda^*(t), -1 \right\}, \tag{4.247}$$

see an illustration in Fig. 4.44.

(III) In case $\xi_1 \in (-2, 0)$, by (4.239), the costate $\lambda^*(t)$ will decrease from ξ_1 at initial time 0, and may be less than -2 after some time $\widehat{t_1}$ before the system reaches the value zero, see the curve (III) in Fig. 4.42; then by applying (4.245), we get the optimal control u^* may be in one of the following forms

$$\left\{ -\frac{1}{2}\lambda^*(t) \right\}, \text{ or } \left\{ -\frac{1}{2}\lambda^*(t), 1 \right\}, \tag{4.248}$$

see an illustration in Fig. 4.45.

(IV) In case $\xi_1 \leq -2$, by (4.239),

$$u^*(t) = 1, \tag{4.249}$$

for all $t \in [0, t_f]$, or equivalently to say u^* is in the form of $\{1\}$, see an illustration in Fig. 4.46.

As analyzed, the controls specified in (4.246) and (4.247) are non-positive valued for all $t \in [0, t_f]$, and it should be applied to positive state values. We can verify this statement as below.

By the given state equation,

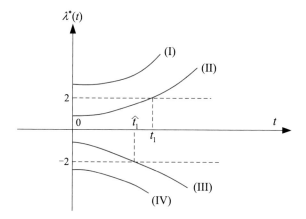

Fig. 4.42 Possible forms for an extremal costate trajectory

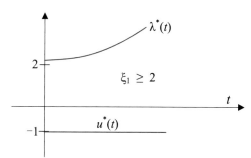

Fig. 4.43 The evolution of the optimal control u^* with respect to dynamics of λ^* given in the curve (I) in Fig. 4.42

$$x(t_f) = 0 = \exp(-a[t_f - t])x(t) + \exp(-at_f) \int_t^{t_f} \exp(at_f)u(\tau)d\tau, \quad (4.250)$$

which implies that

$$- \exp(at_f)x(t) = \int_t^{t_f} \exp(a\tau)u(\tau)d\tau. \quad (4.251)$$

For $u(\tau)$ non-positive when $\tau \in [t, t_f]$, the integral is negative; therefore, $x(t)$ must be positive.

Similarly, it can be claimed that the non-negative controls specified in (4.248) and (4.249), respectively, should be applied to negative values of state $x(t)$.

Since the final time t_f is free, and the Hamiltonian $\mathscr{H}$ does not depend upon the time t explicitly, by (4.31), it is necessary that

$$\mathscr{H}(x^*(t), u^*(t), \lambda^*(t)) = 0, \quad t \in [t_0, t_f]. \quad (4.252)$$

In case the control reaches -1, the lower bound of the control is set at time $t = t_1$, $\lambda^*(t_1) = 2$ by (4.245); then by applying $u^*(t_1) = -1$ and $\lambda^*(t_1) = 2$ into the

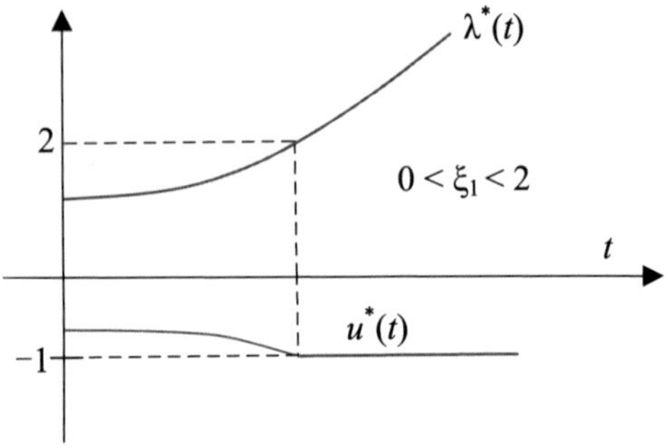

Fig. 4.44 The evolution of
the optimal control u^* with
respect to dynamics of λ^*
given in the curve (II) in
Fig. 4.42

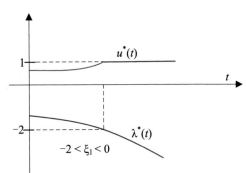

Fig. 4.45 The evolution of
the optimal control u^* with
respect to dynamics of λ^*
given in the curve (III) in
Fig. 4.42

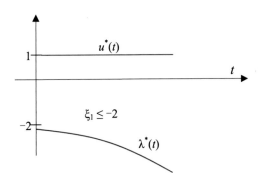

Fig. 4.46 The evolution of
the optimal control u^* with
respect to dynamics of λ^*
given in the curve (IV) in
Fig. 4.42

Hamiltonian $\mathcal{H}$, it can be obtained

$$\mathcal{H}(x^*(t_1), u^*(t_1), \lambda^*(t_1)) = \beta + 1 - 2ax^*(t_1) - 2 = 0, \tag{4.253}$$

which implies that

$$x^*(t_1) = \frac{\beta - 1}{2a}. \tag{4.254}$$

In case the control reaches -1, the lower bound of the control set at time $t = t_1$, by (4.246), we have $u^*(t) = -1$ for all $t \in [t_1, t_f]$, and $x^*(t) < x^*(t_1)$ for $t > t_1$; thus

$$u^*(t) = -1, \text{ with } 0 < x^*(t) < \frac{\beta - 1}{2a}. \tag{4.255}$$

Following the same analysis, we can get that in case the control reaches, 1, the upper bound of the control set at another time $t = \widehat{t_1}$, then

$$x^*(\widehat{t_1}) = \frac{\beta - 1}{-2a}, \tag{4.256}$$

and

$$u^*(t) = 1, \text{ with } \frac{\beta - 1}{-2a} < x^*(t) < 0. \tag{4.257}$$

Notice that, in case $\beta \leq 1, x^*(t_1) \leq 0$ in (4.254), and while $x^*(\widehat{t_1}) \geq 0$ in (4.256). However, as stated, (4.254) applies for positive state values and (4.256) applies for negative state values. Thus the optimal control u^* does not reach the bounds of the control set in case $\beta \leq 1$.

Next, we will analyze the inner region of the control set where the optimal control is given as $u^*(t) = -\frac{1}{2}\lambda^*(t)$.

By applying the necessary condition of (4.252) and $u^*(t) = -\frac{1}{2}\lambda^*(t)$, it can be obtained

$$\mathscr{H}(x^*(t), u^*(t), \lambda^*(t)) = \beta + \frac{1}{4}[\lambda^*(t)]^2 - \lambda^*(t)ax^*(t) - \frac{1}{2}[\lambda^*(t)]^2 = 0, \tag{4.258}$$

by which we get the solution of $\lambda^*(t)$ given as

$$\lambda^*(t) = 2\left[-ax^*(t) \pm \sqrt{a^2[x^*(t)]^2 + \beta}\right]. \tag{4.259}$$

Thus by (4.259) and $u^*(t) = -\frac{1}{2}\lambda^*(t)$, it can be obtained

$$u^*(t) = ax^*(t) \pm \sqrt{a^2[x^*(t)]^2 + \beta}. \tag{4.260}$$

As stated earlier, if $x^*(t) \geq 0$, $u^*(t)$ must be negative valued, thus the minus sign applies for (4.260); otherwise, in case $x^*(t) < 0$, the positive sign in (4.260) applies for $u^*(t)$.

By (4.260) together with (4.257) and (4.255), we get the optimal control with the feedback of the state given as

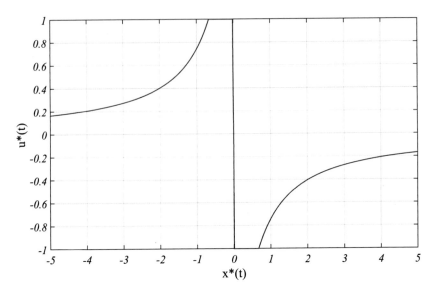

Fig. 4.47 The time-energy optimal control for Example 4.11 with $a = 3$ and $\beta = 5$

Fig. 4.48 An implementation of the time-energy optimal control for Example 4.11

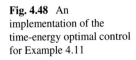

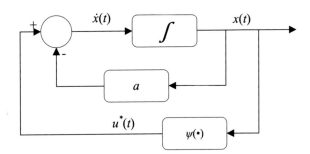

$$u^*(t) = \psi(x^*(t)) \triangleq \begin{cases} ax(t) - \sqrt{a^2[x^*(t)]^2 + \beta}, & \text{in case } 0 < \frac{\beta-1}{2a} < x^*(t) \\ -1, & \text{in case } 0 < x^*(t) \leq \frac{\beta-1}{2a} \\ 1, & \text{in case } -\frac{\beta-1}{2a} \leq x^*(t) < 0 \,, \\ ax^*(t) + \sqrt{a^2[x^*(t)]^2 + \beta}, & \text{in case } x^*(t) < -\frac{\beta-1}{2a} < 0 \\ 0, & \text{in case } x^*(t) = 0 \end{cases}$$

$$(4.261)$$

for all $t \in [0, t_f]$.

Figures 4.47 and 4.48 illustrate the optimal control with respect to the state and its implementation diagram, respectively.   □

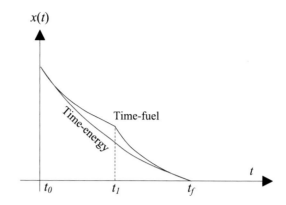

Fig. 4.49 Weighted-time-fuel and weighted-time-energy optimal trajectories

It is worth to emphasize here that distinct from the time-fuel-optimal controls, see an illustration in Fig. 4.28, which are set as either the lower/upper boundary of the control set or zero, the time-energy optimal controls can be set as any values in the control set $[-1, 1]$.

For comparison, Fig. 4.49, displays the typical state processes subject to the optimal controls for minimum time-fuel problems and minimum time-energy problems, respectively.

4.8 Summary

In this chapter, Pontryagin's minimum principle for optimal control problems with constrained control and state is developed; then by applying this result, we study how to implement the optimal controls for minimum time, minimum fuel, and minimum energy problems. For the purpose of comparison, we further study the optimal controls for minimum time-fuel and minimum time-energy problems as well.

4.9 Exercises

Exercise 4.1 Consider a state system given as

$$\dot{x}_1(t) = x_2(t),$$
$$\dot{x}_2(t) = -x_1(t) + x_2(t) + u(t),$$

and the performance cost function is specified as

$$J \triangleq \int_0^1 \left[x_1^2(t) + x_2^2(t) + u^2(t) \right] dt.$$

Suppose that the initial and final state are specified; then

- to implement the costate equation for the underlying problem;
- to implement the optimal controls minimizing the Hamiltonian $\mathscr{H}$ with $u(t)$ unbounded and $|u(t)| \leq 1$, respectively.

Exercise 4.2 Implement the optimal control for the following state system

$$\dot{x}(t) = -2x(t) + u(t),$$

to drive the system from a given initial state x_0 to zero in a minimum time, with $u(t)$ is unbounded and $|u(t)| \leq 1$ respectively.

Exercise 4.3 Implement the optimal control for the following state system

$$\begin{aligned} \dot{x}_1(t) &= x_2(t), \\ \dot{x}_2(t) &= u(t), \end{aligned}$$

to drive the system from a given initial state x_0 to the final state $\begin{bmatrix} -2 \\ 2 \end{bmatrix}$ in a minimum time, with $u(t)$ is unbounded and $|u(t)| \leq 1$ respectively.

Exercise 4.4 Implement the optimal control for the minimum time problem for an unmanned vehicle driving system introduced in Example 1.1.

Exercise 4.5 Implement the optimal control for the state system $\dot{x}(t) = -u(t)$ with the control satisfying the constraint of $|u(t)| \leq 1$, such that the state is driven from an initial state x_0 to the origin at the final time, and the following performance cost function is minimized

$$J = \int_0^{t_f} |u(t)| dt,$$

where the final time t_f is free.

Exercise 4.6 Implement the optimal control for the state system

$$\dot{x}(t) = -a + u(t)$$

with $a > 0$ and $|u(t)| \leq 1$, for all t, such that the state is driven from an initial state x_0 to the origin at the final time, and the following performance cost function is minimized

$$J = \int_0^{t_f} [\beta + |u(t)|] dt$$

with $\beta > 0$, where the final time t_f is free.

Exercise 4.7 Implement the optimal control for the state system defined in Exercise 4.3 such that the system is driven from $\begin{bmatrix} -2 \\ 2 \end{bmatrix}$ at t_0 to the origin, and the following performance cost function

$$J = \int_0^{t_f} |u(t)| dt$$

is minimized.

Exercise 4.8 Implement the optimal control for the state system

$$\dot{x}_1(t) = 2x_2(t),$$
$$\dot{x}_1(t) = u(t)$$

with $|u(t)| \le 1$, for all t, such that the state is driven from an initial state $x_0 = \begin{bmatrix} 1 \\ -1 \end{bmatrix}$ to the origin at the final time, and the following performance cost function is minimized

$$J = \int_0^{t_f} [1 + |u(t)|] dt,$$

where the final time t_f is free.

Exercise 4.9 Analyze the properties of the optimal control for the fuel-minimization problem specified in Example 4.6 with the initial state $x_0 < 0$.

Exercise 4.10 Suppose that the optimal control u^* for the minimum time-fuel problems specified in Example 4.9 is in the form of $u^* = \{-1, 0, 1\}$ and denote by $\widehat{t}_1$ the switching time when the control switches from -1 to 0; then to verify that at the switching time $\widehat{t}_1$, the values of the optimal state x_1^* and x_2^* satisfy (4.198).

Exercise 4.11 Implement the optimal control for the state system $\dot{x}(t) = u(t)$ with the control satisfying the constraint of $|u(t)| \le 1$, such that the state is driven from an initial state x_0 to the origin at the final time, and the following performance cost function is minimized

$$J = \int_0^{t_f} u^2(t) dt,$$

where the final time t_f is free.

Exercise 4.12 Implement the optimal control for the state system

$$\dot{x}(t) = -a + u(t)$$

with $a > 0$ and $|u(t)| \leq 1$, for all t, such that the state is driven from an initial state x_0 to the origin at the final time, and the following performance cost function is minimized

$$J = \int_0^{t_f} \left[\beta + u^2(t)\right] dt,$$

with $\beta > 0$, where the final time t_f is free.

Exercise 4.13 Implement the optimal control for the state system

$$\dot{x}_1(t) = 2x_2(t)$$
$$\dot{x}_1(t) = u(t)$$

with $|u(t)| \leq 1$, for all t, such that the state is driven from an initial state $x_0 = \begin{bmatrix} 1 \\ -1 \end{bmatrix}$ to the origin at the final time, and the following performance cost function is minimized

$$J = \int_0^{t_f} [1 + u^2(t)] dt,$$

where the final time t_f is free.

Exercise 4.14 Considering a DC motor speed control system which is given as

$$\dot{x}(t) = \begin{bmatrix} 0 & 1 \\ -16 & -8 \end{bmatrix} x(t) + \begin{bmatrix} 0 \\ 16 \end{bmatrix} u(t),$$

where $x_1(t)$, $x_2(t)$, and $u(t)$ represent the speed of the motor and the current in the armature circuit, the voltage control to an amplifier of the motor, respectively.

Implement the optimal control for the underlying system to drive the speed of the motor to a given value such that the following performance cost function

$$J = \int_0^{t_f} u^2(t) dt,$$

with $u(t) \leq 1$ and t_f being free, is minimized.

Chapter 5
Dynamic Programming

In Chaps. 2–4, it has been introduced how to implement the optimal control problems via the variational method. This chapter will briefly give an introduction to dynamic programming, the other key branch of optimal control method, which was firstly developed by R. Bellman during the 1950s–1960s.

In Sect. 5.1, the Hamilton–Jacobi–Bellman equation based on the principle of optimality is developed. Section 5.2 gives the necessary condition of the optimal controls for those problems without considering any constraints on the controls in a way different from that given in Chap. 3, and the specification of the optima control is further studied. In Sect. 5.3, the optimal controls for linear-quadratic regulation problems are studied, while in Sects. 5.4 and 5.5, the optimal controls for affine-quadratic regulation and tracking problems are studied, respectively. For the purpose of comparison, Pontryagin's minimum principle is developed via the dynamic program method in Sect. 5.6. A brief summary of this chapter is given in Sect. 5.7. Finally, in Sect. 5.8, exercises for the readers to verify the results introduced in this chapter are given.

5.1 The Hamilton–Jacobi–Bellman Equation

This section firstly presents the principle of optimality for the optimal control problems studied in previous parts in this book, which simply states that any portion of the optimal trajectory is optimal. Alternatively, the optimal control has the property that no matter what the previous controls have been, the remaining decision must constitute an optimal one.

And then it develops the so-called Hamilton–Jacobi–Bellman (HJB) equation based upon the principle of optimality.

Z. Ma and S. Zou, *Optimal Control Theory*,
https://doi.org/10.1007/978-981-33-6292-5_5

In earlier parts, the state system has been considered

$$\dot{x}(t) = f(x(t), u(t), t), \tag{5.1}$$

and the performance cost function during the time interval $[t_0, t_f]$

$$J(x(t_0), t_0, u) \triangleq h(x(t_f), t_f) + \int_{t_0}^{t_f} g(x(t), u(t), t)dt, \tag{5.2}$$

with the state value at initial time t_0 given as $x(t_0)$.

Denote by $\mathscr{U}([t, s])$ the set of admissible controls, such that

$$\mathscr{U}([t, s]) \triangleq \{u(\tau) \in U, \text{ for all } \tau \in [t, s]\}, \tag{5.3}$$

for any pair of t and s, with $t_0 \le t < s \le t_f$.

In the following, the dynamic programming to implement the optimal control solution subject will be introduced to which the above cost function J is minimized.

Notice that since the proposed optimal control is determined by the state, it results in the closed-loop optimal control.

Given the value of $x(t)$, say the state at time t, define the cost function with respect to $(x(t), t)$ over the time interval $[t, t_f]$ subject to a control u, denoted by $J(x(t), t, u)$, such that

$$J(x(t), t, u) \triangleq h(x(t_f), t_f) + \int_{t}^{t_f} g(x(\tau), u(\tau), \tau)d\tau, \tag{5.4}$$

for all $t \in [t_0, t_f]$.

Moreover, denote by $J^*(x(t), t)$ the minimum value of the performance cost function J over the time interval $[t, t_f]$ with the state $x(t)$ at time t, say

$$J^*(x(t), t) \triangleq \min_{u \in \mathscr{U}([t, t_f])} J(x(t), t, u)$$

$$= \min_{u \in \mathscr{U}([t, t_f])} \left\{ h(x(t_f), t_f) + \int_{t}^{t_f} g(x(\tau), u(\tau), \tau)d\tau \right\}. \tag{5.5}$$

That is to say, $J^*(x(t), t)$ represents the value of the performance cost function when evaluated along the optimal trajectory starting at $(x(t), t)$.

In Theorem 5.1, the necessary condition of the optimal control for the underlying problem will be developed by applying the principle of optimality. Before that, it gives an analysis below.

For any control u, we have

$$J(x(t), t, u) = h(x(t_f), t_f) + \int_t^{t_f} g(x(\tau), u(\tau), \tau)d\tau$$

$$= h(x(t_f), t_f) + \int_t^s g(x(\tau), u(\tau), \tau)d\tau + \int_s^{t_f} g(x(\tau), u(\tau), \tau)d\tau$$

$$= \int_t^s g(x(\tau), u(\tau), \tau)d\tau + \left[h(x(t_f), t_f) + \int_s^{t_f} g(x(\tau), u(\tau), \tau)d\tau \right]$$

$$= \int_t^s g(x(\tau), u(\tau), \tau)d\tau + J(x(s), s, u). \tag{5.6}$$

Theorem 5.1 (Necessary Condition of Optimal Control) *Suppose that $J^*(x, t)$ is the minimum cost function; then it satisfies the following HJB equation,*

$$-\frac{\partial J^*(x, t)}{\partial t} = \min_{u \in U} \left\{ g(x, u, t) + \left[\frac{\partial J^*(x, t)}{\partial x} \right]^\top f(x, u, t) \right\}. \tag{5.7}$$

Proof Firstly, we consider the following principle of optimality

$$J^*(x(t), t) = \min_{u(\cdot) \in \mathcal{U} ([t, t+dt])} \left\{ \int_t^{t+dt} g(x(s), u(s), s)ds + J^*(x(t + dt), t + dt) \right\}. \tag{5.8}$$

By considering the Taylor series expansion for $J^*(x(t + dt), t + dt)$, we can obtain

$$J^*(x(t + dt), t + dt) = J^*(x, t) + \left[\frac{\partial J^*(x, t)}{\partial x} \right]^\top dx(t) + \frac{\partial J^*(x, t)}{\partial t} dt + \mathcal{O}(dx, dt), \tag{5.9}$$

where $\mathcal{O}(dx, dt)$ represents the higher-order parts of dx and dt.

Also, for infinitesimal-valued dt, we have

$$\int_t^{t+dt} g(x(s), u(s), s)ds \doteq g(x(t), u(t), t)dt, \tag{5.10}$$

where $\doteq$ represents the equality up to the first-order term.

As a consequence, from the above analysis, for infinitesimal-valued dx and dt, we can get

$$J^*(x,t) = \min_{u(\cdot) \in \mathcal{U}([t,t+dt])} \left\{ g(x(t),u(t),t)dt + J^*(x,t) + \left[\frac{\partial J^*(x,t)}{\partial x} \right]^\top dx(t) \right.$$

$$\left. + \frac{\partial J^*(x,t)}{\partial t} dt + \mathcal{O}(dx,dt) \right\}$$

$$= \min_{u(\cdot) \in \mathcal{U}([t,t+dt])} \left\{ g(x(t),u(t),t)dt + \left[\frac{\partial J^*(x,t)}{\partial x} \right]^\top dx(t) + \mathcal{O}(dx,dt) \right\}$$

$$+ J^*(x,t) + \frac{\partial J^*(x,t)}{\partial t} dt; \qquad (5.11)$$

then as dt goes to zero, by (5.11), together with

$$dx(t) \doteq f(x(t),u(t),t)dt, \qquad (5.12)$$

we obtain

$$0 = \min_{u \in U} \left\{ g(x,u,t) + \left[\frac{\partial J^*(x,t)}{\partial x} \right]^\top \lim_{dt \to 0} \frac{dx(t)}{dt} \right\}$$

$$+ \frac{\partial J^*(x,t)}{\partial t} + \lim_{dt \to 0} \frac{\mathcal{O}(dx,dt)}{dt}$$

$$= \frac{\partial J^*(x,t)}{\partial t} + \min_{u \in U} \left\{ g(x,u,t) + \left[\frac{\partial J^*(x,t)}{\partial x} \right]^\top f(x,u,t) \right\}, \qquad (5.13)$$

which is the conclusion of (5.7). □

Here, apply the Hamiltonian $\mathcal{H}$ such that

$$\mathcal{H}(x,u,\lambda,t) \triangleq g(x,u,t) + \lambda^\top f(x,u,t), \qquad (5.14)$$

which has been defined in previous chapters in this book; then by (5.7), the HJB equation can be written as

$$-\frac{\partial J^*(x,t)}{\partial t} = \min_{u(t) \in U} \mathcal{H}\left(x(t),u(t), \frac{\partial J^*(x,t)}{\partial x}, t \right). \qquad (5.15)$$

In Theorem 5.1, the necessary condition for the optimal control solution has been specified. And in the following, a sufficient condition of the optimal control as well will be given.

Theorem 5.2 (Sufficient Condition of Optimal Control) *Suppose $V(x,t)$ is continuously differentiable in t and x, and is a solution to the following:*

$$-\frac{\partial V(x,t)}{\partial t} = \min_{u \in U} \left\{ g(x,u,t) + \left[\frac{\partial V(x,t)}{\partial x}\right]^{\top} f(x,u,t) \right\}, \ \textit{for all } (x,t),$$

(5.16a)

$$V(x,t_f) = h(x,t_f), \ \textit{for all } x;$$

(5.16b)

then $V(x,t) = J^*(x,t)$, say $V(x,t)$ is the minimum cost solution for the optimal control problems.

Proof Consider an admissible control $\widehat{u}$, and denote by $\widehat{x}$ the state trajectory subject to $\widehat{u}$; then by (5.16a), we have, for all t,

$$0 \le g(\widehat{x}(t),\widehat{u}(t),t) + \frac{\partial V(\widehat{x}(t),t)}{\partial t} + \left[\frac{\partial V(\widehat{x}(t),t)}{\partial x}\right]^{\top} f(\widehat{x}(t),\widehat{u}(t),t)$$

$$= g(\widehat{x}(t),\widehat{u}(t),t) + \frac{dV(\widehat{x}(t),t)}{dt},$$

(5.17)

where the last equality holds due to the chain rule of the differentiation and the system state equation given in (5.1).

By the above inequality, we can obtain

$$0 \le \int_{t_0}^{t_f} g(\widehat{x}(t),\widehat{u}(t),t)dt + V(\widehat{x}(t_f),t_f) - V(x(t_0),t_0).$$

(5.18)

Thus by the above equation, together with (5.16b),

$$V(x(t_0),t_0) \le h(\widehat{x}(t_f),t_f) + \int_{t_0}^{t_f} g(\widehat{x}(t),\widehat{u}(t),t)dt.$$

(5.19)

Following a similar analysis given above, we can get

$$V(x(t_0),t_0) = h(x^*(t_f),t_f) + \int_{t_0}^{t_f} g(x^*(t),u^*(t),t)dt.$$

(5.20)

By the above relationships, we can claim that u^* is optimal and

$$V(x,t) = J^*(x,t), \ \text{for all } (x,t),$$

(5.21)

i.e., $V(x,t)$ is the minimum cost for the optimal control problems over the interval $[t,t_f]$ with the initial value as (x,t). $\qquad \square$

In the following, the optimal control solution will be studied by applying Theorem 5.2.

Example 5.1 Consider a state system such that

$$\dot{x}(t) = u(t),$$

with the constraint $|u(t)| \le 1$, for all $t \in [t_0, t_f]$.

Implement the optimal control u^* such that the cost function

$$J(x, u) = x^2(t_f)$$

is minimized.

Solution. By the specification of the state equation and the cost function, it is obvious that the promising optimal control is to move the state to zero as quickly as possible, and remain at zero as it is reached, that is to say,

$$v^*(x, t) = -\mathrm{sgn}(x) \equiv \begin{cases} -1, & \text{in case } x > 0 \\ 0, & \text{in case } x = 0 , \\ 1, & \text{in case } x < 0 \end{cases} \tag{5.22}$$

for all $t \in [t_0, t_f]$.

Thus, we can obtain the cost subject to $v^*(x, t)$ specified above as

$$J^*(x, t) = \left[\max\{0, |x| - [t_f - t]\}\right]^2, \tag{5.23}$$

for any t and x; see an illustration in Fig. 5.1.

Actually, by the specification $J^*(x, t)$, we have

$$\frac{\partial J^*(x, t)}{\partial t} = 2\max\{0, |x| - [t_f - t]\}, \tag{5.24a}$$

$$\frac{\partial J^*(x, t)}{\partial x} = 2\mathrm{sgn}(x)\max\{0, |x| - [t_f - t]\}; \tag{5.24b}$$

then the HJB equation is

Fig. 5.1 The cost function
subject to optimal control u^*

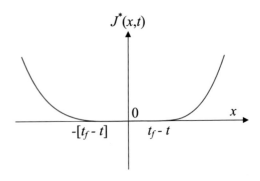

$$0 = \min_{|u| \le 1} 2[1 + \text{sgn}(x)u] \max\{0, |x| - [t_f - t]\}, \tag{5.25}$$

which holds for all t and x.

In the above equation, it has been verified that the specific function satisfies the HJB equation (5.16); then by applying Theorem 5.2, we can claim that the specified $v^*(x, t)$ is optimal. $\qquad\square$

5.2 Analysis on Optimal Control

In the last section, the conditions of the optimal control considering the constraints on the controls has been developed. Here, the necessary condition of the optimal control for those problems will be given without considering any constraints on the controls in a different way, and the specification of the optima control studied further.

Before further analysis, for notational simplicity, in this part, we consider

$$J_x^*(x, t) \equiv \frac{\partial J^*(x, t)}{\partial x}, \tag{5.26a}$$

$$J_t^*(x, t) \equiv \frac{\partial J^*(x, t)}{\partial t}. \tag{5.26b}$$

Thus, by the chain rule of the differentiation and the state equation, we have

$$
\begin{aligned}
\frac{dJ^*(x^*(t), t)}{dt} &= \left[\frac{\partial J^*(x^*(t), t)}{\partial x}\right]^\top \dot{x}^*(t) + \frac{\partial J^*(x^*(t), t)}{\partial t}, \\
&= \left[\frac{\partial J^*(x^*(t), t)}{\partial x}\right]^\top f(x^*(t), u^*(t), t) + \frac{\partial J^*(x^*(t), t)}{\partial t} \\
&\equiv \left[J_x^*(x^*(t), t)\right]^\top f(x^*(t), u^*(t), t) + J_t^*(x^*(t), t), \tag{5.27}
\end{aligned}
$$

by (5.26a) and (5.26b).

By the specification of $J^*(x^*(t), t)$ defined in (5.5), we have

$$\frac{dJ^*(x^*(t), t)}{dt} = -g(x^*(t), u^*(t), t). \tag{5.28}$$

Using (5.27) and (5.28), we get

$$J_t^*(x^*(t), t) + g(x^*(t), u^*(t), t) + [J_x^*(x^*(t), t)]^\top f(x^*(t), u^*(t), t) = 0. \tag{5.29}$$

By (5.29) together with the Hamiltonian $\mathcal{H}(x, u, \lambda, t)$ given in (5.14), we have

$$J_t^*(x^*(t), t) + \mathcal{H}(x^*(t), u^*(t), J_x^*(x^*(t), t), t) = 0, \tag{5.30}$$

for all $t \in [t_0, t_f)$, with the boundary condition from (5.5) as

$$J^*(x^*(t_f), t_f) = h(x^*(t_f), t_f).$$ (5.31)

Equation (5.30) is equivalent to the HJB equation which is firstly verified in (5.7), Theorem 5.1, for the optimal control problems with constraints on the control.

Comparing the Hamiltonian (5.14) with that defined in (3.12) in Chap. 3, the costate function $\lambda^*(t)$ defined earlier is given as the following:

$$\lambda^*(t) = J_x^*(x^*(t), t).$$ (5.32)

Also, by Theorem 3.1 in Chap. 3 that the state and costate satisfy the following

$$\dot{\lambda}^*(t) = -\frac{\partial \mathcal{H}(x^*(t), u^*(t), J_x^*(x^*(t), t), t)}{\partial x},$$ (5.33)

and the optimal control $u^*(t)$ is obtained from

$$\frac{\partial \mathcal{H}(x^*(t), u^*(t), J_x^*(x^*(t), t), t)}{\partial u} = 0,$$ (5.34)

since here it does not consider any constraints on the control u, which implies that

$$u^*(t) = \psi(x^*(t), J_x^*(x^*(t), t), t),$$ (5.35)

for some function $\psi(\cdot)$, that is to say, $u^*(t)$ is determined by $x^*(t)$ and t, hence it is a state-dependent optimal control.

Here, comparing (5.32) and (5.33), we get

$$\frac{dJ_x^*(x^*(t), t)}{dt} = \dot{\lambda}^*(t) = -\frac{\partial \mathcal{H}(x^*(t), u^*(t), J_x^*(x^*(t), t), t)}{\partial x}.$$ (5.36)

This equation, in general, is a nonlinear partial differential equation in J^*, which can be solved. Thus suppose that J^* is specified, then J_x^*, the gradient of J^*, can be calculated and the optimal control $u^*(t)$ can be obtained by applying (5.35).

Usually, it is challenging to solve the HJB equation. Now the proposed results with a simple numerical example are illustrated in the following.

Example 5.2 Considering a one-dimensional state system such that

$$\dot{x}(t) = -\frac{1}{2}x(t) + u(t),$$ (5.37)

with the following performance cost function

$$J(x_0, 0, u) = \frac{1}{2}x^2(t_f) + \frac{1}{2}\int_0^{t_f} \left[x^2(t) + u^2(t)\right] dt. \tag{5.38}$$

Determine the optimal control for the underlying problem.

Solution. For the state system and cost function given in (5.1) and (5.2), respectively, we have

$$g(x(t), u(t), t) = \frac{1}{2}u^2(t) + \frac{1}{2}x^2(t), \tag{5.39a}$$

$$h(x(t_f), t_f) = \frac{1}{2}x^2(t_f), \tag{5.39b}$$

$$f(x(t), u(t), t) = -\frac{1}{2}x(t) + u(t). \tag{5.39c}$$

Hence, by (5.14), the Hamiltonian $\mathscr{H}$ is defined as

$$\mathscr{H}(x(t), u(t), J_x(x(t), t, u), t) \tag{5.40}$$

$$= g(x(t), u(t), t) + J_x(x(t), t, u) f(x(t), u(t), t) \tag{5.41}$$

$$= \frac{1}{2}u^2(t) + \frac{1}{2}x^2(t) + J_x(x(t), t, u)\left[-\frac{1}{2}x(t) + u(t)\right]. \tag{5.42}$$

For an unconstrained control, a necessary condition for the optimization is

$$\frac{\partial \mathscr{H}(x^*(t), u^*(t), J_x^*(x^*(t), t), t)}{\partial u} = 0, \tag{5.43}$$

which implies that

$$u^*(t) + J_x^*(x^*(t), t) = 0. \tag{5.44}$$

Hence, we obtain the optimal control as

$$u^*(t) = -J_x^*(x^*(t), t). \tag{5.45}$$

By (5.40) and (5.45), we get

$$\mathscr{H}(x^*(t), u^*(t), J_x^*(x^*(t), t), t) = \frac{1}{2}[-J_x^*]^2 + \frac{1}{2}[x^*(t)]^2 + J_x^*\left[-\frac{1}{2}x^*(t) - J_x^*\right]$$

$$= -\frac{1}{2}[J_x^*]^2 + \frac{1}{2}[x^*(t)]^2 - \frac{1}{2}x^*(t)J_x^*, \tag{5.46}$$

with $J_x^* \equiv J_x^*(x^*(t), t)$.

Hence, we can obtain the HJB equation (5.30) as

$$J_t^* - \frac{1}{2}[J_x^*]^2 + \frac{1}{2}[x^*(t)]^2 - \frac{1}{2}x^*(t)J_x^* = 0, \tag{5.47}$$

with $J_t^* \equiv J_t^*(x^*(t), t)$ and the boundary condition (5.31) given as

$$J^*(x^*(t_f), t_f) = h(x^*(t_f), t_f) = \frac{1}{2}[x^*(t_f)]^2. \tag{5.48}$$

To solve the HJB equation (5.47) with the boundary condition (5.48), in the following, we firstly assume an optimal solution and then check if this assumed solution satisfies the equation.

For the underlying problem, since we would like to specify the optimal control (5.45) which is dependent upon the state, and the cost function is a quadratic function of state x and control u, we may suppose that the optimal solution is in a quadratic form of the state

$$J^*(x^*(t), t) = \frac{1}{2}\kappa(t)[x^*(t)]^2, \tag{5.49}$$

where the coefficient parameter $\kappa(t)$ shall be specified later.

Due to the boundary condition given in (5.48), we get

$$J^*(x^*(t_f), t_f) = \frac{1}{2}[x^*(t_f)]^2 = \frac{1}{2}\kappa(t_f)[x^*(t)]^2, \tag{5.50}$$

which implies that

$$\kappa(t_f) = 1. \tag{5.51}$$

Thus by (5.49), we can get

$$J_x^* = \kappa(t)x^*(t), \tag{5.52a}$$

$$J_t^* = \frac{1}{2}\dot{\kappa}(t)[x^*(t)]^2, \tag{5.52b}$$

which implies that the closed-loop optimal control (5.45) is given as

$$u^*(t) = -\kappa(t)x^*(t). \tag{5.53}$$

By applying the optimal control (5.53) together with (5.52) to the HJB equation (5.47), we obtain

$$\left[\frac{1}{2}\dot{\kappa}(t) - \frac{1}{2}\kappa^2(t) - \frac{1}{2}\kappa(t) + \frac{1}{2}\right][x^*(t)]^2 = 0, \tag{5.54}$$

which holds for any $x^*(t)$ for all $t \in [t_0, t_f]$.

Thus, for any $x^*(t)$, the previous relation is equivalent with the following:

$$\frac{1}{2}\dot{\kappa}(t) - \frac{1}{2}\kappa^2(t) - \frac{1}{2}\kappa(t) + \frac{1}{2} = 0, \tag{5.55}$$

which upon solving with the boundary condition (5.51) becomes

$$\kappa(t) = -\frac{\sqrt{5}}{2}\tanh\left(\text{artanh}\left(-\frac{3}{\sqrt{5}}\right) + \frac{\sqrt{5}}{2}[t - t_f]\right) - \frac{1}{2}, \tag{5.56}$$

for all $t \in [t_0, t_f]$ with the boundary condition $\kappa(t_f) = 1$.

By applying (5.56), the closed-loop optimal control (5.53) has been established.

□

Notice that in Example 5.2, the relation (5.55) is the scalar version of the matrix Riccati differential equation (3.115) for the finite-time LQR system in Chap. 3.

Furthermore, as the final time t_f goes to infinity, $\kappa(t)$ in (5.56) converges to $\bar{\kappa} = \frac{1}{2}\left[\sqrt{5} - 1\right]$, and then the optimal control (5.53) is given as

$$u^*(t) = -\frac{1}{2}\left[\sqrt{5} - 1\right]x^*(t). \tag{5.57}$$

5.3 Linear-Quadratic Regulation Problems

In this section, we will revisit the LQR problems, which have been analyzed in Sect. 3.3 by applying the variational method.

Here, recall that the following time-variant linear state system is considered

$$\dot{x}(t) = A(t)x(t) + B(t)u(t), \tag{5.58}$$

such that the performance cost function given below is to be minimized as

$$J(x(t_0), t_0, u(t)) \triangleq \frac{1}{2}x^\top(t_f)Fx(t_f)$$
$$+ \frac{1}{2}\int_{t_0}^{t_f} \left[[x(t)]^\top Q(t)x(t) + [u(t)]^\top R(t)u(t)\right]dt, \tag{5.59}$$

where, as defined earlier, F and $Q(t)$ are real, symmetric, positive semidefinite matrices, respectively, and $R(t)$ is a real, symmetric, positive-definite matrix.

Different from the analysis given in Sect. 3.3, Chap. 3, in Theorem 5.3 below, the optimal control for the underlying LQR problems is specified by applying the dynamic programming method which has been introduced in earlier parts of this chapter.

Theorem 5.3 *The optimal control for the LQR problems is linear and time-variant with respect to the state, such that*

$$u^*(t) = -R^{-1}(t)[B(t)]^\top P(t)x^*(t), \tag{5.60}$$

for all $t \in [t_0, t_f]$, where $P(\cdot)$ is given in the following Riccati differential equation:

$$\dot{P}(t) = -P(t)A(t) - [A(t)]^\top P(t) + P(t)B(t)R^{-1}(t)[B(t)]^\top P(t) - Q(t). \tag{5.61}$$

Proof As usual, firstly we define the Hamiltonian for the optimal control problems as follows:

$$\mathcal{H}(x(t), u(t), J_x^*(x(t), t), t) = \frac{1}{2}[x(t)]^\top Q(t)x(t) + \frac{1}{2}[u(t)]^\top R(t)u(t)$$
$$+ [J_x^*(x(t), t)]^\top [A(t)x(t) + B(t)u(t)]. \tag{5.62}$$

By (5.34), a necessary condition for optimization of $\mathcal{H}$ with respect to $u(t)$ is

$$\frac{\partial \mathcal{H}(x^*(t), u^*(t), J_x^*(x^*(t), t), t)}{\partial u} = 0; \tag{5.63}$$

then by this, together with (5.62), we can obtain

$$R(t)u^*(t) + [B(t)]^\top J_x^*(x^*(t), t) = 0. \tag{5.64}$$

It implies that

$$u^*(t) = -R^{-1}(t)[B(t)]^\top J_x^*(x^*(t), t). \tag{5.65}$$

Furthermore, due to the assumption that $R(t)$ is symmetric positive definite and $\frac{\partial^2 \mathcal{H}}{\partial u^2} = R(t)$, it is sufficient to state that the $u^*(t)$ given in (5.65) above is the optimal control for the underlying LQR problems.

With the optimal control given in (5.65), the Hamiltonian (5.62) becomes

$$\mathcal{H}(x^*(t), u^*(t), J_x^*, t)$$
$$= \frac{1}{2}[x^*(t)]^\top Q(t)x^*(t) + \frac{1}{2}[J_x^*]^\top B(t)R^{-1}(t)[B(t)]^\top J_x^*$$
$$+ [J_x^*]^\top A(t)x^*(t) - [J_x^*]^\top B(t)R^{-1}(t)[B(t)]^\top J_x^*$$
$$= \frac{1}{2}[x^*(t)]^\top Q(t)x^*(t) - \frac{1}{2}[J_x^*]^\top B(t)R^{-1}(t)[B(t)]^\top J_x^* + [J_x^*]^\top A(t)x^*(t), \tag{5.66}$$

where, as earlier, for notational simplicity, we consider that $J_x^* \equiv J_x^*(x^*(t), t)$.

The HJB equation is

$$J_t^* + \mathscr{H}(x^*(t), u^*(t), J_x^*, t) = 0. \tag{5.67}$$

Thus, with (5.66), the HJB equation (5.67) becomes

$$J_t^* + \frac{1}{2}[x^*(t)]^\top Q(t)x^*(t) - \frac{1}{2}[J_x^*]^\top B(t)R^{-1}(t)[B(t)]^\top J_x^* + [J_x^*]^\top A(t)x^*(t) = 0, \tag{5.68}$$

with the boundary condition as

$$J^*(x^*(t_f), t_f) = \frac{1}{2}[x^*(t_f)]^\top Fx^*(t_f). \tag{5.69}$$

Since the cost function $J(x(t), t, u)$ is a quadratic function of the state, as considered in an example analyzed earlier, it may be supposed that the minimum cost function subject to the optimal control still possesses a quadratic form with respect to the state $x(t)$, such that

$$J^*(x^*(t), t) = \frac{1}{2}[x^*(t)]^\top P(t)x^*(t), \tag{5.70}$$

where $P(t)$ is assumed to be a real, symmetric, positive-definite matrix, with

$$J_t^* \equiv \frac{\partial J^*(x^*(t), t)}{\partial t} = \frac{1}{2}[x^*(t)]^\top \dot{P}(t)x^*(t), \tag{5.71a}$$

$$J_x^* \equiv \frac{\partial J^*(x^*(t), t)}{\partial x} = P(t)x^*(t), \tag{5.71b}$$

and using the assumed performance cost function (5.70) in the HJB equation (5.68), we get

$$\frac{1}{2}[x^*(t)]^\top \dot{P}(t)x^*(t) + \frac{1}{2}[x^*(t)]^\top Q(t)x^*(t)$$
$$- \frac{1}{2}[x^*(t)]^\top P(t)B(t)R^{-1}(t)[B(t)]^\top P(t)x^*(t)$$
$$+ [x^*(t)]^\top P(t)A(t)x^*(t) = 0. \tag{5.72}$$

Also, we can easily show that since all the terms, except for the last term on the right-hand side of (5.72), are symmetric.
Moreover, we have

$$[x^*(t)]^\top P(t)A(t)x^*(t) = [P(t)x^*(t)]^\top A(t)x^*(t) = [A(t)x^*(t)]^\top P(t)x^*(t)$$
$$= [x^*(t)]^\top [A(t)]^\top P(t)x^*(t). \tag{5.73}$$

Thus by (5.72) and (5.73), we get

$$\frac{1}{2}[x^*(t)]^\top \dot{P}(t)x^*(t) + \frac{1}{2}[x^*(t)]^\top Q(t)x^*(t)$$

$$- \frac{1}{2}[x^*(t)]^\top P(t)B(t)R^{-1}(t)[B(t)]^\top P(t)x^*(t)$$

$$+ \frac{1}{2}[x^*(t)]^\top P(t)A(t)x^*(t) + \frac{1}{2}[x^*(t)]^\top [A(t)]^\top P(t)x^*(t) = 0, \qquad (5.74)$$

which should hold for any $x^*(t)$; then we obtain

$$\dot{P}(t) + Q(t) - P(t)B(t)R^{-1}(t)[B(t)]^\top P(t) + P(t)A(t) + [A(t)]^\top P(t) = 0. \tag{5.75}$$

By reorganizing the above equation, we can obtain the differential Riccati equation as specified in (5.61).

Also by (5.69) and (5.70), we obtain

$$\frac{1}{2}[x^*(t_f)]^\top P(t_f)x^*(t_f) = \frac{1}{2}[x^*(t_f)]^\top F x^*(t_f); \tag{5.76}$$

then we have the final condition for $P(t)$ as

$$P(t_f) = F. \tag{5.77}$$

Consequently, by (5.65) and (5.71b), we obtain the optimal control of the underlying LQR problems in the form of the state as given in (5.60). □

Notice that the $n \times n$ matrix $P(t)$ is determined by numerically integrating backward from t_f to t_0, and due to the symmetry property; instead of n^2 equations, we need to solve only $\frac{1}{2}n[n+1]$ ones.

Suppose that the optimal control for the underlying LQR problems is closed-loop, linear, and time-variant with respect to the state, then it is natural to assume the solution of the quadratic form (5.70).

It may be feasible to specify the solution to the HJB equation for the linear, time-variant system with a quadratic performance cost function. However in general, it is challenging to find the solution, and it may be implemented by numerical techniques.

In the following, the optimal control for the numerical LQR problems studied in Example 5.2 will be specified by directly applying the results developed in Theorem 5.3.

Example 5.3 Implement the optimal control to Example 5.2 by applying Theorem 5.3.

Solution. Firstly by the specification of Example 5.2,

$$A(t) = -\frac{1}{2}, \; B(t) = F = Q(t) = R(t) = 1, \tag{5.78}$$

for all $t \in [t_0, t_f]$; then by applying (5.75),

$$\dot{P}(t) + 1 - P^2(t) - P(t) = 0, \tag{5.79}$$

with the boundary $P(t_f) = 1$, which is the same as (5.55).
Thus,

$$P(t) = -\frac{\sqrt{5}}{2}\tanh\left(\text{artanh}\left(-\frac{3}{\sqrt{5}}\right) + \frac{\sqrt{5}}{2}[t - t_f]\right) - \frac{1}{2}, \tag{5.80}$$

for all $t \in [t_0, t_f]$, with the boundary condition $P(t_f) = 1$.
Also by (5.60) in Theorem 5.3,

$$u^*(t) = -R^{-1}(t)[B(t)]^\top P(t)x^*(t) = -P(t)x^*(t), \tag{5.81}$$

with the solution of $P(t)$ specified in (5.80).   □

Here, consider the following time-invariant linear state system

$$\dot{x}(t) = Ax(t) + Bu(t), \tag{5.82}$$

such that the performance cost function over the infinite-horizon time interval is to be minimized as

$$J(x(0), 0, u(t)) \triangleq \frac{1}{2}\int_0^\infty \left[[x(t)]^\top Qx(t) + [u(t)]^\top Ru(t)\right]dt, \tag{5.83}$$

where Q is a real, symmetric, positive semidefinite matrix, while R is a real, symmetric, positive-definite matrix.

Theorem 5.4 *The optimal control for the infinite-horizon time-invariant LQR problems is linear and time-invariant with respect to the state, such that*

$$u^*(t) = -R^{-1}B^\top Px^*(t), \tag{5.84}$$

where P is given in the algebra equation

$$PA + A^\top P - PBR^{-1}B^\top P + Q = 0. \tag{5.85}$$

□

Notice that Theorem 5.4 can be verified by applying the same technique used for the proof of Theorem 5.3.

We study an example below to implement the optimal control solution for the infinite-horizon LQR problems by applying the results developed in Theorem 5.4.

Example 5.4 Implement the optimal control for the one-dimensional state system

$$\dot{x}(t) = -x(t) + u(t), \tag{5.86}$$

with the infinite-horizon performance cost function as

$$J(x(0), 0, u) = \int_0^\infty \left[x^2(t) + u^2(t) \right] dt. \tag{5.87}$$

Solution. Firstly, we assume that $J^*(x(t), t) = \eta x^2(t)$ for some coefficient η; then we have

$$J_x^*(x(t), t) = 2\eta x(t). \tag{5.88}$$

The Hamiltonian $\mathcal{H}$ is specified as

$$
\begin{aligned}
\mathcal{H}(x(t), u(t), J_x^*(x(t), t), t) &= g(x(t), u(t), t) + J_x^*(x(t), t) f(x(t), u(t)) \\
&= x^2(t) + u^2(t) + 2\eta x(t)[-x(t) + u(t)] \\
&= x^2(t) + u^2(t) - 2\eta x^2(t) + 2\eta x(t)u(t). \quad (5.89)
\end{aligned}
$$

The optimal control $u^*(t)$ can be given by minimizing the $\mathcal{H}$ with respect to the control u as follows:

$$\frac{\partial \mathcal{H}(x^*(t), u^*(t), J_x^*(x^*(t), t), t)}{\partial u} = 2u^*(t) + 2\eta x^*(t) = 0, \tag{5.90}$$

which implies that

$$u^*(t) = -\eta x^*(t). \tag{5.91}$$

Thus, we can obtain the optimal Hamiltonian as

$$\mathcal{H}(x^*(t), u^*(t), J_x^*(x^*(t), t), t) = [x^*(t)]^2 - 2\eta[x^*(t)]^2 - \eta^2[x^*(t)]^2. \tag{5.92}$$

Then, we have the HJB equation

$$\mathcal{H}(x^*(t), u^*(t), J_x^*(x^*(t), t), t) + J_t^*(x^*(t), t), t) = [x^*(t)]^2 - 2\eta[x^*(t)]^2 - \eta^2[x^*(t)]^2 = 0, \tag{5.93}$$

since $J_t^*(x(t), t) = \dfrac{\partial(\eta x^2(t))}{\partial t} = 0$.

Equation (5.93) should hold for any $x^*(t)$; then we can obtain the following:

$$\eta^2 + 2\eta - 1 = 0, \tag{5.94}$$

which implies that

$$\eta = -1 \pm \sqrt{2}. \tag{5.95}$$

Since by the specification of the cost function given in (5.87), the underlying cost function has to be positive valued, we can get the minimum cost function in the following by selecting the positive value of η implemented in (5.95), say $\eta = -1 + \sqrt{2}$; then

$$J^*(x^*(t), t) = \eta \left[x^*(t) \right]^2 = \left[\sqrt{2} - 1 \right] [x^*(t)]^2. \tag{5.96}$$

Finally, by applying the value of η selected in (5.96), we can obtain the optimal control as

$$u^*(t) = -\eta x^*(t) = - \left[\sqrt{2} - 1 \right] x^*(t). \tag{5.97}$$

$\square$

5.4 Affine-Quadratic Regulation Problems

In this section, we will study a class of affine QR problems based upon the results developed in Sect. 5.3.

Here, consider the following time-variant affine state system

$$\dot{x}(t) = A(t)x(t) + B(t)u(t) + C(t), \tag{5.98}$$

such that the performance cost function specified in (5.59) is minimized.

In Theorem 5.5 below, the optimal control for the underlying affine QR problems is specified by applying the dynamic programming method.

Theorem 5.5 *The optimal control for the affine QR problems is linear and time-variant with respect to the state, such that*

$$u^*(t) = -R^{-1}(t)[B(t)]^\top \left[P(t)x^*(t) + \eta(t) \right], \tag{5.99}$$

for all $t \in [t_0, t_f]$, and

$$J^*(x^*(t), t) = \frac{1}{2}[x^*(t)]^\top P(t)x^*(t) + [\eta(t)]^\top x^*(t) + \tau(t), \tag{5.100}$$

where $P(t)$, $\eta(t)$, and $\tau(t)$ are given in the following, respectively:

$$\dot{P}(t) = -P(t)A(t) - [A(t)]^\top P(t) + P(t)B(t)R^{-1}(t)[B(t)]^\top P(t) - Q(t),$$
$$\tag{5.101a}$$

$$\dot{\eta}(t) = \big[P(t)B(t)R^{-1}(t)[B(t)]^\top - [A(t)]^\top\big]\eta(t) - P(t)C(t), \tag{5.101b}$$

$$\dot{\tau}(t) = \frac{1}{2}[\eta(t)]^\top B(t)R^{-1}(t)[B(t)]^\top \eta(t) - [\eta(t)]^\top C(t), \tag{5.101c}$$

with the boundary conditions as

$$P(t_f) = F, \tag{5.102a}$$
$$\eta(t_f) = 0, \tag{5.102b}$$
$$\tau(t_f) = 0. \tag{5.102c}$$

Proof As usual, firstly define the Hamiltonian for the optimal control problems as follows:

$$\mathcal{H}(x(t), u(t), J_x^*(x(t), t), t)$$
$$= \frac{1}{2}[x(t)]^\top Q(t)x(t) + \frac{1}{2}[u(t)]^\top R(t)u(t)$$
$$+ [J_x^*(x(t), t)]^\top [A(t)x(t) + B(t)u(t) + C(t)]. \tag{5.103}$$

Following the same analysis given in Theorem 5.3, the Hamiltonian (5.103) becomes

$$\mathcal{H}(x^*(t), u^*(t), J_x^*, t)$$
$$= \frac{1}{2}[x^*(t)]^\top Q(t)x^*(t) + \frac{1}{2}[J_x^*]^\top B(t)R^{-1}(t)[B(t)]^\top J_x^*$$
$$+ [J_x^*]^\top A(t)x^*(t) - [J_x^*]^\top B(t)R^{-1}(t)[B(t)]^\top J_x^* + [J_x^*]^\top C(t)$$
$$= \frac{1}{2}[x^*(t)]^\top Q(t)x^*(t) - \frac{1}{2}[J_x^*]^\top B(t)R^{-1}(t)[B(t)]^\top J_x^*$$
$$+ [J_x^*]^\top A(t)x^*(t) + [J_x^*]^\top C(t). \tag{5.104}$$

Thus, by (5.104), the HJB equation (5.67) becomes

$$J_t^* + \frac{1}{2}[x^*(t)]^\top Q(t)x^*(t) - \frac{1}{2}[J_x^*]^\top B(t)R^{-1}(t)[B(t)]^\top J_x^*$$
$$+ [J_x^*]^\top A(t)x^*(t) + [J_x^*]^\top C(t) = 0, \tag{5.105}$$

with the boundary condition given as

$$J^*(x^*(t_f), t_f) = \frac{1}{2}[x^*(t_f)]^\top F x^*(t_f). \tag{5.106}$$

As already considered in Theorem 5.3, since the cost function $J(x(t), t, u(t))$ is a quadratic function of the state, here it is still supposed that J^* is in a quadratic form of $x(t)$, such that

$$J^*(x^*(t), t) = \frac{1}{2}[x^*(t)]^T P(t)x^*(t) + [\eta(t)]^T x^*(t) + \tau(t),$$

which is (5.100), where $P(t)$ is a real, symmetric, positive-definite matrix; then

$$J_t^* \equiv \frac{\partial J^*(x^*(t), t)}{\partial t} = \frac{1}{2}[x^*(t)]^T \dot{P}(t)x^*(t) + [\dot{\eta}(t)]^T x^*(t) + \dot{\tau}(t), \quad (5.107a)$$

$$J_x^* \equiv \frac{\partial J^*(x^*(t), t)}{\partial x} = P(t)x^*(t) + \eta(t), \quad (5.107b)$$

and using the performance cost function (5.100) in the HJB equation (5.105), we get

$$\frac{1}{2}[x^*(t)]^T \dot{P}(t)x^*(t) + [\dot{\eta}(t)]^T x^*(t) + \dot{\tau}(t) + \frac{1}{2}[x^*(t)]^T Q(t)x^*(t)$$
$$- \frac{1}{2}[x^*(t)]^T P(t)B(t)R^{-1}(t)[B(t)]^T P(t)x^*(t)$$
$$- [\eta(t)]^T B(t)R^{-1}(t)[B(t)]^T P(t)x^*(t) - \frac{1}{2}[\eta(t)]^T B(t)R^{-1}(t)[B(t)]^T \eta(t)$$
$$+ [x^*(t)]^T P(t)A(t)x^*(t) + [\eta(t)]^T A(t)x^*(t)$$
$$+ [x^*(t)]^T P(t)C(t) + [\eta(t)]^T C(t) = 0. \quad (5.108)$$

Thus by (5.108) and (5.73), we obtain

$$\frac{1}{2}[x^*(t)]^T \left[\dot{P}(t) + Q(t) - P(t)B(t)R^{-1}(t)[B(t)]^T P(t) + P(t)A(t) + [A(t)]^T P(t) \right] x^*(t)$$
$$+ \left[[\dot{\eta}(t)]^T - [\eta(t)]^T B(t)R^{-1}(t)[B(t)]^T P(t) + [\eta(t)]^T A(t) + [C(t)]^T P(t) \right] x^*(t)$$
$$+ \left[\dot{\tau}(t) - \frac{1}{2}[\eta(t)]^T B(t)R^{-1}(t)[B(t)]^T \eta(t) + [\eta(t)]^T C(t) \right] = 0, \quad (5.109)$$

which should hold for any $x^*(t)$; then we obtain

$$\dot{P}(t) + Q(t) - P(t)B(t)R^{-1}(t)[B(t)]^T P(t) + P(t)A(t) + [A(t)]^T P(t) = 0,$$
$$(5.110a)$$

$$\dot{\eta}(t) - P(t)B(t)R^{-1}(t)[B(t)]^T \eta(t) + [A(t)]^T \eta(t) + P(t)C(t) = 0, \quad (5.110b)$$

$$\dot{\tau}(t) - \frac{1}{2}[\eta(t)]^T B(t)R^{-1}(t)[B(t)]^T \eta(t) + [\eta(t)]^T C(t) = 0. \quad (5.110c)$$

By reorganizing the above equation, $P(t)$, $\eta(t)$, and $\tau(t)$ are specified in the collection of differential equations (5.101).

By (5.106) and (5.100), we obtain

$$J^*(x^*(t_f), t_f) = \frac{1}{2}[x^*(t_f)]^\top P(t_f)x^*(t_f) + [\eta(t_f)]^\top x^*(t_f) + \tau(t_f)$$

$$= \frac{1}{2}[x^*(t_f)]^\top F x^*(t_f); \tag{5.111}$$

then we have the final boundary conditions as

$$P(t_f) = F, \eta(t_f) = 0, \text{ and } \tau(t_f) = 0,$$

which is (5.102).

Consequently, by (5.65) and (5.107b), we obtain the optimal control of the underlying affine QR problems in the form of the state as given in (5.99). □

5.5   Affine-Quadratic Tracking Problems

In this section, we will study a class of affine-quadratic tracking (QT) problems based upon the results developed in Sect. 5.3.

Here, we consider the following time-variant affine state system

$$\dot{x}(t) = A(t)x(t) + B(t)u(t) + C(t), \tag{5.112}$$

such that the following performance cost function is minimized:

$$J(x(t_0), t_0, u) \triangleq \frac{1}{2}[x(t_f) - r(t_f)]^\top F[x(t_f) - r(t_f)]$$

$$+ \frac{1}{2}\int_{t_0}^{t_f} \left[[x(t) - r(t)]^\top Q(t)[x(t) - r(t)] + [u(t)]^\top R(t)u(t) \right] dt, \tag{5.113}$$

where, as defined earlier, F and $Q(t)$ are real, symmetric, positive semidefinite matrices, respectively, and $R(t)$ is a real, symmetric, positive-definite matrix, and $r(\cdot)$ represents a vector-valued function to be tracked by the state trajectory.

In Theorem 5.6 below, the optimal control for the underlying affine QT problems is specified by applying the dynamic programming method.

Theorem 5.6 *The optimal control for the affine QT problems is linear and time-variant with respect to the state, such that*

$$u^*(t) = -R^{-1}(t)[B(t)]^\top \left[P(t)x^*(t) - P(t)r(t) + \eta(t) \right], \tag{5.114}$$

for all $t \in [t_0, t_f]$.

Proof It firstly introduces a new state denoted by $\widehat{x}$, such that

$$\widehat{x}(t) \triangleq x(t) - r(t), \tag{5.115}$$

for all $t \in [t_0, t_f]$; then by (5.112), we get

$$\dot{\widehat{x}}(t) = A(t)\widehat{x}(t) + B(t)u(t) + \widehat{C}(t), \tag{5.116}$$

with $\widehat{C}(t)$ given as follows:

$$\widehat{C}(t) \equiv A(t)r(t) + C(t) - \dot{r}(t). \tag{5.117}$$

Also define a performance cost function with respect to $\widehat{x}$ such that

$$J(\widehat{x}(t_0), t_0, u(t)) \triangleq \frac{1}{2}[\widehat{x}(t_f)]^\top F\widehat{x}(t)$$

$$+ \frac{1}{2}\int_{t_0}^{t_f} \left[[\widehat{x}(t)]^\top Q(t)[\widehat{x}(t)] + [u(t)]^\top R(t)u(t) \right] dt. \tag{5.118}$$

By the above analysis, the underlying affine LT problems are converted to affine LR problems. Thus by (5.99) in Theorem 5.5, we can obtain

$$u^*(t) = -R^{-1}(t)[B(t)]^\top \left[P(t)\widehat{x}^*(t) + \eta(t) \right], \tag{5.119}$$

for all $t \in [t_0, t_f]$, which implies the conclusion of (5.114).

Also by (5.101) and (5.102), we get

$$\dot{P}(t) = -P(t)A(t) - [A(t)]^\top P(t) + P(t)B(t)R^{-1}(t)[B(t)]^\top P(t) - Q(t), \tag{5.120a}$$

$$\dot{\eta}(t) = \left[P(t)B(t)R^{-1}(t)[B(t)]^\top - [A(t)]^\top \right]\eta(t) - P(t)\widehat{C}(t), \tag{5.120b}$$

$$\dot{\tau}(t) = \frac{1}{2}[\eta(t)]^\top B(t)R^{-1}(t)[B(t)]^\top \eta(t) - [\eta(t)]^\top \widehat{C}(t), \tag{5.120c}$$

with the following final boundary conditions

$$P(t_f) = F, \tag{5.121a}$$

$$\eta(t_f) = 0, \tag{5.121b}$$

$$\tau(t_f) = 0. \tag{5.121c}$$

$\square$

In the following, we will apply the result developed in Theorem 5.6 to the EV charging problems firstly introduced in Example 2.12 in Chap. 2.

Example 5.5 (*An EV Charging Control Problem*) An EV is charged in a microgrid system with the state equation given as

$$\dot{x}(t) = \beta u(t), \tag{5.122}$$

for all $t \in [t_0, t_f]$, with β as the charging efficiency.

Consider the following cost function:

$$J(x(t_0), t_0, u) \triangleq \frac{1}{2}\left[x(t_f) - x_{\text{ref}}\right]^2 + \frac{1}{2}\int_{t_0}^{t_f}\left[\gamma u^2(t) + [u(t) + D(t)]^2\right]dt, \tag{5.123}$$

where x_{ref} and $D(t)$ denote the referred SoC value of the EV and the inelastic base demand in the system, respectively.

Notice that $\gamma u^2(t)$ and $[u(t) + D(t)]^2$ may represent the battery degradation cost with respect to charging rate $u(t)$ and the electricity cost in the system, respectively.

Solution. By (5.123), we can get

$$J(x(t_0), t_0, u) = \frac{1}{2}\left[x(t_f) - x_{\text{ref}}\right]^2 + \frac{1}{2}\int_{t_0}^{t_f}[\gamma + 1]\left[u(t) + \frac{D(t)}{\gamma + 1}\right]^2 dt$$
$$+ \frac{\gamma}{6[\gamma + 1]}\left[D^3(t_f) - D^3(t_0)\right]. \tag{5.124}$$

Based upon the above cost function, it defines a new control denoted by $\widehat{u}$ such that

$$\widehat{u}(t) = u(t) + \frac{D(t)}{\gamma + 1}; \tag{5.125}$$

then the state equation becomes

$$\dot{x}(t) = \beta\widehat{u}(t) - \frac{\beta}{\gamma + 1}D(t), \tag{5.126}$$

and the cost function (5.124) subject to $\widehat{u}$ becomes

$$J(x(t_0), t_0, \widehat{u}) = \frac{1}{2}\left[x(t_f) - x_{\text{ref}}\right]^2 + \frac{1}{2}\int_{t_0}^{t_f}[\gamma + 1]\widehat{u}^2(t)dt$$
$$+ \frac{\gamma}{6[\gamma + 1]}\left[D^3(t_f) - D^3(t_0)\right]. \tag{5.127}$$

Thus for the above charging control problems, we have

$$A(t) = 0, B(t) = \beta, C(t) = -\frac{\beta}{\gamma+1}D(t), F = 1, r(t) = x_{\text{ref}}, Q(t) = 0, R(t) = \gamma + 1.$$
(5.128)

Hence by applying Theorem 5.6,

$$\widehat{u}^*(t) = -R^{-1}(t)[B(t)]^\top \left[P(t)x^*(t) - P(t)r(t) + \eta(t)\right]$$

$$= -\frac{\beta}{\gamma+1}\left[P(t)x^*(t) - x_{\text{ref}}P(t) + \eta(t)\right],$$
(5.129)

for all $t \in [t_0, t_f]$; then

$$u^*(t) = -\frac{\beta}{\gamma+1}\left[P(t)x^*(t) - x_{\text{ref}}P(t) + \eta(t)\right] - \frac{D(t)}{\gamma+1},$$
(5.130)

with $P(\cdot), \eta(\cdot),$ and $\tau(\cdot)$ given as

$$\dot{P}(t) = \frac{\beta^2}{\gamma+1}P^2(t),$$
(5.131a)

$$\dot{\eta}(t) = \frac{\beta^2}{\gamma+1}P(t)\eta(t) - P(t)\widehat{C}(t),$$
(5.131b)

$$\dot{\tau}(t) = \frac{\beta^2}{2[\gamma+1]}\eta^2(t) - \eta(t)\widehat{C}(t),$$
(5.131c)

where, by (5.117),

$$\widehat{C}(t) = A(t)r(t) + C(t) - \dot{r}(t) = C(t) = -\frac{\beta}{\gamma+1}D(t),$$

and the boundary conditions are given as $P(t_f) = 1, \eta(t_f) = 0$ and $\tau(t_f) = 0.$ □

5.6 Development of Pontryagin's Minimum Principle via Dynamic Programming

In this part, it will develop Pontryagin's minimum principle by applying the dynamic programming method.

Consider the time-invariant state system in the following:

$$\dot{x}(t) = f(x(t), u(t)),$$
(5.132)

and the performance cost function given as

$$J(x(t_0), t_0, u) \triangleq h(x(t_f)) + \int_{t_0}^{t_f} g(x(t), u(t))dt, \qquad (5.133)$$

with the state value at initial time t_0 given as $x(t_0)$.

Suppose that x^* is the solution of the HJB equation specified in (5.16), and u^* is the control associated with the state trajectory x^*; then by Theorem 5.2, u^* is the optimal control for the underlying problems, say

$$u^*(t) = \arg\min_{u \in U} \left\{ g(x^*(t), u) + J_t^*(x^*(t), t) + \left[J_x^*(x^*(t), t) \right]^\top f(x^*(t), u) \right\}.$$
$$(5.134)$$

Before dealing with the analysis further, we will firstly give a lemma below.

Lemma 5.1 *Consider a function denoted by $F(x, u, t)$ which is continuously differentiable on its arguments, and suppose that U is a convex set. Assume that $v^*(x, t)$ is continuously differentiable on x and t such that*

$$v^*(x, t) = \arg\min_{u \in U} F(x, u, t); \qquad (5.135)$$

then the following holds:

$$\frac{\partial \min_{u \in U} F(x, u, t)}{\partial t} = \frac{\partial F(x, v^*(x, t), t)}{\partial t}, \qquad (5.136a)$$

$$\frac{\partial \min_{u \in U} F(x, u, t)}{\partial x} = \frac{\partial F(x, v^*(x, t), t)}{\partial x}. \qquad (5.136b)$$

$\square$

In Theorem 5.7 below, we redevelop Pontryagin's minimum principle by applying the dynamic programming method.

Theorem 5.7 *Suppose that u^* is an optimal control for the problems defined earlier in this section; then we can show that (4.8), in Theorem 4.1, holds by applying the dynamic programming method.*

Proof Firstly suppose that

$$v^*(x, t) = \arg\min_{u \in U} \left\{ g(x, u) + \left[\frac{\partial J^*(x, t)}{\partial x} \right]^\top f(x, u) \right\}; \qquad (5.137)$$

then by applying Lemma 5.1 and and by differentiating both sides of the HJB equation (5.7), with respect to x and t, respectively, we can obtain

$$0 = g_x(x, v^*(x, t)) + J^*_{xt}(x, t)$$
$$+ J^*_{xx}(x, t) f(x, v^*(x, t)) + f_x(x, v^*(x, t)) J^*_x(x, t), \tag{5.138a}$$
$$0 = J^*_{tt}(x, t) + \left[J^*_{xt}(x, t)\right]^\top f(x, v^*(x, t)), \tag{5.138b}$$

for all x and t, where

$$J^*_{xt}(x, t) \equiv \frac{\partial^2 J^*(x, t)}{\partial x \, \partial t},$$

$$J^*_{xx}(x, t) \equiv \frac{\partial^2 J^*(x, t)}{\partial x \, \partial x},$$

$$J^*_{tt}(x, t) \equiv \frac{\partial^2 J^*(x, t)}{\partial t \, \partial t},$$

$$f_x(x, v^*(x, t)) \equiv \begin{bmatrix} \frac{\partial f_1(x, v^*(x,t))}{\partial x_1} & \cdots & \frac{\partial f_n(x, v^*(x,t))}{\partial x_1} \\ \cdots & \cdots & \cdots \\ \frac{\partial f_1(x, v^*(x,t))}{\partial x_n} & \cdots & \frac{\partial f_n(x, v^*(x,t))}{\partial x_n} \end{bmatrix}.$$

Thus, by specifying the optimal control as $u^*(t) = v^*(x^*(t), t)$ for all $t \in [t_0, t_f]$, we have

$$\dot{x}^*(t) = f(x^*(t), u^*(t)). \tag{5.140}$$

Hence we can obtain

$$\frac{dJ^*_x(x^*(t), t)}{dt} = J^*_{xt}(x^*(t), t) + J^*_{xx}(x^*(t), t) f(x^*(t), u^*(t)), \tag{5.141a}$$

$$\frac{dJ^*_t(x^*(t), t)}{dt} = J^*_{tt}(x^*(t), t) + [J^*_{xt}(x^*(t), t)]^\top f(x^*(t), u^*(t)). \tag{5.141b}$$

Consider

$$\lambda^*(t) \equiv J^*_x(x^*(t), t), \tag{5.142a}$$

$$\widehat{\lambda}^*(t) \equiv J^*_t(x^*(t), t); \tag{5.142b}$$

then by the specification of $\lambda^*(t)$, we have

$$\dot{\lambda}^*(t) = \frac{dJ^*_x(x^*(t), t)}{dt}$$
$$= J^*_{xt}(x^*(t), t) + J^*_{xx}(x^*(t), t) f(x^*(t), u^*(t))$$
$$= -f_x(x^*(t), u^*(t)) \lambda^*(t) - g_x(x^*(t), u^*(t)), \tag{5.143}$$

for all $t \in [t_0, t_f]$, where the last equality holds by (5.138a), which is equivalent to the following:

$$\dot{\lambda}^*(t) = -\frac{\partial \mathcal{H}(x^*(t), u^*(t), \lambda^*(t), t)}{\partial x}, \tag{5.144}$$

by the specification of the Hamiltonian $\mathcal{H}(x, u, \lambda, t)$ given in (5.14).

Furthermore, by the boundary condition $J^*(x, t_f) = h(x, t_f)$, we can obtain

$$J_x^*(x, t_f) = h_x(x, t_f), \text{ for all } x; \tag{5.145}$$

then by the specification of $\lambda^*(t)$, we obtain the boundary condition of the costate equation $\lambda^*(\cdot)$ as

$$\lambda^*(t_f) = h_x(x^*(t_f), t_f). \tag{5.146}$$

By the specification of $\widehat{\lambda}^*(t)$, we have

$$\dot{\widehat{\lambda}}^*(t) = \frac{d J_t^*(x^*(t), t)}{dt} = 0, \tag{5.147}$$

for all $t \in [t_0, t_f]$, where the last equality holds by (5.138b), which implies that

$$\widehat{\lambda}^*(t) \equiv J_t^*(x^*(t), t) = \xi, \tag{5.148}$$

for some constant-valued ξ; then by Theorem 5.1, the Hamiltonian $\mathcal{H}$ satisfies the following property:

$$\mathcal{H}(x^*(t), u^*(t), \lambda^*(t), t) = \min_{u \in U} \mathcal{H}(x^*(t), u, \lambda^*(t), t) = -J_t^*(x^*(t), t) = -\xi, \tag{5.149}$$

for all t, say the optimal Hamiltonian is constant valued.

Also by (5.134), the specification of the Hamiltonian $\mathcal{H}$ and (5.148), we obtain the following:

$$u^*(t) = \arg\min_{u \in U} \left\{ \mathcal{H}(x^*(t), u, \lambda^*(t), t) \right\}. \tag{5.150}$$

Moreover, by the specification of $\mathcal{H}$, the state equation subject to the optimal control u^* is given as

$$\dot{x}^*(t) = f(x^*(t), u^*(t)) = \frac{\partial \mathcal{H}(x^*(t), u^*(t), \lambda^*(t), t)}{\partial \lambda}, \tag{5.151}$$

with the initial condition $x^*(0) = x_0$.

In summary we obtain (5.151), (5.143), and (5.150) which are Pontryagin's minimum principle developed in Theorem 4.1, Chap. 4.

Thus we can get the conclusion that Pontryagin's minimum principle is redeveloped here by applying the dynamic programming method. □

Notice that Pontryagin's minimum principle provides necessary optimality conditions, so all optimal control trajectories satisfy these conditions, but if a control trajectory satisfies these conditions, it is not necessarily optimal. Further analysis is needed to guarantee the optimality.

One method that often works is to prove that an optimal control trajectory exists, and to verify that there is only one control trajectory satisfying the conditions of Pontryagin's minimum principle (or that all control trajectories satisfying these conditions have equal cost). Another possibility to conclude optimality arises when the state system function f is linear in x and u, the constraint set U is convex, and the cost functions h and g are convex. Then it can be shown that the conditions of Pontryagin's minimum principle are both necessary and sufficient for optimality.

5.7 Summary

In this chapter, we briefly introduce another key branch of the optimal control method, say the dynamic programming method, to implement the optimal control solution. More specially, we firstly develop the HJB equation based on the principle of optimality, and then study the optimal controls for certain classes of optimal control problems, such as linear-quadratic regulation problems and affine-quadratic regulation and tracking problems. Pontryagin's minimum principle is also developed via the dynamic program method for comparison.

5.8 Exercises

Exercise 5.1 Implement the optimal control for a state system given by

$$\dot{x} = -2x(t) + u(t),$$

with the boundary conditions $x(0) = 2$ and $x(2) = 1$, such that the performance cost function

$$J(u) = \frac{1}{2} \int_0^2 \left[x^2(t) + u^2(t) \right] dt$$

is minimized.

Exercise 5.2 Implement the optimal control for a state system given by

$$\dot{x} = -2x(t) + u(t),$$

with the boundary condition $x(0) = 2$, such that the performance cost function

$$J(u) = \frac{1}{2}[x(2) - 1]^2 + \frac{1}{2}\int_0^2 \left[x^2(t) + u^2(t)\right]dt$$

is minimized.

Exercise 5.3 Consider a state system given as

$$\dot{x}_1(t) = x_2(t),$$
$$\dot{x}_2(t) = -x_1(t) + x_2(t) + u(t),$$

and the performance cost function is specified as

$$J \triangleq \int_0^1 \left[x_1^2(t) + x_2^2(t) + u^2(t)\right]dt.$$

Suppose that the initial and final states are specified; then specify the HJB equation for the underlying optimal control problem.

Exercise 5.4 Specify the HJB equation for the system

$$\dot{x}_1(t) = x_2(t),$$
$$\dot{x}_2(t) = -x_2(t) - x_1^2(t) + 2u(t),$$

with the performance function given as

$$J = \frac{1}{2}\int_{t_0}^{t_f} \left[x_1^2(t) + u^2(t)\right]dt.$$

Exercise 5.5 Consider a one-dimensional state system such that

$$\dot{x}(t) = -2x(t) + u(t),$$

with the following performance cost function

$$J = x^2(t_f) + \int_{t_0}^{t_f} u^2(t)dt.$$

Find the optimal control solution for the underlying problem.

Exercise 5.6 Implement the optimal control for a state system given by

$$\dot{x} = -2x(t) + u(t),$$

with the boundary condition $x(0) = 2$, such that the performance cost function

$$J(u) = \frac{1}{2} \int_0^\infty \left[x^2(t) + u^2(t) \right] dt$$

is minimized.

Exercise 5.7 Solve the minimum energy consumption problem in an electric circuit system formulated in Example 1.2 by applying dynamic programming.

Exercise 5.8 Formulate the vehicle suspension system introduced in Example 1.10 in Chap. 1 as an optimal control problem and implement its optimal control solution by applying dynamic programming.

Exercise 5.9 For the DC motor speed control system described in Exercise 4.14 in Chap. 4, specify the HJB equation and hence the closed-loop optimal control to keep the speed at a constant value.

Exercise 5.10 For the liquid-level control system described in Example 1.11 in Chap. 1, specify the HJB equation and hence the closed-loop optimal control to keep the liquid-level constant at a particular value.

Chapter 6
Differential Games

In this chapter, differential games are studied.

In Sect. 6.1, a class of noncooperative differential games is formulated, the properties of the Nash equilibrium of the underlying differential games are developed, and affine-quadratic differential games are further studied. In Sect. 6.2, a class of zero-sum differential games is formulated and the saddle point of two-person zero-sum differential games is studied and implemented via the dynamic programming method. Moreover, in this section, the two-person linear-quadratic zero-sum differential games are studied.

A brief summary of this chapter is given in Sect. 6.3. Finally, in Sect. 6.4, exercises for the readers to verify the results introduced in this chapter are given.

6.1 Noncooperative Differential Games

6.1.1 Formulation of Noncooperative Differential Games

In this part, a class of noncooperative differential games is introduced. Basically, in each of these games, there are a set of individual players denoted by $\mathcal{N} \equiv \{1, 2, \ldots, N\}$.

Moreover, it is considered that each of these players tries to maximize his own payoff function or minimize his own cost function.

We will formulate this class of differential games in the following.

Denote by $u_n(\cdot)$ the control strategy of player n over the time interval $[t_0, t_f]$.

Further, define $\mathcal{U}_n([t_0, t_f])$ as *the set of control strategies* for player n, such that

$$\mathcal{U}_n([t_0, t_f]) \triangleq \{u_n(\cdot) : [t_0, t_f] \to U_n, \text{ s.t. } u_n(\cdot) \text{ is measurable}\}, \tag{6.1}$$

where U_n is *the admissible control set* of player n.

Consider a state system denoted by $x(\cdot)$ which is driven by a collection of control strategies $u_n(\cdot)$, with $n \in \mathcal{N}$, such that

$$\dot{x}(t) = f(x(t), u_1(t), \ldots, u_N(t), t), \tag{6.2}$$

for all $t \in [t_0, t_f]$, with an initial value of $x(t_0) = x_0$.

Define a cost function of individual player n, denoted by $J_n(u_n, u_{-n})$, with $u_{-n} \equiv (u_1, \ldots, u_{n-1}, u_{n+1}, \ldots, u_N)$, such that

$$J_n(u_n, u_{-n}) \triangleq h_n(x(t_f)) + \int_{t_0}^{t_f} g_n(x(t), u_1(t), \ldots, u_N(t), t)dt, \tag{6.3}$$

where $h_n(x(t_f))$ represents the terminal cost with respect to the state of agent n at final time t_f.

It is called an N-person noncooperative differential game if the objective of each of the individual players is to minimize the cost function (6.3), respectively.

In the following, a noncooperative differential game will be formulated for the EV charging control problems introduced in Example 2.12 in Chap. 2.

Example 6.1 (*An EV Charging Differential Game*) A collection of EVs are charged in a power grid system with the state equation given as follows:

$$\dot{x}_n(t) = \beta_n u_n(t), \tag{6.4}$$

for all $t \in [t_0, t_f]$, with $n \in \mathcal{N}$, and β_n as the charging efficiency of EV n.

Consider the following local cost function for each EV n such that

$$J_n(u_n, u_{-n}) \triangleq \frac{1}{2} Q_n^f \left[x_n(t_f) - x_{n,\max} \right]^2 + \int_{t_0}^{t_f} \left[\gamma_n[u_n(t)]^2 + p(u(t))u_n(t) \right] dt, \tag{6.5}$$

where $p(u(t))$ represents the (real-time) electricity price at time t with respect to the charging strategy of EVs at that time.

And it is considered that

$$U_n \triangleq [0, u_{n,\max}], \quad \text{for all } n \in \mathcal{N}, \tag{6.6}$$

i.e., each EV n is to be charged with a maximum allowed power rate $u_{n,\max}$. □

Notice that, the local cost function given in (6.5),

- The term $\gamma_n[u_n(t)]^2$ represents the battery degradation cost subject to the charging rate u_n.
- In the literature, it may be considered that

$$p(u(t)) = \rho_1 \left[\sum_{n \in \mathcal{N}} u_n(t) + D(t) \right] + \rho_2, \tag{6.7}$$

that is to say, the (real-time) electricity price at time t is a linear function of the total demand at t.

Definition 6.1 (*Nash Equilibrium (NE) of Noncooperative Differential Game*) A strategy denoted by $u^*(\cdot) \equiv (u_n^*(\cdot); n \in \mathcal{N})$ is called a Nash equilibrium (NE) of the noncooperative differential game, if the following holds:

$$J_n(u_n^*, u_{-n}^*) \le J_n(u_n, u_{-n}^*), \tag{6.8}$$

for all $u_n \in \mathcal{U}_n([t_0, t_f])$ and all $n \in \mathcal{N}$.   □

6.1.2 Nash Equilibrium of Noncooperative Differential Games

Theorem 6.1 *Suppose that*

- $f(x, u_1, \dots, u_N, t)$ *is continuous on* $t \in [t_0, t_f]$, *for each* $x \in \mathcal{X}$,
- $f(x, u_1, \dots, u_N, t)$ *is uniformly Lipschitz continuous on* $x, u_1, \dots, u_N$, *such that*

$$|f(x, u_1, \dots, u_N, t) - f(\widehat{x}, \widehat{u}_1, \dots, \widehat{u}_N, t)|$$

$$\le \xi \max_{t \in [t_0, t_f]} \left\{ |x(t) - \widehat{x}(t)| + \sum_{n \in \mathcal{N}} |u_n(t) - \widehat{u}_n(t)| \right\}, \tag{6.9}$$

for some constant positive-valued ξ, *for any pair of state trajectories* $x(\cdot), \widehat{x}(\cdot) \in \mathcal{C}([t_0, t_f])$ *which represents the set of continuously differentiable functions over the interval* $[t_0, t_f]$, *and any pair of admissible control strategies* $u_n(\cdot), \widehat{u}_n(\cdot) \in \mathcal{U}_n([t_0, t_f])$,

- *Consider* $v_n(\cdot) \equiv \{v_n(x(t), t); t \in [t_0, t_f]\}$, *with* $v_n(\cdot) \in \mathcal{U}_n([t_0, t_f])$, *such that* $v_n(\cdot)$ *is continuous on* t *for each* $x(\cdot) \in \mathcal{C}([t_0, t_f])$ *and is uniformly Lipschitz continuous on* $x(\cdot) \in \mathcal{C}([t_0, t_f])$;

then the system state equation (6.2) *has a unique continuous solution for any* $v_n(\cdot)$ *given above, so* $u_n(t) = v_n(x(t), t)$, *for all* $t \in [t_0, t_f]$.

Proof The condition given above is a sufficient condition for the existence of the unique continuous solution for differential equations.   □

In Theorem 6.2 below, a necessary condition for an NE of a noncooperative differential game will be given. Before that, define a local Hamiltonian for an individual player n, denoted by $\mathcal{H}_n(x(t), u_1(t), \dots, u_N(t), \lambda_n(t), t)$, with $n \in \mathcal{N}$, such that

$$\mathcal{H}_n(x(t), u_1(t), \ldots, u_N(t), \lambda_n(t), t)$$
$$\triangleq g_n(x(t), u_1(t), \ldots, u_N(t), t) + [\lambda_n(t)]^\top f(x(t), u_1(t), \ldots, u_N(t), t), \quad (6.10)$$

for each time $t \in [t_0, t_f]$, where λ_n represents the costate associated with agent n of the noncooperative differential game system.

Theorem 6.2 *Suppose that u^* is an NE strategy of a noncooperative differential game, and x^* is the corresponding state trajectory subject to this NE strategy, say*

$$\dot{x}^*(t) = f\left(x^*(t), u_1^*(t), \ldots, u_N^*(t), t\right), \quad (6.11)$$

for all $t \in [t_0, t_f]$, with an initial value of $x(t_0) = x_0$.

Also consider that the following hold for this noncooperative differential game,

- $f(\cdot, u_1, \ldots, u_N, t)$ *is continuously differentiable on $\mathbb{R}^n$,*
- $g_n(\cdot, u_1, \ldots, u_N, t)$ *is continuously differentiable on $\mathbb{R}^n$,*
- $h_n(\cdot)$ *is continuously differentiable on $\mathbb{R}^n$,*

for all $t \in [t_0, t_f]$; then there exists a costate trajectory for each player $n \in \mathcal{N}$, denoted by $\lambda_n^*(\cdot) : [t_0, t_f] \to \mathbb{R}^{mN}$, such that

$$u_n^*(t) = \arg\min \mathcal{H}_n(x^*(t), u_n, u_{-n}^*(t), \lambda_n^*(t), t), \quad (6.12a)$$
$$\dot{\lambda}_n^*(t) = -\frac{\partial \mathcal{H}_n(x^*(t), u_1^*(t), \ldots, u_N^*(t), \lambda_n^*(t), t)}{\partial x}, \quad (6.12b)$$

with the boundary condition

$$\lambda_n^*(t_f) = \frac{\partial h_n(x^*(t_f))}{\partial x}. \quad (6.13)$$

Proof Suppose that $u^* \equiv \left(u_1^*, \ldots, u_N^*\right)$ is an NE for the underlying dynamic game; then by Definition 6.1, the following holds:

$$J_n\left(u_n^*, u_{-n}^*\right) \le J_n\left(u_n, u_{-n}^*\right), \quad (6.14)$$

for each individual player n, that is to say, for each given controls of other players $u_{-n} \equiv (u_1, \ldots, u_{n-1}, u_{n+1}, \ldots, u_N)$, u_n^* is an optimal control for $J_n\left(u_n, u_{-n}^*\right)$, which is the cost function of player n, defined in (6.3) subject to the the state equation

$$\dot{x}(t) = f\left(x(t), u_1^*(t), \ldots, u_{n-1}^*(t), u_{n+1}^*(t), \ldots, u_N^*(t), t\right),$$

for all $t \in [t_0, t_f]$.

Then following Theorem 4.1 in Chap. 4, we can obtain the result by applying Pontryagin's minimum principle to implement the optimal control problems with a fixed final time and unspecified final state over a time interval $[t_0, t_f]$. $\qquad\square$

In the following, we will apply Theorem 6.2 with a numerical example.

Example 6.2 Give a necessary condition for the NE strategy of the differential game specified in Example 6.1.

Solution. By the specifications of EV charging differential game given in Example 6.1, and by applying Theorem 6.2, the NE strategy u^* satisfies (6.12a).

And by (6.12b), the associated costate λ_n^* satisfies the following:

$$\dot{\lambda}_n^*(t) = -\frac{\partial \mathscr{H}_n(x^*(t), u^*(t), \lambda_n^*(t), t)}{\partial x} = 0, \tag{6.15}$$

since $\mathscr{H}_n(x^*(t), u^*(t), \lambda_n^*(t), t)$ is unrelated to the state x, which implies that

$$\lambda_n^*(t) = \xi_n, \tag{6.16}$$

for some constant value ξ_n, for all $t \in [t_0, t_f]$.

By (6.13), λ_n^* satisfies the following boundary condition:

$$\lambda_n^*(t_f) = \frac{\partial h_n(x^*(t_f))}{\partial x} = \begin{bmatrix} 0 \\ \cdots \\ 0 \\ Q_n^f \left[x_n^*(t_f) - x_{n,\max} \right] \\ 0 \\ \cdots \\ 0 \end{bmatrix}. \tag{6.17}$$

Thus,

$$\lambda_n^*(t) = \lambda_n^*(t_f), \tag{6.18}$$

with $\lambda_n^*(t_f)$ given in (6.17), that is, the n-th component of λ_n^* equals Q_n^f $\left[x_n^*(t_f) - x_{n,\max} \right]$, for all $t \in [t_0, t_f]$, and all the other components remain at zeros over $[t_0, t_f]$. □

Theorem 6.2 specifies a necessary condition of the NE strategy with a general form of the state equation specified in (6.2). In Corollary 6.1 below, a necessary condition for the NE strategy of differential game will be given such that

$$\dot{x}_n(t) = f_n(x_n(t), u_n(t), t), \tag{6.19a}$$

$$J_n(u_n, u_{-n}) = h_n(x_n(t_f)) + \int_{t_0}^{t_f} g_n(x_n(t), u(t), t)dt, \tag{6.19b}$$

that is to say, the evolution of the state and the cost function of agent n is only determined by its local state.

Corollary 6.1 (A Special Form of Theorem 6.2) *Suppose that u^* is an NE of a noncooperative differential game with state equation given in (6.19a), and $x^* \equiv (x_1^*, \ldots, x_N^*)$ is the corresponding state trajectory subject to this NE strategy, say*

$$\dot{x}^*(t) = f_n\left(x_n^*(t), u^*(t), t\right), \tag{6.20}$$

for all $t \in [t_0, t_f]$, with an initial value of $x_n^*(t_0) = x_{n0}$.

Also consider the smooth property specified in Theorem 6.2; then there exists a costate trajectory for each player $n \in \mathcal{N}$, denoted by $\lambda_n^*(\cdot) : [t_0, t_f] \to \mathbb{R}_n$, such that

$$u_n^*(t) = \arg\min_{u_n \in \mathcal{U}_n} \mathcal{H}_n(x_n^*(t), u_n, u_{-n}^*(t), \lambda_n^*(t), t), \tag{6.21a}$$

$$\dot{\lambda}_n^*(t) = -\frac{\partial \mathcal{H}_n(x_n^*(t), u^*(t), \lambda_n^*(t), t)}{\partial x_n}, \tag{6.21b}$$

with the boundary condition

$$\lambda_n^*(t_f) = \frac{\partial h_n(x_n^*(t_f))}{\partial x_n}. \tag{6.22}$$

Proof It can be verified by applying the same statement used in the proof of Theorem 6.2. □

Actually, the differential game formulated in Example 6.1 satisfies the properties considered in Corollary 6.1; then the NE strategy satisfies the following necessary condition.

By the specifications of EV charging differential game given in Example 6.1, and by applying Theorem 6.2, the NE strategy u^* satisfies (6.21a).

And by (6.21b), the associated costate λ_n^* satisfies the following:

$$\dot{\lambda}_n^*(t) = -\frac{\partial \mathcal{H}_n(x_n^*(t), u^*(t), \lambda_n^*(t), t)}{\partial x_n} = 0, \tag{6.23}$$

since $\mathcal{H}_n(x_n^*(t), u^*(t), \lambda_n^*(t), t)$ is unrelated to the state x_{-n}, which implies that

$$\lambda_n^*(t) = \xi_n, \tag{6.24}$$

for some constant value ξ_n, for all $t \in [t_0, t_f]$.

By (6.22), λ_n^* satisfies the following boundary condition:

$$\lambda_n^*(t_f) = \frac{\partial h_n(x^*(t_f))}{\partial x_n} = Q_n^f\left[x_n^*(t_f) - x_{n,\max}\right]. \tag{6.25}$$

6.1.3   Affine-Quadratic Noncooperative Differential Games

Definition 6.2 (*Affine-Quadratic Differential Games*) A game is called an N-person affine-quadratic differential game if $f(x, u_1, \ldots, u_N, t)$ in the state equation (6.2) and $g_n(x, u_1, \ldots, u_N, t)$ in the local cost function of each individual player n specified in (6.3) satisfy the following, respectively:

$$f(x, u_1, \ldots, u_N, t) = A(t)x + \sum_{n \in \mathcal{N}} B_n(t)u_n + C(t), \tag{6.26a}$$

$$g_n(x, u_1, \ldots, u_N, t) = \frac{1}{2} \left[x^\top Q_n(t)x + \sum_{\widehat{n} \in \mathcal{N}} [u_{\widehat{n}}]^\top R_{n\widehat{n}}(t)u_{\widehat{n}} \right], \tag{6.26b}$$

$$h_n(x) = \frac{1}{2} x^\top Q_n^f x, \tag{6.26c}$$

for all $t \in [t_0, t_f]$, where $A(t)$, $B_n(t)$, $C(t)$, $Q_n(t)$, Q_n^f, and $R_{n\widehat{n}}(t)$ are matrices with appropriate dimensions, respectively.

It is further supposed that $Q_n(t)$, Q_n^f are symmetric and positive definite, and $R_{nn}(t)$ is strictly positive definite, for all $t \in [t_0, t_f]$ and all $n \in \mathcal{N}$. $\square$

Theorem 6.3 below shows the existence and uniqueness of NE for the affine-quadratic differential game specified in Definition 6.2 above. Before that, firstly define a notation $\mathfrak{M}_n(\cdot)$, such that it is a solution to the following Riccati differential equations:

$$\dot{\mathfrak{M}}_n(t) = -\mathfrak{M}_n(t)A(t) - [A(t)]^\top \mathfrak{M}_n(t) - Q_n(t)$$
$$+ \mathfrak{M}_n(t) \sum_{\widehat{n} \in \mathcal{N}} B_{\widehat{n}}(t)[R_{\widehat{n}\widehat{n}}(t)]^{-1}[B_{\widehat{n}}(t)]^\top \mathfrak{M}_{\widehat{n}}(t), \tag{6.27}$$

for all $n \in \mathcal{N}$, with the boundary condition $\mathfrak{M}_n(t_f) = Q_n^f$.

Theorem 6.3 (Uniqueness and Existence of NE of Affine-Quadratic Differential Game) *Consider an N-person affine-quadratic differential game; then the game system has a unique NE specified as*

$$u_n^*(t) = -[R_{nn}(t)]^{-1}[B_n(t)]^\top \left[\mathfrak{M}_n(t)x^*(t) + \mathfrak{m}_n(t) \right], \tag{6.28}$$

for all $t \in [t_0, t_f]$ and $n \in \mathcal{N}$, where $\mathfrak{m}_n(\cdot)$ is given as

$$\dot{\mathfrak{m}}_n(t) = - [A(t)]^\top \mathfrak{m}_n(t) - \mathfrak{M}_n(t)C(t)$$
$$+ \mathfrak{M}_n(t) \sum_{\widehat{n} \in \mathcal{N}} B_{\widehat{n}}(t)[R_{\widehat{n}\widehat{n}}(t)]^{-1}[B_{\widehat{n}}(t)]^\top \mathfrak{m}_{\widehat{n}}(t), \tag{6.29}$$

for all $t \in [t_0, t_f]$ and $n \in \mathcal{N}$, with the boundary condition as $\mathfrak{m}_n(t_f) = 0$, and $x^*(\cdot)$, which represents the state trajectory subject to $u^*(\cdot)$, and is given as

$$x^*(t) = \Phi(t, 0)x(t_0)$$

$$+ \int_{t_0}^t \Phi(t, s) \left[C(t) - \sum_{n \in \mathcal{N}} B_n(s)[R_{nn}(s)]^{-1}[B_n(s)]^{\top} \mathfrak{m}_n(s) \right] ds, \quad (6.30)$$

with the initial condition $x^*(t_0) = x_0$ and where $\Phi(\cdot, s)$ is specified with the following differential equation:

$$\dot{\Phi}(t, s) = \left[A(t) - \sum_{n \in \mathcal{N}} B_n(t)[R_{nn}(t)]^{-1}[B_n(t)]^{\top} \mathfrak{M}_n(t) \right] \Phi(t, s), \quad (6.31)$$

such that $\Phi(s, s) = \mathbb{I}$, for all s.

Proof For the underlying affine-quadratic differential game defined in Definition 6.2, the local cost function $J_n(u_1, \ldots, u_N)$ is strictly convex on $u_n(\cdot)$ for any given $u_{-n}(\cdot)$.

Firstly, by the specification of the local Hamiltonian $\mathcal{H}_n$ of player n,

$$\mathcal{H}_n(x, u_1, \ldots, u_N, \lambda, t)$$

$$= \frac{1}{2} \left[x^{\top} Q_n(t)x + \sum_{\widehat{n} \in \mathcal{N}} [u_{\widehat{n}}]^{\top} R_{n\widehat{n}}(t)u_{\widehat{n}} \right] + [\lambda_n(t)]^{\top} \left[A(t)x + \sum_{n \in \mathcal{N}} B_n(t)u_n + C(t) \right]. \quad (6.32)$$

Thus, by applying (3.104) in Theorem 3.6 in Chap. 3, and by the positive-definite properties of $Q_n(\cdot)$ and $R_{nn}(\cdot)$, there exists a unique solution to minimize $\mathcal{H}_n$ which is given as

$$u_n^*(t) = -[R_{nn}(t)]^{-1}[B_n(t)]^{\top}\lambda_n^*(t), \quad (6.33)$$

for all $n \in \mathcal{N}$.

Also, by (3.106) in Theorem 3.6 in Chap. 3, the state and costate trajectories corresponding to the optimal control u^* are given as follows:

$$\dot{x}^*(t) = A(t)x^*(t) - \sum_{n \in \mathcal{N}}[R_{nn}(t)]^{-1}[B_n(t)]^{\top}\lambda_n^*(t), \quad (6.34a)$$

$$\dot{\lambda}_n^*(t) = -Q_n(t)x^*(t) - [A(t)]^{\top}\lambda_n^*(t), \quad (6.34b)$$

with the boundary conditions $\lambda_n^*(t_f) = Q_n^f x^*(t_f)$ and $x^*(t_0) = x_0$, respectively.

As observed, the above-given set of differential equations is a two-point boundary value problem.

Now with $\mathfrak{M}_n$ and $\mathfrak{m}_n$ satisfying (6.27) and (6.29), respectively, and by using

$$\lambda_n^*(t) = \mathfrak{M}_n(t)x^*(t) + \mathfrak{m}_n(t), \quad (6.35)$$

for all $n \in \mathcal{N}$, to the costate equation (6.34b), we can solve the underlying two-point boundary value problem.

Thus, the NE strategy u^* in the form of (6.28) is obtained by using $\lambda_n^*(t) = \mathfrak{M}_n(t)x^*(t) + \mathfrak{m}_n(t)$ again to (6.33). Also, the state trajectory associated with u^* in the form of (6.30) is obtained by applying (6.34a).   □

In the following, we will apply Theorem 6.3 to analyze a specific differential game introduced in Example 6.1 without considering any constraints on the control u.

Example 6.3 Consider a differential game with the state system given as

$$\dot{x}_1(t) = a_1 x_1(t) + b_1 u_1(t), \tag{6.36a}$$
$$\dot{x}_2(t) = a_2 x_2(t) + b_2 u_2(t), \tag{6.36b}$$

with a given performance cost function

$$J_n(u) = \frac{1}{2} q_n^f [x_n(t_f)]^2 + \int_{t_0}^{t_f} \left[\frac{1}{2} q_n [x_n(t)]^2 + \frac{1}{2} r_n [u_n(t)]^2 \right] dt, \tag{6.37}$$

with $r_n > 0$, $n = 1, 2$.

Implement the NE strategy $u^* \equiv (u_1^*, u_2^*)$ for the differential game system.

Solution. Firstly, by (6.26a), (6.26b), and (6.26c), we obtain

$$A(t) = \begin{bmatrix} a_1 & 0 \\ 0 & a_2 \end{bmatrix}, B_1(t) = \begin{bmatrix} b_1 \\ 0 \end{bmatrix}, B_2(t) = \begin{bmatrix} 0 \\ b_2 \end{bmatrix}, C(t) = \begin{bmatrix} 0 \\ 0 \end{bmatrix},$$

$$Q_1(t) = \begin{bmatrix} q_1 & 0 \\ 0 & 0 \end{bmatrix}, Q_2(t) = \begin{bmatrix} 0 & 0 \\ 0 & q_2 \end{bmatrix}, R_{11}(t) = r_1, R_{22}(t) = r_2, R_{12}(t) = R_{21}(t) = 0,$$

$$Q_1^f = \begin{bmatrix} q_1^f & 0 \\ 0 & 0 \end{bmatrix}, Q_2^f = \begin{bmatrix} 0 & 0 \\ 0 & q_2^f \end{bmatrix}, \tag{6.38}$$

for all $t \in [t_0, t_f]$.

By (6.27), we get

$$\dot{\mathfrak{M}}_n(t) = -\mathfrak{M}_n(t) \begin{bmatrix} a_1 & 0 \\ 0 & a_2 \end{bmatrix} - \begin{bmatrix} a_1 & 0 \\ 0 & a_2 \end{bmatrix} \mathfrak{M}_n(t) - Q_n(t)$$

$$+ \mathfrak{M}_n(t) \left[\begin{bmatrix} \frac{|b_1|^2}{r_1} & 0 \\ 0 & 0 \end{bmatrix} \mathfrak{M}_1(t) + \begin{bmatrix} 0 & 0 \\ 0 & \frac{|b_2|^2}{r_2} \end{bmatrix} \mathfrak{M}_2(t) \right], \tag{6.39}$$

for $n = 1, 2$, with the boundary condition $\mathfrak{M}_n(t_f) = Q_n^f$.

Then by Theorem 6.3, the NE strategy u^* is given as

$$u_n^*(t) = -\frac{1}{r_n} [B_n(t)]^\top \left[\mathfrak{M}_n(t)x^*(t) + \mathfrak{m}_n(t) \right], \tag{6.40}$$

for all $t \in [t_0, t_f]$ and $n = 1, 2$, where $\mathfrak{m}_n(\cdot)$ is given as

$$\dot{\mathfrak{m}}_n(t) = -\begin{bmatrix} a_1 & 0 \\ 0 & a_2 \end{bmatrix} \mathfrak{m}_n(t) + \mathfrak{M}_n(t) \left[\begin{bmatrix} \frac{|b_1|^2}{r_1} & 0 \\ 0 & 0 \end{bmatrix} \mathfrak{m}_1(t) + \begin{bmatrix} 0 & 0 \\ 0 & \frac{|b_2|^2}{r_2} \end{bmatrix} \mathfrak{m}_2(t) \right], \quad (6.41)$$

for all $t \in [t_0, t_f]$ and $n = 1, 2$, with the boundary condition as $\mathfrak{m}_n(t_f) = 0$, and $x^*(\cdot)$, which represents the state trajectory subject to $u^*(\cdot)$, and is given as

$$x^*(t) = \Phi(t, 0)x(t_0) - \int_{t_0}^{t} \Phi(t, s) \left[\begin{bmatrix} \frac{|b_1|^2}{r_1} & 0 \\ 0 & 0 \end{bmatrix} \mathfrak{m}_1(s) + \begin{bmatrix} 0 & 0 \\ 0 & \frac{|b_2|^2}{r_2} \end{bmatrix} \mathfrak{m}_2(s) \right] ds,$$
$$(6.42)$$

with the initial condition $x^*(t_0) = x_0$ and where $\Phi(\cdot, s)$ is specified with the following differential equations

$$\dot{\Phi}(t, s) = \left[\begin{bmatrix} a_1 & 0 \\ 0 & a_2 \end{bmatrix} - \begin{bmatrix} \frac{|b_1|^2}{r_1} & 0 \\ 0 & 0 \end{bmatrix} \mathfrak{M}_1(t) - \begin{bmatrix} 0 & 0 \\ 0 & \frac{|b_2|^2}{r_2} \end{bmatrix} \mathfrak{M}_2(t) \right] \Phi(t, s), \quad (6.43)$$

such that $\Phi(s, s) = \mathbb{I}_{2 \times 2}$, for all s. □

6.2 Two-Person Zero-Sum Differential Games

In this part, we will study a class of two-person zero-sum differential games.

6.2.1 Formulation of Two-Person Zero-Sum Differential Games

Basically, in each of these games, two players can control the dynamics of a state system. We will formulate this class of differential games in the following.

Suppose that there are two players, and denote by $u_1(\cdot)$ and $u_2(\cdot)$ the controls of player 1 and player 2, respectively, over the time interval $[t, t_f]$, with $t \geq t_0$.

Consider a state system denoted by $x(\cdot)$ which is driven by the control strategies $u_1(\cdot)$ and $u_2(\cdot)$, such that

$$\dot{x}(t) = f(x(t), u_1(t), u_2(t)), \quad (6.44)$$

for all $t \in [t, t_f]$, with an initial value of $x(0) = x_0$, where $x \in \mathbb{R}^n$.

Define a payoff function $J(x, t, u_1, u_2)$ subject to $u_1(\cdot)$ and $u_2(\cdot)$ the controls of players, such that

$$J(x, t, u_1, u_2) \triangleq h(x(t_f)) + \int_t^{t_f} g(x(s), u_1(s), u_2(s))ds. \qquad (6.45)$$

Suppose that the cost functions of individual players are given as

$$J_n(x, t, u_1, u_2) \triangleq h_n(x(t_f)) + \int_t^{t_f} g_n(x(s), u_1(s), u_2(s))ds, \qquad (6.46)$$

with $n = 1, 2$, such that

$$g(x(s), u_1(s), u_2(s)) = g_1(x(s), u_1(s), u_2(s)) = -g_2(x(s), u_1(s), u_2(s)), \qquad (6.47a)$$

$$h(x(t_f)) = h_1(x(t_f)) = -h_2(x(t_f)). \qquad (6.47b)$$

Thus, for the underlying two-person zero-sum game system, one of the players tries to maximize the payoff function $J(x, t, u_1, u_2)$, while the other one tries to minimize this same performance function.

Denote by U_1 and U_2 the admissible control set for player 1 and player 2, respectively.

Further, define $\mathscr{U}_1([t, t_f])$ and $\mathscr{U}_2([t, t_f])$ as *the sets of control strategies* for player 1 and player 2, respectively, such that

$$\mathscr{U}_1([t, t_f]) \triangleq \{u_1(\cdot) : [t, t_f] \to U_1, \text{ s.t. } u_1(\cdot) \text{ is measurable}\}, \qquad (6.48a)$$

$$\mathscr{U}_2([t, t_f]) \triangleq \{u_2(\cdot) : [t, t_f] \to U_2, \text{ s.t. } u_2(\cdot) \text{ is measurable}\}. \qquad (6.48b)$$

It is called a *two-person zero-sum differential game* if the objectives of player 1 and player 2 are to maximize and minimize the payoff function (6.45), respectively.

For the underlying two-person zero-sum differential game, a pair of control strategies denoted by $u_1^*(\cdot)$ and $u_2^*(\cdot)$ is called *optimal* in case that $(u_1^*(\cdot), u_2^*(\cdot))$ is a *saddle point* for a common payoff function $J(x, t, u_1, u_2)$ for these two players specified in (6.45), say

$$J(x, t, u_1(\cdot), u_2^*(\cdot)) \leq J(x, t, u_1^*(\cdot), u_2^*(\cdot)) \leq J(x, t, u_1^*(\cdot), u_2(\cdot)), \qquad (6.49)$$

for all control strategies $u_1(\cdot)$ and $u_2(\cdot)$.

Notice that we can state that the saddle point $(u_1^*(\cdot), u_2^*(\cdot))$ is an equilibrium since by

$$J(x, t, u_1(\cdot), u_2^*(\cdot)) \leq J(x, t, u_1^*(\cdot), u_2^*(\cdot)),$$

player 1 may prefer staying at $u_1^*(\cdot)$, since he cannot benefit by deviating from this strategy; similarly, by

$$J(x, t, u_1^*(\cdot), u_2^*(\cdot)) \leq J(x, t, u_1^*(\cdot), u_2(\cdot)),$$

player 2 may prefer staying at $u_2^*(\cdot)$, since he cannot benefit by deviating from this strategy as well.

We will demonstrate the developed result with the following example originally designed by R. Isaacs.

Suppose that two players are at war against each other. Denote by $x_i(t)$ the resources for player i, with $i = 1, 2$.

It is considered that at each time, each player i will dedicate some fraction of his efforts to the direct attack against the other player, and the remaining to the attrition.

For analytical simplicity, the control sets of players are defined as

$$U_1 = U_2 = [0, 1]. \tag{6.50}$$

The control of players is specified as

- The control of player 1, denoted by $u_1(t) \in U_1$, represents the fraction of player 1's effort dedicated to the attrition, and $1 - u_1(t)$ is the fraction of player 1's effort dedicated to the attack.
- The control of player 2, denoted by $u_2(t) \in U_2$, represents the fraction of player 2's effort dedicated to the attrition, and $1 - u_2(t)$ is the fraction of player 2's effort dedicated to the attack.

The state process of the system subject to $(u_1(\cdot), u_2(\cdot))$ evolves as follows:

$$\dot{x}_1(t) = m_1 - c_1 u_2(t) x_2(t), \tag{6.51a}$$
$$\dot{x}_2(t) = m_2 - c_2 u_1(t) x_1(t), \tag{6.51b}$$

where m_i, with $i = 1, 2$, represents the production rate of materials for player i, and c_i denotes the effectiveness on player i from his opponent's weapon.

Also without loss of generality, suppose that

$$c_2 > c_1. \tag{6.52}$$

We also specify a payoff function J as

$$J(x, t_0, u_1(\cdot), u_2(\cdot)) \triangleq \int_{t_0}^{t_f} \Big[[1 - u_1(t)]x_1(t) - [1 - u_2(t)]x_2(t)\Big] dt. \tag{6.53}$$

6.2.2 Saddle Point of Two-Person Zero-Sum Differential Games

In Theorem 6.4 below, we will give a necessary condition for a saddle point strategy of a zero-sum differential game; before that, we define the Hamiltonian in the following:

$$\mathcal{H}(x(t), u_1(t), u_2(t), \lambda(t))$$
$$\triangleq g(x(t), u_1(t), u_2(t)) + [\lambda(t)]^\top f(x(t), u_1(t), u_2(t)), \qquad (6.54)$$

for each time $t \in [t_0, t_f]$, where λ represents the costate of the noncooperative differential game system.

Moreover we consider the following assumptions for this noncooperative differential game,

- $f(\cdot, u_1, u_2)$ is continuously differentiable on $\mathbb{R}^n$,
- $g(\cdot, u_1, u_2)$ is continuously differentiable on $\mathbb{R}^n$,
- $h(\cdot)$ is continuously differentiable on $\mathbb{R}^n$.

Theorem 6.4 *Suppose that u^* is a saddle point strategy of a zero-sum differential game, and x^* is the corresponding state trajectory subject to this saddle point strategy, say*

$$\dot{x}^*(t) = f(x^*(t), u_1^*(t), u_2^*(t)), \qquad (6.55)$$

for all $t \in [t_0, t_f]$, with an initial value of $x(t_0) = x_0$.

Then there exists a costate trajectory for each player $n \in \mathcal{N}$, denoted by $\lambda_n^*(\cdot)$: $[t_0, t_f] \to \mathbb{R}^n$, such that

$$\mathcal{H}(x^*(t), u_1^*(t), u_2(t), \lambda^*(t))$$
$$\leq \mathcal{H}(x^*(t), u_1^*(t), u_2^*(t), \lambda^*(t)) \leq \mathcal{H}(x^*(t), u_1(t), u_2^*(t), \lambda^*(t)), \qquad (6.56a)$$
$$\dot{\lambda}^*(t) = -\frac{\partial \mathcal{H}(x^*(t), u_1^*(t), u_2^*(t), \lambda_n^*(t))}{\partial x}$$
$$= -\frac{\partial g(x^*(t), u_1^*(t), u_2^*(t))}{\partial x} - [\lambda^*(t)]^\top \frac{\partial f(x^*(t), u_1^*(t) u_2^*(t))}{\partial x}, \qquad (6.56b)$$

for all $t \in [t_0, t_f]$, with the boundary condition

$$\lambda(t_f) = \frac{\partial h(x^*(t_f))}{\partial x}. \qquad (6.57)$$

Moreover, we have

$$\min_{u_1 \in U_1} \mathcal{H}(x^*(t), u_1, u_2^*(t), \lambda^*(t)) = \mathcal{H}(x^*(t), u_1^*(t), u_2^*(t), \lambda^*(t))$$
$$= \max_{u_2 \in U_2} \mathcal{H}(x^*(t), u_1^*(t), u_2, \lambda^*(t)), \qquad (6.58)$$

for all $t \in [t_0, t_f]$.

Proof The conclusion of (6.56) is obtained by applying Theorem 6.2 with $N \equiv |\mathcal{N}| = 2$, $g_n(\cdot)$ and $h_n(\cdot)$ specified in (6.47), respectively, and the Hamiltonian given as

$$\mathcal{H}(x(t), u_1(t), u_2(t), \lambda(t)) = \mathcal{H}_1(x(t), u_1(t), u_2(t), \lambda(t))$$
$$= -\mathcal{H}_2(x(t), u_1(t), u_2(t), \lambda(t)). \tag{6.59}$$

Considering a fixed strategy u_2^*, the state equation becomes

$$\dot{x}(t) = f(x(t), u_1(t), u_2^*(t)), \tag{6.60}$$

for all $t \in [t, t_f]$, and the objective of the cost function degenerates into the following:

$$J(x, t, u_1, u_2^*) = h(x(t_f)) + \int_t^{t_f} g(x(s), u_1(s), u_2^*(s)) ds. \tag{6.61}$$

Also by the definition of the saddle point of $u^* \equiv (u_1^*, u_2^*)$, we have

$$\min_{u_1 \in \mathcal{U}_1[t, t_f]} J(x, t, u_1, u_2^*) = J(x, t, u_1^*, u_2^*), \tag{6.62}$$

say u_1^* is the optimal control strategy of player 1 for the controlled process defined above. And x^* is the associated optimal state trajectory. Then by applying Theorem 4.1 in Chap. 4, the first equality of (6.58) holds.

Following the same analysis, the second equality of (6.58) holds as well. $\qquad \square$

Example 6.4 A necessary control of the saddle point of the two-person zero-sum differential game formulated in Sect. 6.2.1.

Solution. Firstly for the example designed in Sect. 6.2.1, the Hamiltonian $\mathcal{H}(x, u_1, u_2, \lambda)$, with $\lambda \equiv (\lambda_1, \lambda_2)$, is specified as follows:

$$\mathcal{H}(x(t), u_1(t), u_2(t), \lambda(t))$$
$$\triangleq [\lambda(t)]^\top f(x(t), u_1(t), u_2(t)) + g(x(t), u_1(t), u_2(t))$$
$$= [m_1 - c_1 u_2(t) x_2(t)] \lambda_1(t) + [m_2 - c_2(t) u_1(t) x_1(t)] \lambda_2(t)$$
$$+ [1 - u_1(t)] x_1(t) - [1 - u_2(t)] x_2(t)$$
$$= [-x_1(t) - c_2 x_1(t) \lambda_2(t)] u_1(t) + [x_2(t) - c_1 x_2(t) \lambda_1(t)] u_2(t)$$
$$+ [m_1 \lambda_1(t) + m_2 \lambda_2(t) + x_1(t) - x_2(t)]. \tag{6.63}$$

Thus by Theorem 6.4, the costate equation is given as

$$\dot{\lambda}_1^*(t) = -\frac{\partial \mathcal{H}(x^*(t), u_1^*(t), u_2^*(t), \lambda^*(t))}{\partial x_1} = u_1(t) - 1 + \lambda_2^*(t) c_2 u_1(t), \tag{6.64a}$$

$$\dot{\lambda}_2^*(t) = -\frac{\partial \mathcal{H}(x^*(t), u_1^*(t), u_2^*(t), \lambda^*(t))}{\partial x_2} = 1 - u_2(t) + \lambda_1^*(t) c_1 u_2(t), \tag{6.64b}$$

with the boundary condition

$$\lambda_1^*(t_f) = \lambda_2^*(t_f) = 0, \tag{6.65}$$

due to $h(x^*(t_f)) = 0$.   □

Example 6.5 Consider a zero-sum differential game with the state process given as

$$\dot{x}_1(t) = x_2(t), \tag{6.66a}$$
$$\dot{x}_2(t) = u_1(t) - u_2(t), \tag{6.66b}$$

for all $t \in [t_0, t_f]$ with $x_1(t_0) = x_2(t_0) = 0$, $|u_1(t)| \leq 3$, and $|u_2(t)| \leq 1$.
The cost function is specified as

$$J(x_0, t_0, u_1, u_2) \triangleq \int_{t_0}^{t_f} [-x_2(t)]dt. \tag{6.67}$$

Implement the saddle point u^* of the above zero-sum differential game.
Solution. As usual, firstly define the Hamiltonian $\mathcal{H}$ such that

$$\mathcal{H}(x(t), u_1(t), u_2(t), \lambda(t)) = -x_2(t) + [\lambda(t)]^{\top} \begin{bmatrix} x_2(t) \\ u_1(t) - u_2(t) \end{bmatrix}. \tag{6.68}$$

Then we have

$$\dot{\lambda}_1^*(t) = -\frac{\partial \mathcal{H}(x(t), u_1^*(t), u_2^*(t), \lambda^*(t))}{\partial x_1} = 0, \tag{6.69a}$$

$$\dot{\lambda}_2^*(t) = -\frac{\partial \mathcal{H}(x(t), u_1^*(t), u_2^*(t), \lambda^*(t))}{\partial x_2} = 1 - \lambda_1^*(t), \tag{6.69b}$$

with the boundary condition $\lambda^*(t_f) = 0$. Thus we get

$$\lambda_1^*(t) = 0, \tag{6.70a}$$
$$\lambda_2^*(t) = t - t_f, \tag{6.70b}$$

which implies that $\lambda_2^*(t) < 0$ for all $t \in [t_0, t_f)$.
    Hence, we can get $u_1^*(t) = 3$ and $u_2^*(t) = 1$, by which the associated state trajectories are given as

$$x_1^*(t) = t^2, \tag{6.71a}$$
$$x_2^*(t) = 2t, \tag{6.71b}$$

for all $t \in [t_0, t_f]$.

Consequently, we obtain that at the saddle point u^*, the cost function is given as

$$J(x_0, t_0, u_1^*, u_2^*) = \int_{t_0}^{t_f} -x_2^*(t)dt = -x_1^*(t_f) = -t_f^2. \tag{6.72}$$

$\square$

6.2.3 Implementation of Saddle Point of Two-Person Zero-Sum Differential Games via Dynamic Programming

Besides the implementation of the underlying differential games by applying Pontryagin's minimum principle in Sect. 6.2.2, in this section, we will analyze how to implement them by applying the dynamic programming method.

Firstly, define some notations in the following.

Definition 6.3 (*Control Strategies of Individual Players*)

- A mapping $\Phi : \mathscr{U}_2([t, t_f]) \to \mathscr{U}_1([t, t_f])$ is called a strategy of player 1 if the following holds:

$$\Phi(u_2(\tau)) = \Phi(\widehat{u}_2(\tau)), \ \forall \tau \in [t, s], \tag{6.73}$$

in case

$$u_2(\tau) = \widehat{u}_2(\tau), \ \forall \tau \in [t, s], \tag{6.74}$$

for all instants $s \in [t, t_f]$.
Denote by $\mathscr{U}_1([t, t_f])$ *the set of strategies* of player 1.
- A mapping $\Psi : \mathscr{U}_1([t, t_f]) \to \mathscr{U}_2([t, t_f])$ is called a strategy of player 2 if the following holds:

$$\Psi(u_1(\tau)) = \Psi(\widehat{u}_1(\tau)), \ \forall \tau \in [t, s], \tag{6.75}$$

in case

$$u_1(\tau) = \widehat{u}_1(\tau), \ \forall \tau \in [t, s], \tag{6.76}$$

for all instants $s \in [t, t_f]$.
Denote by $\mathscr{U}_2([t, t_f])$ *the set of strategies* of player 2. $\square$

Also, the value functions for the differential games are introduced in the following.

Definition 6.4 (*Value Functions for the Differential Games*)

- Denote by $\bar{v}(x, t)$ the upper value function such that

$$\bar{v}(x, t) \triangleq \inf_{\Psi \in \mathcal{U}_2([t, t_f])} \sup_{u_1(\cdot) \in \mathcal{U}_1([t, t_f])} J(x, t, u_1(\cdot), \Psi(u_1(\cdot))). \qquad (6.77)$$

- Denote by $\underline{v}(x, t)$ the lower value function such that

$$\underline{v}(x, t) \triangleq \sup_{\Phi \in \mathcal{U}_1([t, t_f])} \inf_{u_2(\cdot) \in \mathcal{U}_2([t, t_f])} J(x, t, u_2(\cdot), \Phi(u_2(\cdot))). \qquad (6.78)$$

□

Theorem 6.5 *Suppose that $\bar{v}(x, t)$ and $\underline{v}(x, t)$ are continuously differentiable; then*

- $\bar{v}(x, t)$ *solves the following so-called upper Isaacs's equation*

$$\frac{\partial \bar{v}(x, t)}{\partial t} + \min_{u_2 \in U_2} \max_{u_1 \in U_1} \left\{ \left[\frac{\partial \bar{v}(x, t)}{\partial x} \right]^{\top} f(x, u_1, u_2) + g(x, u_1, u_2) \right\} = 0, \quad (6.79)$$

with the boundary condition $\bar{v}(x, t_f) = h(x)$, and

- $\underline{v}(x, t)$ *solves the following so-called lower Isaacs's equation*

$$\frac{\partial \underline{v}(x, t)}{\partial t} + \max_{u_1 \in U_1} \min_{u_2 \in U_2} \left\{ \left[\frac{\partial \underline{v}(x, t)}{\partial x} \right]^{\top} f(x, u_1, u_2) + g(x, u_1, u_2) \right\} = 0, \quad (6.80)$$

with the boundary condition $\underline{v}(x, t_f) = h(x)$.

Proof • *Verification of (6.79).*

Firstly, we consider the following Isaacs–Bellman optimality principle

$$\bar{v}(x, t) = \inf_{u_2(\cdot) \in \mathcal{U}_2([t, t+dt])} \sup_{u_1(\cdot) \in \mathcal{U}_1([t, t+dt])}$$
$$\left\{ \int_t^{t+dt} g(x(s), u_1(s), u_2(s)) ds + \bar{v}(x(t+dt), t+dt) \right\}. \qquad (6.81)$$

Considering the Taylor series expansion for $u(x(t + dt), t + dt)$, we can obtain

$$\bar{v}(x(t+dt), t+dt) = \bar{v}(x, t) + \frac{\partial \bar{v}(x, t)}{\partial x} dx(t) + \frac{\partial \bar{v}(x, t)}{\partial t} dt + \mathcal{O}(dx, dt).$$
$$(6.82)$$

Also by the state equation specified in (6.48), we get

$$dx(t) = f(x(t), u_1(t), u_2(t)) dt. \qquad (6.83)$$

As a consequence, by the above analysis, for infinitesimal-valued dx and dt, we can get

$$\bar{v}(x,t) = \inf_{u_2(\cdot)\in\mathscr{U}_2([t,t+dt])} \sup_{u_1(\cdot)\in\mathscr{U}_1([t,t+dt])}$$

$$\left\{ g(x(t), u_1(t), u_2(t))dt + \bar{v}(x,t) + \left[\frac{\partial\bar{v}(x,t)}{\partial x}\right]^{\mathsf{T}} dx(t) + \frac{\partial\bar{v}(x,t)}{\partial t}dt + \mathscr{O}(dx,dt) \right\}$$

$$= \inf_{u_2(\cdot)\in\mathscr{U}_2([t,t+dt])} \sup_{u_1(\cdot)\in\mathscr{U}_1([t,t+dt])}$$

$$\left\{ g(x(t), u_1(t), u_2(t))dt + \left[\frac{\partial\bar{v}(x,t)}{\partial x}\right]^{\mathsf{T}} dx(t) + \mathscr{O}(dx,dt) \right\}$$

$$+ \bar{v}(x,t) + \frac{\partial\bar{v}(x,t)}{\partial t}dt; \tag{6.84}$$

then as dt goes to zero, by the above analysis, we have

$$0 = \min_{u_2\in U_2} \max_{u_1\in U_1} \left\{ g(x, u_1, u_2) + \left[\frac{\partial\bar{v}(x,t)}{\partial x}\right]^{\mathsf{T}} \lim_{dt\to 0} \frac{dx(t)}{dt} \right\}$$

$$+ \frac{\partial\bar{v}(x,t)}{\partial t} + \lim_{dt\to 0} \frac{\mathscr{O}(dx,dt)}{dt}$$

$$= \frac{\partial\bar{v}(x,t)}{\partial t} + \min_{u_2\in U_2} \max_{u_1\in U_1} \left\{ \left[\frac{\partial\bar{v}(x,t)}{\partial x}\right]^{\mathsf{T}} f(x, u_1, u_2) + g(x, u_1, u_2) \right\}, \tag{6.85}$$

which is the conclusion of (6.79).

- **Verification of (6.80).**
  Similarly, we also consider the following Isaacs–Bellman optimality principle

$$\underline{v}(x,t) = \sup_{u_1(\cdot)\in\mathscr{U}_1([t,t+dt])} \inf_{u_2(\cdot)\in\mathscr{U}_2([t,t+dt])}$$

$$\left\{ \int_t^{t+dt} g(x(s), u_1(s), u_2(s))ds + \underline{v}(x(t+dt), t+dt) \right\}. \tag{6.86}$$

Considering the Taylor series expansion for $\underline{v}(x(t+dt), t+dt)$, we can obtain

$$\underline{v}(x(t+dt), t+dt) = \underline{v}(x,t) + \left[\frac{\partial\underline{v}(x,t)}{\partial x}\right]^{\mathsf{T}} dx(t) + \frac{\partial\underline{v}(x,t)}{\partial t}dt + \mathscr{O}(dx,dt). \tag{6.87}$$

Also by the state equation specified in (6.48), we get

$$dx(t) = f(x(t), u_1(t), u_2(t))dt. \tag{6.88}$$

As a consequence, by the above analysis, for infinitesimal-valued dx and dt, we can get

$$\underline{v}(x, t) = \sup_{u_1(\cdot) \in \mathcal{U}_1([t, t+dt])} \inf_{u_2(\cdot) \in \mathcal{U}_2([t, t+dt])}$$

$$\left\{ g(x(t), u_1(t), u_2(t))dt + \underline{v}(x, t) + \left[\frac{\partial \underline{v}(x, t)}{\partial x} \right]^{\top} dx(t) + \frac{\partial \underline{v}(x, t)}{\partial t} dt + \mathcal{O}(dx, dt) \right\}$$

$$= \sup_{u_1(\cdot) \in \mathcal{U}_1([t, t+dt])} \inf_{u_2(\cdot) \in \mathcal{U}_2([t, t+dt])}$$

$$\left\{ g(x(t), u_1(t), u_2(t))dt + \left[\frac{\partial \underline{v}(x, t)}{\partial x} \right]^{\top} dx(t) + \mathcal{O}(dx, dt) \right\}$$

$$+ \underline{v}(x, t) + \frac{\partial \underline{v}(x, t)}{\partial t} dt; \tag{6.89}$$

then as dt goes to zero, by the above analysis, we have

$$0 = \max_{u_1 \in U_1} \min_{u_2 \in U_2} \left\{ g(x, u_1, u_2) + \left[\frac{\partial \underline{v}(x, t)}{\partial x} \right]^{\top} \lim_{dt \to 0} \frac{dx(t)}{dt} \right\}$$

$$+ \frac{\partial \underline{v}(x, t)}{\partial t} + \lim_{dt \to 0} \frac{\mathcal{O}(dx, dt)}{dt}$$

$$= \frac{\partial \underline{v}(x, t)}{\partial t} + \max_{u_1 \in U_1} \min_{u_2 \in U_2} \left\{ \left[\frac{\partial \underline{v}(x, t)}{\partial x} \right]^{\top} f(x, u_1, u_2) + g(x, u_1, u_2) \right\}, \tag{6.90}$$

which is the conclusion of (6.80). □

Notice that the upper Isaacs's equation given in (6.79) and the lower Isaacs's equation given in (6.80) developed in Theorem 6.5 are the counterpart of the HJB equation in the framework of two-person zero-sum differential games.

Define a notation of Hamiltonian for the underlying two-person zero-sum differential game, $\mathcal{H}(x, u_1, u_2, \lambda)$, such that

$$\mathcal{H}(x, u_1, u_2, \lambda) \triangleq g(x, u_1, u_2) + \lambda^{\top} f(x, u_1, u_2), \tag{6.91}$$

for any state x, p, player 1's control $u_1 \in U_1$, and player 2's control $u_2 \in U_2$.

- The upper Isaacs's equation (6.79) is rewritten in the following compact form

$$\frac{\partial \bar{v}(x, t)}{\partial t} + \mathcal{H}^+ \left(x, \frac{\partial \bar{v}(x, t)}{\partial x} \right) = 0, \tag{6.92}$$

with *the upper Hamiltonian* $\mathcal{H}^+(x, \lambda)$ defined as

$$\mathcal{H}^+(x, \lambda) \triangleq \min_{u_2 \in U_2} \max_{u_1 \in U_1} \mathcal{H}(x, u_1, u_2, \lambda). \tag{6.93}$$

- The lower Isaacs's equation (6.80) is rewritten in the following compact form

$$\frac{\partial \underline{v}(x,t)}{\partial t} + \mathscr{H}^-\left(x, \frac{\partial \underline{v}(x,t)}{\partial x}\right) = 0, \qquad (6.94)$$

with *the lower Hamiltonian* $\mathscr{H}^-(x, \lambda)$ defined as

$$\mathscr{H}^-(x, \lambda) \triangleq \max_{u_1 \in U_1} \min_{u_2 \in U_2} \mathscr{H}(x, u_1, u_2, \lambda). \qquad (6.95)$$

It can be shown that

$$\mathscr{H}^-(x, \lambda) \le \mathscr{H}^+(x, \lambda), \qquad (6.96)$$

for all x and p; thus in general, the upper Isaacs's equation (6.79) and the lower Isaacs's equation (6.80) are different from each other, and hence the upper value function $\bar{v}(x, t)$ and the lower value function $\underline{v}(x, t)$ are not equal to each other.

Nevertheless, in case that

$$\mathscr{H}^-(x, \lambda) = \mathscr{H}^+(x, \lambda), \qquad (6.97)$$

for all x and p, then we claim that the game satisfies *the minimax condition*, or it may be called *the Isaacs's condition*.

Suppose that (6.97) holds; then we can claim that the game has *value*, and the Isaacs's equations (6.79) and (6.80) can be solved by applying the dynamic programming method. And the optimal controls for individual players can be specified as well.

For the purpose of demonstration, in the following example, we will study the saddle point for the example introduced in Sect. 6.2.1.

Example 6.6 Specify the two-person zero-sum differential game formulated in Sect. 6.2.1 by applying the dynamic programming method.

Solution. Firstly, by the Hamiltonian for the designed game specified in (6.63), it is straightforward to verify that the minimax condition of (6.97) holds for the underlying game system, say

$$\mathscr{H}(x, \lambda) = \mathscr{H}^-(x, \lambda) = \mathscr{H}^+(x, \lambda).$$

Thus we have $\bar{v}(x, t) = \underline{v}(x, t)$ for all x and t, and it satisfies

$$\frac{\partial \underline{v}(x,t)}{\partial t} + \mathscr{H}\left(x, \frac{\partial \underline{v}(x,t)}{\partial x}\right) = 0. \qquad (6.98)$$

Considering $\lambda \equiv [\lambda_1; \lambda_2] = \dfrac{\partial \underline{v}(x,t)}{\partial x} \equiv \left[\dfrac{\partial \underline{v}(x,t)}{\partial x_1}, \dfrac{\partial \underline{v}(x,t)}{\partial x_2}\right]^\top$, we have

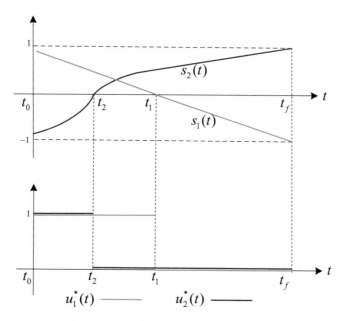

Fig. 6.1 An illustration of evolutions of $u_1^*(\cdot)$ and $u_2^*(\cdot)$ with respect to $s_1(\cdot)$ and $s_2(\cdot)$, respectively

$$\mathscr{H}(x, u_1, u_2, \lambda) = \left[-1 - c_2 \frac{\partial \underline{v}(x, t)}{\partial x_2}\right] x_1 u_1 + \left[1 - c_1 \frac{\partial \underline{v}(x, t)}{\partial x_1}\right] x_2 u_2 + \eta,$$

$$(6.99)$$

with $\eta \equiv m_1 \dfrac{\partial \underline{v}(x, t)}{\partial x_1} + m_2 \dfrac{\partial \underline{v}(x, t)}{\partial x_2} + x_1 - x_2.$

As a consequence, the optimal control solution is specified as

$$u_1^*(t) = \begin{cases} 1, & \text{in case } s_1(t) \equiv -1 - c_2 \frac{\partial \underline{v}(x,t)}{\partial x_2} \geq 0 \\ 0, & \text{otherwise} \end{cases},$$

$$(6.100a)$$

$$u_2^*(t) = \begin{cases} 0, & \text{in case } s_2(t) \equiv 1 - c_1 \frac{\partial \underline{v}(x,t)}{\partial x_1} \geq 0 \\ 1, & \text{otherwise} \end{cases},$$

$$(6.100b)$$

for all $t \in [t_0, t_f]$.

See an illustration in Fig. 6.1 for the dynamics of the optimal control $u_1^*(\cdot)$ and $u_2^*(\cdot)$ with respect to the evolutions of $s_1(\cdot)$ and $s_2(\cdot)$, respectively. □

As proposed in Example 6.6, the optimal feedback control strategies for players are designed with respect to the evolution of the value $\underline{v}(x, t)$. However, it is challenging to solve Isaacs's equation for $\underline{v}(x, t)$.

By (6.100), the optimal control is determined in case that the dynamics of processes $s_1(t)$ and $s_2(t)$ is specified.

Next, continue to implement the example designed earlier in this chapter.

Example 6.7 Implement the equilibrium strategy for the two-person zero-sum differential game formulated in Sect. 6.2.1.

Solution. By (6.100) and (6.64) developed in Examples 6.6 and 6.4, respectively, the processes of $s_1(t)$ and $s_2(t)$ are specified as follows:

$$\dot{s}_1(t) = -c_2\dot{\lambda}_2^*(t) = c_2[u_2(t) - 1 - \lambda_1^*(t)c_1u_2(t)] = c_2[-1 + u_2(t)s_2(t)],$$
$$\tag{6.101a}$$

$$\dot{s}_2(t) = -c_1\dot{\lambda}_1^*(t) = c_1[1 - u_1(t) - \lambda_2^*(t)c_2u_1(t)] = c_1[1 + u_1(t)s_1(t)], \tag{6.101b}$$

with the boundary condition

$$\begin{bmatrix} s_1(t_f) \\ s_2(t_f) \end{bmatrix} = \begin{bmatrix} -1 \\ 1 \end{bmatrix}. \tag{6.102}$$

Next, discuss the dynamics of $s_1(t)$ and $s_2(t)$ backward from the final time t_f.

As given, $s_1(t_f) = -1 < 0$ and $s_2(t_f) = 1 > 0$, and due to the smoothness property of $s(\cdot)$, we can claim that

$$s_1(t) < 0 \text{ and } s_2(t) > 0, \tag{6.103}$$

with t close to the final time t_f.

Thus by (6.100), we have

$$u_1^*(t) = u_2^*(t) = 0, \tag{6.104}$$

with t close to the final time t_f, i.e., the optimal controls for both players are to attack against each other at these instants near the final time t_f.

Suppose that t_1, with $t_1 < t_f$, represents the first instant, backward from the final time t_f, at which one of the players switches their strategies. That is to say, the players' strategies switch to $u_1^* = u_2^* = 0$ at t_1, and

$$u_1^*(t) = u_2^*(t) = 0, \tag{6.105}$$

for all $t \in [t_1, t_f]$.

Thus by (6.101), we get

$$\dot{s}_1(t) = -c_2, \tag{6.106a}$$
$$\dot{s}_2(t) = c_1, \tag{6.106b}$$

for all $t \in [t_1, t_f]$, with the boundary conditions $s_1(t_f) = -1$ and $s_2(t_f) = 1$.

Hence, we have

$$s_1(t) = -1 + c_2[t_f - t], \tag{6.107a}$$
$$s_2(t) = 1 + c_1[t - t_f], \tag{6.107b}$$

for all $t \in [t_1, t_f]$; then s_1 reaches zero at time $t_f - \frac{1}{c_2}$, and s_2 reaches zero at time $t_f - \frac{1}{c_1}$.

Since as assumed in (6.52), it is considered that $c_2 > c_1$, then $t_f - \frac{1}{c_1} < t_f - \frac{1}{c_2}$. Thus by the definition of t_1, we obtain

$$t_1 = t_f - \frac{1}{c_2}. \tag{6.108}$$

Similarly, define t_2 to be the 2nd time backward, from the final time t_f, at which one of the players switches. Hence $u_1^*(t) = 1$ and $u_2^*(t) = 0$ for all $t \in [t_2, t_1]$, and then

$$\dot{s}_1(t) = -c_2, \tag{6.109}$$
$$\dot{s}_2(t) = c_1[1 + s_1(t)], \tag{6.110}$$

for all $t \in [t_2, t_1]$, with $s_1(t_1) = 0$ and $s_2(t_1) = 1 - \frac{c_1}{c_2}$.

$$s_1(t) = -1 + c_2[t_f - t], \tag{6.111a}$$
$$s_2(t) = 1 - \frac{c_1}{2c_2} - \frac{c_1 c_2}{2}[t - t_f]^2, \tag{6.111b}$$

for all $t \in [t_2, t_1]$.

However, we have $s_1(t) > 0$ for all $t \in [t_2, t_1]$, and $s_2(t_2) = 0$ with

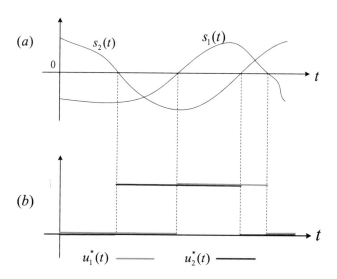

Fig. 6.2 An illustration of $u_1^*(\cdot)$ and $u_2^*(\cdot)$ with respect to $s_1(\cdot)$ and $s_2(\cdot)$, respectively

$$t_2 = t_f - \frac{1}{c_2}\sqrt{\frac{2c_2}{c_1} - 1}. \tag{6.112}$$

We can get that s_1 and s_2 will not change the sign any longer, if it can solve (6.101) over the interval $[t_0, t_2]$ subject to $u_1^*(t) = u_2^*(t) = 1$.

See an illustration in Fig. 6.2 for the dynamics of the optimal control $u_1^*(\cdot)$ and $u_2^*(\cdot)$ with respect to the evolutions of $s_1(\cdot)$ and $s_2(\cdot)$, respectively.   □

6.2.4 Linear-Quadratic Two-Person Zero-Sum Differential Games

Definition 6.5 (*Linear-Quadratic Two-Person Zero-Sum Differential Games*) A game is called a linear-quadratic two-person zero-sum differential game if $f(x, u_1, u_2, t)$ in the state equation (6.44), and $g(x, u_1, u_2, t)$ and $h(x)$ in the functions (6.47a) and (6.47b) satisfy the following, respectively:

$$f(x, u_1, u_2, t) = A(t)x + B_1(t)u_1 + B_2(t)u_2, \tag{6.113a}$$

$$g(x, u_1, u_2, t) = \frac{1}{2}\left[x^\top Q(t)x + [u_1]^\top u_1 - [u_2]^\top u_2\right], \tag{6.113b}$$

$$h(x) = \frac{1}{2}x^\top Q^f x, \tag{6.113c}$$

for all $t \in [t_0, t_f]$, where $A(t)$, $B_1(t)$, $B_2(t)$, $Q(t)$, and Q^f are matrices with appropriate dimensions, respectively.

It is further supposed that $Q(t)$ and Q^f are symmetric and positive definite for all $t \in [t_0, t_f]$.   □

Theorem 6.6 shows the existence and uniqueness of the saddle point solution for the linear-quadratic two-person zero-sum differential game specified in Definition 6.5 above. Before that, in Lemma 6.1 below, firstly give a sufficient condition for the strict concavity on u_2 of the underlying two-person zero-sum game.

Lemma 6.1 (Sufficient Condition of the Strict Concavity on u_2 in Two-Person Linear-Quadratic Zero-Sum Differential Game) *The performance cost function $J(u_1, u_2)$ of a linear-quadratic two-person zero-sum differential game is strictly concave on, u_2, the control of player 2, for all given u_1, if and only if there exists a unique bounded solution to the following Riccati equation:*

$$\dot{S}(t) = -Q(t) - [A(t)]^\top S(t) - S(t)A(t) - S(t)B_2(t)[B_2(t)]^\top S(t), \tag{6.114}$$

for all $t \in [t_0, t_f]$, with the the boundary condition $S(t_f) = Q^f$.

Proof By applying the same method used in the proof of Lemma 7.1, the proof of the strict concavity of $J(u_1, u_2)$ given in (6.45) is is equivalent to the existence of a unique solution of the following optimal control problems:

$$\min_{\widehat{u}(\cdot)} \left\{ -[\widehat{x}(t_f)]^\top Q^f \widehat{x}(t_f) - \int_{t_0}^{t_f} \left[[\widehat{x}(t)]^\top Q(t)\widehat{x}(t) - [\widehat{u}(t)]^\top \widehat{u}(t) \right] dt \right\}, \quad (6.115)$$

where $\widehat{x}(\cdot)$ satisfies the following state equation

$$\dot{\widehat{x}}(t) = A(t)\widehat{x}(t) + B_2(t)\widehat{u}, \quad (6.116)$$

with the boundary condition $\widehat{x}(t_0) = x_0$. □

Further define a notation $\mathfrak{M}_n(\cdot)$, such that it is a solution to the following Riccati differential equations:

$$\dot{\mathfrak{M}}(t) = -\mathfrak{M}(t)A(t) - [A(t)]^\top \mathfrak{M}(t) - Q(t)$$
$$+ \mathfrak{M}(t) \left[B_1(t)[B_1(t)]^\top - B_2(t)[B_2(t)]^\top \right] \mathfrak{M}(t), \quad (6.117)$$

with the boundary condition $\mathfrak{M}(t_f) = Q^f$.

Theorem 6.6 (Uniqueness and Existence of Saddle Point of Linear-Quadratic Two-Person Zero-Sum Differential Game) *Consider a linear-quadratic two-person zero-sum differential game such that the matrices given in (7.262a) are invertible; then the game system has a unique saddle point as specified below*

$$u_n^*(t) = [-1]^n [B_n(t)]^\top \left[\mathfrak{M}(t)x^*(t) + \mathfrak{m}(t) \right], \quad (6.118)$$

for all $t \in [t_0, t_f]$ and $n = 1, 2$, where $\mathfrak{m}(\cdot)$ is specified as

$$\dot{\mathfrak{m}}(t) = -[A(t)]^\top \mathfrak{m}(t) + [\mathfrak{M}(t)]^\top \left[B_1(t)[B_1(t)]^\top - B_2(t)[B_2(t)]^\top \right] \mathfrak{m}(t), \quad (6.119)$$

for all $t \in [t_0, t_f]$, with the boundary condition as $\mathfrak{m}(t_f) = 0$, and $x^*(\cdot)$, which represents the state trajectory subject to $u^*(\cdot)$, and is given as

$$x^*(t) = \Phi(t, 0)x(t_0)$$
$$- \int_{t_0}^t \Phi(t, s) \left[B_1(s)[B_1(s)]^\top - B_2(s)[B_2(s)]^\top \right] \mathfrak{m}(s)ds, \quad (6.120)$$

with the initial condition $x^*(t_0) = x_0$ and where $\Phi(\cdot, s)$ is specified with the following differential equations:

$$\dot{\Phi}(t, s) = \left[A(t) - \left[B_1(s)[B_1(s)]^\top - B_2(s)[B_2(s)]^\top \right] \mathfrak{M}(t) \right] \Phi(t, s), \quad (6.121)$$

such that $\Phi(s, s) = \mathbb{I}$, *for all* s.   □

Notice that the result stated in Theorem 6.6 is a special case developed in Theorem 6.14, Dynamic Noncooperative Game Theory by Tamer Basar and Geert Olsder, 1995, where it shows the uniqueness and existence of the saddle point of the affine-quadratic two-person zero-sum differential game.

6.3 Summary

In this chapter, noncooperative differential games are introduced, including affine-quadratic differential games, and some properties of the NE for these games are developed. A class of zero-sum differential games is also formulated and the saddle point of two-person zero-sum differential games is studied.

6.4 Exercises

Exercise 6.1 Consider a differential game with the state system given as

$$\dot{x}_1(t) = x_1(t) + u_1(t), \tag{6.122a}$$
$$\dot{x}_2(t) = -x_2(t) + 2u_2(t), \tag{6.122b}$$

with $t_0 = 0$, $t_f = 2$, the boundary condition as $x(t_0) = x_0 = \begin{bmatrix} 2 \\ 2 \end{bmatrix}$, and a given performance cost function as

$$J_n(x_0, t_0, u) = \frac{1}{2}[x_n(t_f)]^2 + \int_{t_0}^{t_f} \left[\frac{1}{2}[x_n(t)]^2 + \frac{n}{2}[u_n(t)]^2 \right] dt, \tag{6.123}$$

with $r_n > 0$, for $n = 1, 2$.

Implement the NE strategy $u^* \equiv (u_1^*, u_2^*)$ for the differential game system, and illustrate the trajectories of individual strategies with respect to time t.

Exercise 6.2 Consider a two-person zero-sum differential game with the state system given as

$$\dot{x}_1(t) = x_2(t), \tag{6.124a}$$
$$\dot{x}_2(t) = u_1(t) - u_2(t), \tag{6.124b}$$

for all $t \in [t_0, t_f]$ with $x(t_0) = x_0 = \begin{bmatrix} 2 \\ 0 \end{bmatrix}$, $|u_1(t)| \leq 2$, and $|u_2(t)| \leq 1$.

The cost function is specified as

$$J(x_0, t_0, u_1, u_2) \triangleq \int_{t_0}^{t_f} [-x_1(t)]dt. \tag{6.125}$$

Specify the saddle point u^* of the above zero-sum differential game.

Chapter 7
Discrete-Time Optimal Control Problems

In earlier parts, we have studied the optimal control for continuous-time state systems which are specified with differential equations, such that certain performance costs in the forms of integrals are minimized. This chapter will introduce optimal control problems for discrete-time systems which are characterized by difference equations, and the objective is to minimize some performance costs which are in the forms of the summations of functions.

More specially, this chapter is organized as follows. In Sect. 7.1, the necessary condition for the optimization of discrete-time problems is developed based upon the variational method. Based upon the results given in Sect. 7.1, the optimal control for discrete-time problems in Sect. 7.2 is studied. Linear-quadratic regulation and tracking problems in discrete-time cases are studied in Sects. 7.3 and 7.4, respectively. In Sect. 7.5, Pontryagin's minimum principle for discrete-time optimal control problems is developed. For the purpose of comparison, in Sect. 7.6, the optimal control solutions of discrete-time problems are developed by applying the dynamic programming method. In Sects. 7.7 and 7.8, noncooperative dynamic games and two-person zero-sum dynamic games in discrete-time case are studied, respectively. A brief summary of this chapter is given in Sect. 7.9. Finally, in Sect. 7.10, exercises for the readers to verify the results introduced in this chapter are given.

7.1 Variational Calculus for Discrete-Time Systems

In this section, the necessary conditions for the optimization of performance costs which are in the form of summations are given, such that

$$J(x) \triangleq \sum_{k=k_0}^{k_f-1} g(x(k), x(k+1), k), \tag{7.1}$$

with the initial state at k_0 given as $x(k_0) = x_0$, where $x(k)$ denotes the state at discrete time k.

Firstly we study the extreme of the performance costs defined in (7.1) such that the final time and state are fixed with $x(t_f) = x_f$ in Sect. 7.1.1.

7.1.1 Optimum of Performance Functions with Fixed Final Time and Fixed Final Value

For the extremization (maximization or minimization) of performance functions, analogous to the case of continuous-time systems studied in Chap. 2, we apply the fundamental theorem of the calculus of variations which states that the variation is equal to zero.

Denote by x^* the state such that the performance function specified in (7.1) reaches an extrema. We denote

$$J(x^*) = \sum_{k=k_0}^{k_f-1} g(x^*(k), x^*(k+1), k). \tag{7.2}$$

Let $x(k)$ take on a variation $\delta x(k)$ from the optimal value $x^*(k)$, with $k = k_0, \cdots, k_f$, such that

$$x(k) = x^*(k) + \delta x(k). \tag{7.3}$$

Now with the above variations, the performance function given in (7.1) becomes

$$J(x^* + \delta x) = \sum_{k=k_0}^{k_f-1} g(x^*(k) + \delta x(k), x^*(k+1) + \delta x(k+1), k). \tag{7.4}$$

The increment of the performance function is defined as

$$\Delta J(x^*, \delta x) \triangleq J(x) - J(x^*). \tag{7.5}$$

The variation, denoted by $\delta J(x^*, \delta x)$, is the first-order approximation of the increment $\Delta J(x^*, \delta x)$. Thus, using the Taylor series expansion for (7.2) and (7.4), we have

$$\delta J(x^*, \delta x) = \sum_{k=k_0}^{k_f-1} \left[\left[\frac{\partial g(x^*(k), x^*(k+1), k)}{\partial x(k)} \right]^\mathsf{T} \delta x(k) \right.$$

$$+ \left. \left[\frac{\partial g(x^*(k), x^*(k+1), k)}{\partial x(k+1)} \right]^\mathsf{T} \delta x(k+1) \right]$$

$$= \sum_{k=k_0}^{k_f-1} \left[\left[\frac{\partial g^*(k)}{\partial x(k)} \right]^\mathsf{T} \delta x(k) + \left[\frac{\partial g^*(k)}{\partial x(k+1)} \right]^\mathsf{T} \delta x(k+1) \right], \qquad (7.6)$$

where for notational simplicity, we consider

$$g^*(k) \equiv g(x^*(k), x^*(k+1), k).$$

Now in order to express the coefficient $\delta x(k+1)$ in terms of $\delta x(k)$, reorganize the second expression in (7.6) as follows:

$$\sum_{k=k_0}^{k_f-1} \left[\frac{\partial g^*(k)}{\partial x(k+1)} \right]^\mathsf{T} \delta x(k+1)$$

$$= \sum_{k=k_0+1}^{k_f} \left[\frac{\partial g^*(k-1)}{\partial x(k)} \right]^\mathsf{T} \delta x(k)$$

$$= \sum_{k=k_0+1}^{k_f-1} \left[\frac{\partial g^*(k-1)}{\partial x(k)} \right]^\mathsf{T} \delta x(k) + \left[\frac{\partial g^*(k_f-1)}{\partial x(k_f)} \right]^\mathsf{T} \delta x(k_f). \qquad (7.7)$$

Theorem 7.1 *Suppose that x^* is an extremal solution to the performance function (7.1) of discrete-time systems with given final time and final state; then x^* satisfies the following equation:*

$$\frac{\partial g^*(k)}{\partial x(k)} + \frac{\partial g^*(k-1)}{\partial x(k)} = 0, \ \text{with } g^*(k) \equiv g(x^*(k), x^*(k+1), k), \qquad (7.8)$$

for all $k \in \{k_0+1, k_0+2, \cdots, k_f-1\}$.

Proof Substituting (7.7) in (7.6) and noting that the variation should be zero, we have

$$\sum_{k=k_0+1}^{k_f-1} \left[\frac{\partial g^*(k)}{\partial x(k)} + \frac{\partial g^*(k-1)}{\partial x(k)} \right]^\mathsf{T} \delta x(k)$$

$$+ \left[\frac{\partial g^*(k_0)}{\partial x(k_0)} \right]^\mathsf{T} \delta x(k_0) + \left[\frac{\partial g^*(k_f-1)}{\partial x(k_f)} \right]^\mathsf{T} \delta x(k_f) = 0. \qquad (7.9)$$

For (7.9) to be satisfied for arbitrary variations $\delta x(k)$, we have the condition that the coefficient of $\delta x(k)$ in the first term in (7.9) should be zero. That is,

$$\frac{\partial g^*(k)}{\partial x(k)} + \frac{\partial g^*(k-1)}{\partial x(k)} = 0, \tag{7.10}$$

with $k = k_0 + 1, k_0 + 2, \cdots, k_f - 1$.

This may be called the discrete-time version of the Euler–Lagrange equation.

The boundary or transversality condition is obtained by setting the second term in (7.9) equal to zero. That is,

$$\left[\frac{\partial g^*(k_0)}{\partial x(k_0)}\right]^\top \delta x(k_0) + \left[\frac{\partial g^*(k_f - 1)}{\partial x(k_f)}\right]^\top \delta x(k_f) = 0. \tag{7.11}$$

For a fixed end point system, we have the boundary conditions $x(k_0)$ and $x(k_f)$ fixed and hence $\delta x(k_0) = \delta x(k_f) = 0$. The additional (or derived) boundary condition (7.11) does not exist. □

Notice that in the Euler–Lagrange equation (7.8), we have the following:

- The first term involves taking the partial derivative of the given function $g(x^*(k)$, $x^*(k+1), k)$ with respect to $x(k)$;
- The second term considers taking the partial derivative of $g(x^*(k-1), x^*(k)$, $k-1)$, which is one time step behind, with respect to the same function $x(k)$.

Notice that the function $g(x^*(k-1), x^*(k), k-1)$ can be easily obtained from the given function $g(x^*(k), x^*(k+1), k)$ just by replacing k by $k-1$.

Also, compare the previous results with the corresponding results for continuous-time systems in Chap. 2.

Now illustrate the application of the Euler–Lagrange equation for discrete-time functions with a simple example.

Example 7.1 Consider the minimization of a function

$$J(x) = \sum_{k=0}^{2} \left[x(k)x(k+1) + x^2(k) \right], \tag{7.12}$$

with the boundary conditions $x(0) = 2$ and $x(3) = 4$.

Solution. Specify $g(x(k), x(k+1))$ in (7.12) as

$$g(x(k), x(k+1)) = x(k)x(k+1) + x^2(k); \tag{7.13}$$

then by applying the Euler–Lagrange equation (7.8) in Theorem 7.1, we obtain

$$x^*(k+1) + 2x^*(k) + x^*(k-1) = 0, \tag{7.14}$$

together with the boundary conditions $x(0) = 2$ and $x(3) = 4$, say

$$x^*(2) + 2x^*(1) + 2 = 0, \tag{7.15a}$$
$$4 + 2x^*(2) + x^*(1) = 0. \tag{7.15b}$$

Thus the extrema solution is given as

$$x^*(1) = 0, \ x^*(2) = -2, \tag{7.16}$$

and the associated performance function $J(x^*) = 0$. □

7.1.2 Optimum with Fixed Final Time and Free Final Value

In this part, consider the case that the final value at t_f is unspecified or free. Thus, besides the summation cost given in the last section, we may formulate the performance function including a terminal cost as well, such that

$$J(x) \triangleq h(x(k_f), k_f) + \sum_{k=k_0}^{k_f-1} g(x(k), x(k+1), k), \tag{7.17}$$

with the initial state $x(k_0)$ and the final time k_f being fixed, and the final state $x(k_f)$ being free.

Theorem 7.2 *Suppose that x^* is an extremal solution to the performance function (7.17) of discrete-time systems with given final time and final state being free; then x^* satisfies the following equation:*

$$\frac{\partial g^*(k)}{\partial x(k)} + \frac{\partial g^*(k-1)}{\partial x(k)} = 0, \ \text{with } g^*(k) \equiv g(x^*(k), x^*(k+1), k), \tag{7.18}$$

for all $k \in \{k_0 + 1, k_0 + 2, \cdots, k_f - 1\}$, with the boundary condition

$$\frac{\partial g(x^*(k_f - 1), x^*(k_f), k_f - 1)}{\partial x(k_f)} + \frac{\partial h(x^*(k_f), k_f)}{\partial x(k_f)} = 0. \tag{7.19}$$

Proof Following the same procedure as given earlier for a function without terminal cost, it is straightforward to get the variation for the performance function given in (7.17) as

$$\sum_{k=k_0+1}^{k_f-1} \left[\frac{\partial g^*(k)}{\partial x(k)} + \frac{\partial g^*(k-1)}{\partial x(k)} \right]^\top \delta x(k)$$

$$+ \left[\frac{\partial g^*(k_0)}{\partial x(k_0)} \right]^\top \delta x(k_0) + \left[\frac{\partial g^*(k_f-1)}{\partial x(k_f)} + \frac{\partial h(x^*(k_f), k_f)}{\partial x(k_f)} \right]^\top \delta x(k_f) = 0.$$

(7.20)

For extremization, the variation δJ must be zero. Hence, from (7.20) the Euler–Lagrange equation becomes

$$\frac{\partial g^*(k)}{\partial x(k)} + \frac{\partial g^*(k-1)}{\partial x(k)} = 0,$$

(7.21)

with $k = k_0 + 1, \cdots, k_f - 1$, and the boundary condition becomes

$$\left[\frac{\partial g^*(k_0)}{\partial x(k_0)} \right]^\top \delta x(k_0) + \left[\frac{\partial g^*(k_f-1)}{\partial x(k_f)} + \frac{\partial h(x^*(k_f), k_f)}{\partial x(k_f)} \right]^\top \delta x(k_f) = 0.$$

(7.22)

For a free final point system, given the initial condition $x(k_0)$ we can have $\delta x(k_0) = 0$ in (7.11). Next, at the final point, k_f is specified, and $x(k_f)$ is free, and hence $\delta x(k_f)$ is arbitrary.

Thus in case the coefficient of $\delta x(k_f)$ at $k = k_f$ is zero, the condition (7.22) reduces to the boundary condition (7.19). □

Example 7.2 Consider the minimization of a function

$$J(x) = \frac{1}{2}[x(k_f) - 2]^2 + \sum_{k=0}^{k_f-1} \left[x(k)x(k+1) + x^2(k) \right],$$

(7.23)

with the boundary conditions $x(0) = 2$ and $x(k_f)$ being free with $k_f = 2$.
Solution. Firstly ,by (7.21) and (7.14) given in Example 7.1, we get

$$x^*(2) + 2x^*(1) + x^*(0) = 0,$$

(7.24)

and by (7.19) in Theorem 7.2, we obtain

$$x^*(1) + x^*(2) - 2 = 0.$$

(7.25)

Thus, we have $x^*(0) = 2$, $x^*(1) = -4$, and $x^*(2) = 6$, with the extrema of the performance function $J(u^*) = -4$.

□

7.2 Discrete-Time Optimal Control via Variational Method

By applying a similar technique specified for continuous-time optimal control problems in previous chapters, we can establish the necessary condition for the optimal control for discrete-time optimal control systems based upon the variational method. This part is analogous to that of continuous-time optimal control systems developed in Chap. 3.

Consider a linear, time-variant, discrete-time control system described by

$$x(k+1) = f(x(k), u(k), k),\qquad(7.26)$$

with $k = k_0, \cdots, k_f - 1$, and a fixed initial state as

$$x(k_0) = x_0,\qquad(7.27)$$

where $x(k)$ is an n-dimensional state vector and $u(k)$ is an r-dimensional control vector, respectively.

Also we specify a general performance function with terminal cost as

$$J(x, u) \triangleq h(x(k_f), k_f) + \sum_{k=k_0}^{k_f-1} g(x(k), x(k+1), u(k), k),\qquad(7.28)$$

where $h(x(k_f), k_f)$ represents the terminal cost with respect to the final state $x(k_f)$ and the final time k_f.

Theorem 7.3 will specify the optimal control for the discrete-time optimal control problems.

Before that, we firstly formulate an augmented performance cost by adjoining the original performance cost (7.28) with (7.26) using the Lagrange multiplier, denoted by $\lambda(k+1)$ which will be called a costate function as before,

$$J_a(x, u, \lambda) \triangleq h(x(k_f), k_f)$$
$$+ \sum_{k=k_0}^{k_f-1} \left[g(x(k), x(k+1), u(k), k) + [\lambda(k+1)]^\top [f(x(k), u(k), k) - x(k+1)] \right].$$
$$(7.29)$$

Certainly, the minimization of the augmented performance cost (7.29) is the same as that of the original performance cost (7.28), since $J(x, u) = J_a(x, u, \lambda)$ in case that the state equation (7.26) is satisfied.

Now define a Lagrangian function as follows:

$$\mathscr{L}(x(k), u(k), x(k+1), \lambda(k+1))$$
$$= g(x(k), x(k+1), u(k), k) + [\lambda(k+1)]^\top [f(x(k), u(k), k) - x(k+1)]. \quad(7.30)$$

Notice that, for the purpose of simplicity related to the analysis below, we use $\lambda(k+1)$ instead of $\lambda(k)$ in the specification of $J_a(x, u, \lambda)$.

Moreover, we also define the Hamiltonian as follows:

$$\mathscr{H}(x(k), x(k+1), u(k), \lambda(k+1))$$
$$\triangleq g(x(k), x(k+1), u(k), k) + [\lambda(k+1)]^\top f(x(k), u(k), k). \quad (7.31)$$

Theorem 7.3 *A necessary condition of the optimal control u^* for the discrete-time control problems is given as*

$$\lambda^*(k) = \frac{\partial \mathscr{H}^*(k)}{\partial x(k)} + \frac{\partial \mathscr{H}^*(k-1)}{\partial x(k)}, \quad (7.32a)$$

$$0 = \frac{\partial \mathscr{H}^*(k)}{\partial u(k)}, \quad (7.32b)$$

$$x^*(k+1) = f(x^*(k), u^*(k), k), \quad (7.32c)$$

with $\mathscr{H}^*(k) \equiv \mathscr{H}(x^*(k), x^*(k+1), u^*(k), \lambda^*(k+1))$ and the boundary condition as

$$\left[\frac{\partial g^*(k_f - 1)}{\partial x(k_f)} - \lambda^*(k_f) + \frac{\partial h(x^*(k_f), k_f)}{\partial x(k_f)} \right]^\top \delta x(k_f) = 0. \quad (7.33)$$

Proof We will apply the Euler–Lagrange equation (7.8) to the Lagrangian function $\mathscr{L}$ with respect to the variables $x(k)$, $u(k)$, and $\lambda(k+1)$, respectively.

For notational simplicity, here we consider

$$\mathscr{L}^*(k) \equiv \mathscr{L}(x^*(k), x^*(k+1), u^*(k), \lambda^*(k+1)); \quad (7.34)$$

then we get

$$\frac{\partial \mathscr{L}^*(k)}{\partial x(k)} + \frac{\partial \mathscr{L}^*(k-1)}{\partial x(k)} = 0, \quad (7.35a)$$

$$\frac{\partial \mathscr{L}^*(k)}{\partial u(k)} + \frac{\partial \mathscr{L}^*(k-1)}{\partial u(k)} = 0, \quad (7.35b)$$

$$\frac{\partial \mathscr{L}^*(k)}{\partial \lambda(k)} + \frac{\partial \mathscr{L}^*(k-1)}{\partial \lambda(k)} = 0, \quad (7.35c)$$

and the boundary condition (7.19) becomes

$$\left[\frac{\partial \mathscr{L}^*(k-1)}{\partial x(k)} + \frac{\partial h(x^*(k), k)}{\partial x(k)} \right]^\top \delta x(k) \bigg|_{k=k_f} = 0, \quad (7.36)$$

since $\delta x(k_0) = 0$ with the fixed initial state.

Thus by the specifications of Lagrangian function and Hamiltonian given in (7.30) and (7.31), respectively, we get

$$\mathscr{L}^*(k) \equiv \mathscr{L}(x^*(k), x^*(k+1), u^*(k), \lambda^*(k+1))$$
$$= \mathscr{H}(x^*(k), x^*(k+1), u^*(k), \lambda^*(k+1)) - \left[\lambda^*(k+1)\right]^\top x^*(k+1).$$
(7.37)

Now, by (7.37), together with (7.35), we obtain the conditions for the extremum of optimal control problems in terms of the Hamiltonian $\mathscr{H}(x^*(k), x^*(k+1), u^*(k), \lambda^*(k+1))$ as specified in (7.32) and with the boundary condition given in (7.33).

Consequently, we can state that (7.32) is a necessary condition for the discrete-time control systems with the state equation of (7.26) and the performance cost of (7.28). □

7.2.1 Optimal Control with Fixed Final Time and Fixed Final State

In this part, we study the optimal control with a fixed final state such that

$$x(k_f) = x_f.$$
(7.38)

Notice that in the case of fixed final state, we shall set $h(x(k_f), k_f) = 0$, since it does not make any sense involving a terminal cost term in the performance cost (7.28).

Also, with the fixed final state of (7.38), the variation $\delta x(k_f) = 0$ and hence the boundary condition (7.33) do not exist for this case.

In the following, we will apply the result to a simple example below.

Example 7.3 Given a discrete-time state system

$$x(k+1) = x(k) + u(k),$$
(7.39)

implement the optimal solution with the following cost function:

$$J(x, u) = \sum_{k=k_0}^{k_f-1} \left[x(k)x(k+1) + u^2(k)\right],$$
(7.40)

with the boundary conditions $x(k_0) = 2$ and $x(k_f) = 0$ with $k_0 = 0$ and $k_f = 2$.

Solution. Firstly, the Hamiltonian $\mathscr{H}$ is defined as

$$\mathscr{H}(x(k), x(k+1), u(k), \lambda(k+1)) = x(k)x(k+1) + u^2(k) + \lambda(k+1)[x(k) + u(k)];$$
(7.41)

then by applying Theorem 7.3, the optimality should satisfy the following necessary condition:

$$\frac{\partial \mathcal{H}^*(k)}{\partial u(k)} = 2u^*(k) + \lambda^*(k+1) = 0,$$

which implies that

$$u^*(k) = -\frac{1}{2}\lambda^*(k+1). \tag{7.42}$$

By (7.42) and (7.32) in Theorem 7.3, we also obtain the state and costate difference equations as follows:

$$x^*(k+1) = f(x^*(k), u^*(k), k) = x^*(k) - \frac{1}{2}\lambda^*(k+1), \tag{7.43a}$$

$$\lambda^*(k) = \frac{\partial \mathcal{H}^*(k)}{\partial x(k)} + \frac{\partial \mathcal{H}^*(k-1)}{\partial x(k)} = \lambda^*(k+1) + x^*(k+1) + x^*(k-1), \tag{7.43b}$$

with the boundary conditions $x^*(0) = 2$ and $x^*(2) = 0$.

Thus, the optimal solution is given as

$$\begin{bmatrix} x^*(0) \\ x^*(1) \\ x^*(2) \end{bmatrix} = \begin{bmatrix} 2 \\ \frac{1}{2} \\ 0 \end{bmatrix}, \quad \begin{bmatrix} \lambda^*(1) \\ \lambda^*(2) \end{bmatrix} = \begin{bmatrix} 3 \\ 1 \end{bmatrix}, \quad \begin{bmatrix} u^*(0) \\ u^*(1) \end{bmatrix} = \begin{bmatrix} -\frac{3}{2} \\ -\frac{1}{2} \end{bmatrix}, \tag{7.44}$$

and the associated minimum cost $J(u^*) = \frac{7}{2}$.   □

7.2.2 Optimal Control with Fixed Final Time and Free Final State

In this section, consider a free final state condition such that

$$x(k_f) \text{ is free, and } k_f \text{ is fixed.} \tag{7.45}$$

Firstly, by the final condition specified in (7.33), under (7.45) we obtain

$$\frac{\partial g^*(k_f - 1)}{\partial x(k_f)} - \lambda^*(k_f) + \frac{\partial h(x^*(k_f), k_f)}{\partial x(k_f)} = 0, \tag{7.46}$$

which implies that

$$\lambda^*(k_f) = \frac{\partial g^*(k_f - 1)}{\partial x(k_f)} + \frac{\partial h(x^*(k_f), k_f)}{\partial x(k_f)}. \tag{7.47}$$

We revisit below the optimal control problems specified in Example 7.3. such that the final state $x(t_f)$ is free and t_f is fixed.

Example 7.4 Given a discrete-time state system

$$x(k+1) = x(k) + u(k), \tag{7.48}$$

implement the optimal solution with the following cost function:

$$J(x, u) = \frac{1}{2}[x(k_f) - 2]^2 + \sum_{k=k_0}^{k_f - 1} \left[x(k)x(k+1) + u^2(k) \right], \tag{7.49}$$

where $x(k_0) = 2$ and $x(k_f)$ is unspecified, with $k_0 = 0$ and $k_f = 2$.
  Solution. By Example 7.3, we obtain

$$x^*(k+1) = x^*(k) - \frac{1}{2}\lambda^*(k+1),$$

$$\lambda^*(k) = \lambda^*(k+1) + x^*(k+1) + x^*(k-1),$$

with the boundary condition $x(0) = 2$, and by (7.47)

$$\lambda^*(k_f) = \frac{\partial g^*(k_f - 1)}{\partial x(k_f)} + \frac{\partial h(x^*(k_f), k_f)}{\partial x(k_f)} = x^*(k_f - 1) + x^*(k_f) - 2. \tag{7.51}$$

Thus, the optimal solution is given as

$$\begin{bmatrix} x^*(0) \\ x^*(1) \\ x^*(2) \end{bmatrix} = \begin{bmatrix} 2 \\ \frac{8}{11} \\ \frac{10}{11} \end{bmatrix}, \quad \begin{bmatrix} \lambda^*(1) \\ \lambda^*(2) \end{bmatrix} = \begin{bmatrix} \frac{28}{11} \\ -\frac{4}{11} \end{bmatrix}, \quad \begin{bmatrix} u^*(0) \\ u^*(1) \end{bmatrix} = \begin{bmatrix} -\frac{14}{11} \\ \frac{2}{11} \end{bmatrix}, \tag{7.52}$$

and the associated minimum cost $J(u^*) = \frac{48}{11}$. ☐

7.3 Discrete-Time Linear-Quadratic Regulation Problems

This part is analogous to that of continuous-time optimal control systems developed in Chap. 3.

Consider a linear, time-variant, discrete-time control system described by

$$x(k+1) = A(k)x(k) + B(k)u(k), \qquad (7.53)$$

with $k = k_0, \cdots, k_f - 1$, and with an fixed initial state given as

$$x(k_0) = x_0, \qquad (7.54)$$

where $x(k)$ is an n-dimensional state vector, $u(k)$ is an r-dimensional control vector, and $A(k)$ and $B(k)$ are matrices of $n \times n$ and $n \times r$ dimensions, respectively.

Also we specify a quadratic cost with terminal cost as

$$\begin{aligned}
J(x, u) &\triangleq \frac{1}{2}[x(k_f)]^\top F(k_f)x(k_f) \\
&\quad + \frac{1}{2}\sum_{k=k_0}^{k_f-1} \left[[x(k)]^\top Q(k)x(k) + [u(k)]^\top R(k)u(k) \right],
\end{aligned} \qquad (7.55)$$

where $F(k_f)$ and $Q(k)$ are each $n \times n$, symmetric, positive semidefinite matrices, and $R(k)$ is an $r \times r$ symmetric, positive-definite matrix for each k.

Theorem 7.4 will specify the optimal control for the linear-quadratic optimal control problems.

Before that, we firstly formulate an augmented performance cost by adjoining the original performance cost (7.55) with (7.53) using the Lagrange multiplier, denoted by $\lambda(k+1)$ which will be called a costate function as before

$$\begin{aligned}
J_a(x, u, \lambda) &\triangleq \frac{1}{2}[x(k_f)]^\top F(k_f)x(k_f) \\
&\quad + \frac{1}{2}\sum_{k=k_0}^{k_f-1} \left[[x(k)]^\top Q(k)x(k) + [u(k)]^\top R(k)u(k) \right] \\
&\quad + [\lambda(k+1)]^\top [A(k)x(k) + B(k)u(k) - x(k+1)].
\end{aligned} \qquad (7.56)$$

Certainly, the minimization of the augmented performance cost (7.56) is the same as that of the original performance cost (7.55), since $J(x, u) = J_a(x, u, \lambda)$ in case that the state equation (7.53) is satisfied.

Notice that, for the purpose of simplicity related to the analysis below, we use $\lambda(k+1)$ instead of $\lambda(k)$ in the specification of $J_a(x, u, \lambda)$.

Theorem 7.4 *The optimal control u^* for the discrete-time LQR problems is given as*

$$u^*(k) = -R^{-1}(k)[B(k)]^\top \lambda^*(k+1). \qquad (7.57)$$

Proof Next, we will apply the Euler–Lagrange equation (7.21) to the Lagrangian function $\mathscr{L}$ with respect to the variables $x(k)$, $u(k)$, and $\lambda(k + 1)$, respectively.

For notational simplicity, here consider that

$$\mathscr{L}^*(k) \equiv \mathscr{L}(x^*(k), x^*(k + 1), u^*(k), \lambda(k + 1)); \tag{7.58}$$

then (7.35) with the boundary condition (7.22) becomes (7.33) where by (7.55), $h(x(k_f), k_f)$ is defined as

$$h(x(k_f), k_f) = \frac{1}{2}[x(k_f)]^\top F(k_f)x(k_f). \tag{7.59}$$

Next, we define the Hamiltonian as follows:

$$\mathscr{H}(x^*(k), u^*(k), \lambda^*(k + 1)) = \frac{1}{2}[x^*(k)]^\top Q(k)x^*(k) + \frac{1}{2}[u^*(k)]^\top R(k)u^*(k)$$
$$+ [\lambda^*(k + 1)]^\top \left[A(k)x^*(k) + B(k)u^*(k)\right]. \tag{7.60}$$

Thus by (7.30) and (7.60), we get

$$\mathscr{L}^*(k) \equiv \mathscr{L}(x^*(k), x^*(k + 1), u^*(k), \lambda^*(k + 1))$$
$$= \mathscr{H}(x^*(k), u^*(k), \lambda^*(k + 1)) - \lambda^*(k + 1)x^*(k + 1). \tag{7.61}$$

Now, by (7.61), together with (7.35), we obtain the conditions for the extremum of optimal control in terms of the Hamiltonian $\mathscr{H}(x^*(k), u^*(k), \lambda^*(k + 1))$ as follows:

$$\lambda^*(k) = \frac{\partial \mathscr{H}(x^*(k), u^*(k), \lambda^*(k + 1))}{\partial x(k)}; \tag{7.62a}$$

$$0 = \frac{\partial \mathscr{H}(x^*(k), u^*(k), \lambda^*(k + 1))}{\partial u(k)}; \tag{7.62b}$$

$$x^*(k) = \frac{\partial \mathscr{H}(x^*(k - 1), u^*(k - 1), \lambda^*(k))}{\partial \lambda(k)}. \tag{7.62c}$$

Moreover, (7.62c) can also be rewritten as

$$x^*(k + 1) = \frac{\partial \mathscr{H}(x^*(k), u^*(k), \lambda^*(k + 1))}{\partial \lambda(k + 1)}. \tag{7.63}$$

Consequently, for the optimal control systems with the state equation of (7.53) and the performance cost of (7.55), by (7.62a), (7.62b), and (7.63), we get

$$x^*(k + 1) = A(k)x^*(k) + B(k)u^*(k), \tag{7.64a}$$
$$\lambda^*(k) = Q(k)x^*(k) + [A(k)]^\top \lambda^*(k + 1), \tag{7.64b}$$
$$0 = R(k)u^*(k) + [B(k)]^\top \lambda^*(k + 1). \tag{7.64c}$$

Due to the positive property of $R(k)$, it is invertible; then by (7.64c), the optimal control is given as (7.57).

□

By applying the optimal control (7.57) in the state equation (7.64a), we get

$$
\begin{aligned}
x^*(k+1) &= A(k)x^*(k) - B(k)R^{-1}(k)[B(k)]^\top \lambda^*(k+1) \\
&= A(k)x^*(k) - E(k)\lambda^*(k+1),
\end{aligned} \tag{7.65}
$$

where $E(k) = B(k)R^{-1}(k)[B(k)]^\top$.

Thus the state and costate systems of (7.65) and (7.64b) become

$$
\begin{bmatrix} x^*(k+1) \\ \lambda^*(k) \end{bmatrix} = \begin{bmatrix} A(k) & -E(k) \\ Q(k) & [A(k)]^\top \end{bmatrix} \begin{bmatrix} x^*(k) \\ \lambda^*(k+1) \end{bmatrix}. \tag{7.66}
$$

7.3.1 Linear-Quadratic Regulation Problems with Fixed Final Time and Fixed Final State

In this part, we study the optimal control with a fixed final state given in (7.38).

Notice that in the case of the fixed final state, we shall set $F(k_f) = 0$, since it does not make any senses involving a terminal cost term in the performance cost (7.55).

Also, with the fixed final state of (7.38), the variation $\delta x(k_f) = 0$ and hence the boundary condition (7.33) do not exist for this case.

Thus, the state and costate system (7.66) along with the initial condition (7.54) and the fixed final condition (7.38) constitute a so-called two-point boundary value problem.

The solution of this problem gives $x^*(k)$ and $\lambda^*(k)$ or $\lambda^*(k+1)$ which along with the control relation (7.57) leads to the so-called open-loop optimal control.

In the following, we will apply the result to a simple example below.

Example 7.5 Given a discrete-time state system

$$
x(k+1) = x(k) + 2u(k), \tag{7.67}
$$

implement the optimal solution with the following cost function:

$$
J(x, u) = \frac{1}{2} \sum_{k=k_0}^{k_f-1} u^2(k), \tag{7.68}
$$

with the boundary conditions $x(k_0) = 1$ and $x(k_f) = 0$ with $k_0 = 0$ and $k_f = 5$.

Solution. By the specifications of the state equation (7.53) and the cost function (7.55), we can get

$$A(k) = 1, \ B(k) = 2, \ F(k_f) = 0, \ Q(k) = 0, \ R(k) = 1, \tag{7.69}$$

for the problems defined in (7.67) and (7.68).

Firstly, the Hamiltonian $\mathscr{H}$ is defined as

$$\mathscr{H}(x(k), u(k), \lambda(k+1)) = \frac{1}{2}u^2(k) + \lambda(k+1)[x(k) + 2u(k)]; \tag{7.70}$$

then by applying Theorem 7.4, the optimality should satisfy the following necessary condition:

$$\frac{\partial \mathscr{H}(x^*(k), u^*(k), \lambda^*(k+1))}{\partial u(k)} = u^*(k) + 2\lambda^*(k+1) = 0,$$

which implies that

$$u^*(k) = -2\lambda^*(k+1). \tag{7.71}$$

Subject to the optimal control specified in (7.71), $\mathscr{H}^*$ is specified as follows:

$$\begin{aligned}
\mathscr{H}^*\left(x^*(k), \lambda^*(k+1)\right) &\triangleq \mathscr{H}\left(x^*(k), u^*(k), \lambda^*(k+1)\right) \\
&= x^*(k)\lambda^*(k+1) - 2\left[\lambda^*(k+1)\right]^2.
\end{aligned} \tag{7.72}$$

We also obtain the state and costate difference equations as follows:

$$x^*(k+1) = \frac{\partial \mathscr{H}^*(x^*(k), \lambda^*(k+1))}{\partial \lambda(k+1)} = x^*(k) - 4\lambda^*(k+1), \tag{7.73a}$$

$$\lambda^*(k) = \frac{\partial \mathscr{H}^*(x^*(k), \lambda^*(k+1))}{\partial x(k)} = \lambda^*(k+1), \tag{7.73b}$$

with the solution given as

$$x^*(k) = x_0 - \frac{k}{k_f}\left[x_0 - x_f\right],$$

$$\lambda^*(k) = \frac{1}{4k_f}\left[x_0 - x_f\right],$$

for all k.

Consequently, by the boundary conditions $x(0) = 1$ and $x(5) = 0$, we can obtain

$$x^*(k) = 1 - 0.2k, \tag{7.74a}$$

$$\lambda^*(k+1) = 0.05. \tag{7.74b}$$

Thus by the above equation, the optimal control $u^*(k)$ specified in (7.71), we have

$$u^*(k) = -0.1, \tag{7.75}$$

for all k.   □

7.3.2 Linear-Quadratic Regulation Problems with Fixed Final Time and Free Final State

We consider the LQR problems with fixed final time and a free final state condition as defined in (7.45). Thus by (7.33) in Theorem 7.3, the specification of the quadratic cost function given in (7.55), we have

$$
\begin{aligned}
\lambda^*(k_f) &= \frac{\partial g^*(k_f - 1)}{\partial x(k_f)} + \frac{\partial h\left(x^*(k_f), k_f\right)}{\partial x(k_f)} \\
&= \frac{\partial h\left(x^*(k_f), k_f\right)}{\partial x(k_f)} = \frac{\partial}{\partial x(k_f)}\left(\frac{1}{2}\left[x^*(k_f)\right]^\top F(k_f)x^*(k_f)\right),
\end{aligned}
\tag{7.76}
$$

since, by (7.56),

$$
\frac{\partial g^*(k_f - 1)}{\partial x(k_f)} = \frac{\partial}{\partial x(k_f)}\left([x(k)]^\top Q(k)x(k) + [u(k)]^\top R(k)u(k)\right)\Big|_{k=k_f-1} = 0,
\tag{7.77}
$$

which implies

$$\lambda^*(k_f) = F(k_f)x^*(k_f). \tag{7.78}$$

The state and costate system (7.66) along with the initial condition (7.54) and the final condition (7.78) constitute a two-point boundary value problem. The solution to this problem is difficult to implement because of the coupled property of the resulting solutions. The state $x^*(k)$ is solved forward from its initial condition $x(k_0)$, while the costate $\lambda^*(k)$ is solved backward from its final condition $\lambda^*(k_f)$. This leads to the open-loop optimal control.

In the following, we will apply the result to a simple example below.

Example 7.6 Given a discrete-time state system

$$x(k + 1) = x(k) + u(k), \tag{7.79}$$

implement the optimal solution with the following cost function:

$$J(x, u) = \frac{1}{2}x^2(k_f) + \frac{1}{2}\sum_{k=k_0}^{k_f-1}\left[x^2(k) + u^2(k)\right], \tag{7.80}$$

with the boundary condition $x(k_0) = 1$ with $k_0 = 0$ and $k_f = 2$, and the final state $x(k_f)$ is unspecified.

Solution. By the specifications of the state equation (7.53) and the cost function (7.55), we can get

$$A(k) = 1, B(k) = 1, F = 1, Q(k) = 1, R(k) = 1, \tag{7.81}$$

for the problem specified in (7.79) and (7.80).

By applying Theorem 7.4, the optimality should satisfy the following necessary condition:

$$u^*(k) = -R^{-1}(k)[B(k)]^\top \lambda^*(k + 1) = -\lambda^*(k + 1). \tag{7.82}$$

By (7.64a) and (7.64b), we can obtain that the state and costate difference equations are given as follows:

$$x^*(k + 1) = A(k)x^*(k) + B(k)u^*(k) = x^*(k) - \lambda^*(k + 1), \tag{7.83a}$$
$$\lambda^*(k) = Q(k)x^*(k) + [A(k)]^\top \lambda^*(k + 1) = x^*(k) + \lambda^*(k + 1), \tag{7.83b}$$

with the initial condition $x_0 = 1$, and by (7.78), the following final boundary condition given as

$$\lambda^*(k_f) = F(k_f)x^*(k_f) = x^*(k_f). \tag{7.84}$$

Thus, the solution is given as

$$x^* = \begin{bmatrix} 1 \\ 0.4 \\ 0.2 \end{bmatrix} \text{ and } \lambda^* = \begin{bmatrix} 1.6 \\ 0.6 \\ 0.2 \end{bmatrix},$$

and the associated minimum cost $J(u^*) = 0.8$. $\qquad\square$

7.3.3 Optimal Control with Respect to State

Subject to the optimal control given in (7.57) developed in Theorem 7.4, the state and costate equations can be written as

$$x^*(k + 1) = A(k)x^*(k) - E(k)\lambda^*(k + 1), \tag{7.85}$$
$$\lambda^*(k) = Q(k)x^*(k) + [A(k)]^\top \lambda^*(k + 1), \tag{7.86}$$

where for notational simplicity, we consider

$$E(k) \equiv B(k)R^{-1}(k)[B(k)]^{\top}, \tag{7.87}$$

and the final costate relation (7.78) is given by

$$\lambda^*(k_f) = F(k_f)x^*(k_f). \tag{7.88}$$

In order to specify the optimal control with respect to the state $x^*(k)$ and not the costate $\lambda^*(k+1)$, we need to try to express the costate function $\lambda^*(k+1)$ in the optimal control (7.57) with respect to the state $x^*(k)$.

By observing the form of the final condition given in (7.88), it may be reasonable to suppose that

$$\lambda^*(k) = P(k)x^*(k), \tag{7.89}$$

with certain $P(k)$ to be specified.

Thus, by applying (7.89) to the state and costate equations (7.85) and (7.86), we obtain

$$P(k)x^*(k) = Q(k)x^*(k) + [A(k)]^{\top}P(k+1)x^*(k+1), \tag{7.90a}$$
$$x^*(k+1) = A(k)x^*(k) - E(k)P(k+1)x^*(k+1). \tag{7.90b}$$

By (7.90b), we can have

$$x^*(k+1) = [\mathbb{I} + E(k)P(k+1)]^{-1}A(k)x^*(k), \tag{7.91}$$

where $\mathbb{I}$ denotes an identity matrix, by which together with (7.90a), we obtain

$$P(k)x^*(k) = Q(k)x^*(k) + [A(k)]^{\top}P(k+1)[\mathbb{I} + E(k)P(k+1)]^{-1}A(k)x^*(k). \tag{7.92}$$

Since (7.92) must hold for all values of $x^*(k)$, we have

$$P(k) = [A(k)]^{\top}P(k+1)[\mathbb{I} + E(k)P(k+1)]^{-1}A(k) + Q(k), \tag{7.93}$$

which is called the matrix difference Riccati equation (DRE), with the following boundary condition:

$$P(k_f) = F(k_f), \tag{7.94}$$

by (7.88) and (7.89).

Moreover, by assuming that $P(k)$ is invertible for all $k \neq k_f$, (7.93) can be reorganized as the following:

$$P(k) = [A(k)]^{\top} \left[P^{-1}(k+1) + E(k) \right]^{-1} A(k) + Q(k). \qquad (7.95)$$

In the matrix DRE given in (7.93), $P(k)$ needs to be solved backward starting from the final condition (7.94).

Notice that since $Q(k)$ and $F(k_f)$ are assumed to be positive semidefinite, we can show that the Riccati matrix $P(k)$ is positive definite.

Now by (7.57), (7.86), and (7.89), we can obtain

$$u^*(k) = -R^{-1}(k)[B(k)]^{\top}[A(k)]^{-\top}[P(k) - Q(k)]x^*(k), \qquad (7.96)$$

where $A^{-\top}$ represents the inverse of $A^{\top}$ by assuming the invertibility of $A(k)$.

For notational simplicity, consider

$$u^*(k) = -L(k)x^*(k), \qquad (7.97)$$

with

$$L(k) \equiv R^{-1}(k)[B(k)]^{\top}[A(k)]^{-\top}[P(k) - Q(k)]. \qquad (7.98)$$

This is the required relation for the optimal feedback control, and the feedback gain $L(k)$ is called the *Kalman gain*.

Thus, the optimal state $x^*(k)$ can be specified by applying the optimal control $u^*(k)$ given in (7.97) to the state equation (7.53) as

$$x^*(k+1) = [A(k) - B(k)L(k)]x^*(k). \qquad (7.99)$$

In the development of the closed-form optimal control given in (7.97), it is necessary to assume that $A(k)$ is invertible. As a consequence, in the following, we will give another way to implement the optimal control.

Firstly, by the so-called *matrix inversion lemma*, say

$$\left[A_{11} - A_{12} A_{22}^{-1} A_{21} \right]^{-1} = A_{11}^{-1} + A_{11}^{-1} A_{12} \left[A_{22} - A_{21} A_{11}^{-1} A_{12} \right]^{-1} A_{21} A_{11}^{-1}, \qquad (7.100)$$

where A_{11} and A_{22} are assumed to be invertible, respectively; then by (7.95), we have

$$P(k) = [A(k)]^{\top} \left[P(k+1) - P(k+1)B(k)\Gamma(k)[B(k)]^{\top}P(k+1) \right] A(k) + Q(k), \qquad (7.101)$$

with $\Gamma(k) \equiv \left[[B(k)]^{\top}P(k+1)B(k) + R(k) \right]^{-1}$.

Next, by applying (7.57) and (7.89), we get

$$u^*(k) = -R^{-1}(k)[B(k)]^\top P(k+1)x^*(k+1), \qquad (7.102)$$

subject to which the state equation (7.53) becomes

$$u^*(k) = -R^{-1}(k)[B(k)]^\top P(k+1)\left[A(k)x^*(k) + B(k)u^*(k)\right]. \qquad (7.103)$$

Hence, we have

$$\left[\mathbb{I} + R^{-1}(k)[B(k)]^\top P(k+1)B(k)\right]u^*(k) = -R^{-1}(k)[B(k)]^\top P(k+1)A(k)x^*(k),$$

by which, it is implied that

$$u^*(k) = -L_a(k)x^*(k), \qquad (7.104)$$

with $L_a(k)$ given as

$$L_a(k) \equiv \left[R(k) + [B(k)]^\top P(k+1)B(k)\right]^{-1}[B(k)]^\top P(k+1)A(k). \qquad (7.105)$$

Notice, from the optimal feedback control (7.104), that the Kalman gains are dependent on the solution of the matrix DRE (7.101) involving the system state matrices and performance costs.

Finally, the closed-loop, optimal control (7.104) with the state (7.53) gives the optimal system

$$x^*(k+1) = [A(k) - B(k)L_a(k)]x^*(k). \qquad (7.106)$$

Using the gain relation (7.105), an alternate form for the matrix DRE (7.101) becomes

$$P(k) = [A(k)]^\top P(k+1)[A(k) - B(k)L_a(k)] + Q(k). \qquad (7.107)$$

Notice that

- As analyzed above, there are different forms of the matrix DRE given by (7.93), (7.95), (7.101), and (7.107), respectively.
- The Kalman feedback gain matrix has two forms, say (7.98) and (7.105).
- It can be observed that the matrix DRE (7.93) and the associated Kalman feedback gain matrix (7.98) involve the inversion of the matrix $\mathbb{I} + E(k)P(k+1)$, while the matrix DRE (7.101) and the associated Kalman feedback gain matrix (7.105) involve the inversion of the matrix $R(k) + [B(k)]^\top P(k+1)B(k)$.

7.3.4 Optimal Cost Function

For finding the optimal cost function $J^*(k_0)$, we can follow the same procedure as the one used for the continuous-time systems in Chap. 3 to get

$$J^* = \frac{1}{2}[x^*(k_0)]^T P(k_0)x(k_0). \tag{7.108}$$

Notice that the Riccati function $P(k)$ is generated off-line before we obtain the optimal control $u^*(k)$ to be applied to the system. Thus in general for any initial state k, the optimal cost is given as follows:

$$J^*(k) = \frac{1}{2}[x^*(k)]^T P(k)x(k). \tag{7.109}$$

Example 7.7 Consider a two-dimensional state system such that

$$x_1(k+1) = x_1(k) + x_2(k) + u(k), \tag{7.110a}$$
$$x_2(k+1) = x_2(k) + u(k), \tag{7.110b}$$

with the boundary conditions given as

$$x_1(k_0) = 4, x_2(k_0) = 2, k_0 = 0, k_f = 8, \text{ and } x(k_f) \text{ is unspecified.} \tag{7.111}$$

Implement the optimal control subject to the performance cost given as

$$J = \frac{1}{2}\left[x_1(k_f)\right]^2 + \left[x_2(k_f)\right]^2 + \sum_{k=k_0}^{k_f-1}\left[\frac{1}{2}[x_1(k)]^2 + \frac{1}{2}[x_2(k)]^2 + u^2(k)\right]. \tag{7.112}$$

Solution. By (7.53) and (7.55), we have

$$A(k) = \begin{bmatrix} 1 & 1 \\ 0 & 1 \end{bmatrix}, B(k) = \begin{bmatrix} 1 \\ 1 \end{bmatrix}, F(k_f) = \begin{bmatrix} 1 & 0 \\ 0 & 2 \end{bmatrix}, Q(k) = \begin{bmatrix} 1 & 0 \\ 0 & 1 \end{bmatrix}, R(k) = 2,$$
$$\tag{7.113}$$

for all k.

Solve the matrix difference Riccati equation (7.93),

$$\begin{bmatrix} P_{11}(k) & P_{12}(k) \\ P_{12}(k) & P_{22}(k) \end{bmatrix} = \begin{bmatrix} 1 & 0 \\ 1 & 1 \end{bmatrix}\begin{bmatrix} P_{11}(k+1) & P_{12}(k+1) \\ P_{12}(k+1) & P_{22}(k+1) \end{bmatrix}.$$

$$\left[\begin{bmatrix} 1 & 0 \\ 0 & 1 \end{bmatrix} + \begin{bmatrix} 1 \\ 1 \end{bmatrix}[2]^{-1}[1\ 1]\begin{bmatrix} P_{11}(k+1) & P_{12}(k+1) \\ P_{12}(k+1) & P_{22}(k+1) \end{bmatrix}\right]^{-1}\begin{bmatrix} 1 & 1 \\ 0 & 1 \end{bmatrix} + \begin{bmatrix} 1 & 0 \\ 0 & 1 \end{bmatrix},$$
$$\tag{7.114}$$

with the final condition (7.94) as

$$\begin{bmatrix} P_{11}(10) & P_{12}(10) \\ P_{12}(10) & P_{22}(10) \end{bmatrix} = F(k_f) = \begin{bmatrix} 1 & 0 \\ 0 & 2 \end{bmatrix}. \tag{7.115}$$

By applying the implemented $P(k)$ in (7.114), following (7.97), the optimal control $u^*(k)$ can be specified in the following:

$$u^*(k) = -L(k)x^*(k), \tag{7.116}$$

with $L(k)$ as given in (7.98).

Finally subject to the optimal control specified in (7.116), the optimal state can be obtained by solving the state equation (7.110). □

7.3.5 Infinite-Interval Time-Invariant Linear-Quadratic Regulation Problems

In this part, consider the linear time-invariant state system such that

$$x(k+1) = Ax(k) + Bu(k), \tag{7.117}$$

and an infinite-interval-quadratic performance cost defined as

$$J = \frac{1}{2} \sum_{k=k_0}^{\infty} \left[[x^*(k)]^\top Q x(k) + [u^*(k)]^\top R u^*(k) \right]. \tag{7.118}$$

As the final time k_f tends to infinity, the Riccati matrix $P(k)$ converges to a steady value $\widehat{P}$ in (7.93).

For the underlying infinite-interval time-invariant problems specified above, the Riccati equation (7.93) degenerates to the following form:

$$\widehat{P} = A^\top \widehat{P} \left[\mathbb{I} + E\widehat{P} \right]^{-1} A + Q, \tag{7.119}$$

where $E \equiv BR^{-1}B^\top$, and (7.95) becomes the following:

$$\widehat{P} = A^\top \left[\widehat{P}^{-1} + E \right]^{-1} A + Q. \tag{7.120}$$

The feedback optimal control (7.96) becomes

$$u^*(k) = -R^{-1}B^\top A^{-\top}[\widehat{P} - Q]x^*(k) = -\widehat{L}x^*(k), \tag{7.121}$$

where the Kalman gain (7.98) becomes

$$\widehat{L} = R^{-1}B^{\mathsf{T}}A^{-\mathsf{T}}\left[\widehat{P} - Q\right], \tag{7.122}$$

with $A^{-\mathsf{T}}$ representing the inverse of A^{T}.

As earlier, here we give another way to implement $\widehat{P}$ by considering the steady-state form of the DRE (7.101) as

$$\widehat{P} = A^{\mathsf{T}}\left[\widehat{P} - \widehat{P}B\left[B^{\mathsf{T}}\widehat{P}B + R\right]^{-1}B^{\mathsf{T}}\widehat{P}\right]A + Q. \tag{7.123}$$

The optimal feedback control (7.104) becomes

$$u^*(k) = -\widehat{L}_a x^*(k), \tag{7.124}$$

where $\widehat{L}_a$ is given as

$$\widehat{L}_a = \left[B^{\mathsf{T}}\widehat{P}B + R\right]^{-1}B^{\mathsf{T}}\widehat{P}A. \tag{7.125}$$

Thus, subject to the optimal control (7.124), the associated state (7.117) is given as the following system:

$$x^*(k+1) = \left[A - B\widehat{L}_a\right]x^*(k). \tag{7.126}$$

And the minimum cost function (7.109) becomes

$$J^*(k) = [x^*(k)]^{\mathsf{T}}\widehat{P}x^*(k). \tag{7.127}$$

In the following, we will apply the above results to a simple example.

Example 7.8 Consider an infinite-interval two-dimensional state system as specified in (7.110) with $x_1(0) = 4$ and $x_2(0) = 2$.
Implement the optimal control subject to the performance cost given as

$$J = \sum_{k=0}^{\infty}\left[\frac{1}{2}[x_1(k)]^2 + \frac{1}{2}[x_2(k)]^2 + [u(k)]^2\right]. \tag{7.128}$$

Solution. By (7.117) and (7.118), we have

$$A = \begin{bmatrix} 1 & 1 \\ 0 & 1 \end{bmatrix}, B = \begin{bmatrix} 1 \\ 1 \end{bmatrix}, Q = \begin{bmatrix} 1 & 0 \\ 0 & 1 \end{bmatrix}, R = 2, \tag{7.129}$$

for all k.

$$\widehat{P} = A^{\mathsf{T}} \widehat{P} [\mathbb{I} + E]^{-1} A + Q.$$

Solve the matrix DRE (7.119),

$$\begin{bmatrix} \widehat{P}_{11} & \widehat{P}_{12} \\ \widehat{P}_{12} & \widehat{P}_{22} \end{bmatrix} = \begin{bmatrix} 1 & 0 \\ 1 & 1 \end{bmatrix} \begin{bmatrix} \widehat{P}_{11} & \widehat{P}_{12} \\ \widehat{P}_{12} & \widehat{P}_{22} \end{bmatrix} \cdot$$
$$\left[\begin{bmatrix} 1 & 0 \\ 0 & 1 \end{bmatrix} + \begin{bmatrix} 1 \\ 1 \end{bmatrix} [2]^{-1} \begin{bmatrix} 1 & 1 \end{bmatrix} \begin{bmatrix} \widehat{P}_{11} & \widehat{P}_{12} \\ \widehat{P}_{12} & \widehat{P}_{22} \end{bmatrix} \right]^{-1} \begin{bmatrix} 1 & 1 \\ 0 & 1 \end{bmatrix} + \begin{bmatrix} 1 & 0 \\ 0 & 1 \end{bmatrix}. \qquad (7.130)$$

By applying the implemented $\widehat{P}$ in the above equation, following (7.121), the optimal control $u^*(k)$ can be specified in the following:

$$u^*(k) = -\widehat{L} x^*(k), \qquad (7.131)$$

with $\widehat{L}$ given as

$$\widehat{L} = R^{-1} B^{\mathsf{T}} A^{-\mathsf{T}} [\widehat{P} - Q]$$
$$= \frac{1}{2} \begin{bmatrix} 1 & 1 \end{bmatrix} \begin{bmatrix} 1 & 0 \\ 1 & 1 \end{bmatrix}^{-1} \left[\widehat{P} - \begin{bmatrix} 1 & 0 \\ 0 & 1 \end{bmatrix} \right]$$
$$= \frac{1}{2} \begin{bmatrix} \widehat{P}_{12} & \widehat{P}_{22} - 1 \end{bmatrix}, \qquad (7.132)$$

by (7.122).

Finally, subject to the optimal control specified above, the optimal state can be obtained by solving the state equation (7.117). $\square$

7.4 Discrete-Time Linear-Quadratic Tracking Problems

In this section, we study a class of LQT problems for discrete-time systems.

Consider a linear, time-invariant system described by the state equation as follows:

$$x(k+1) = Ax(k) + Bu(k). \qquad (7.133)$$

The performance cost to be minimized is

$$J = \frac{1}{2} [Cx(k_f) - r(k_f)]^{\mathsf{T}} F[Cx(k_f) - r(k_f)]$$
$$+ \frac{1}{2} \sum_{k=k_0}^{k_f - 1} \left[[Cx(k) - r(k)]^{\mathsf{T}} Q[Cx(k) - r(k)] + [u(k)]^{\mathsf{T}} Ru(k) \right], \qquad (7.134)$$

where $x(k)$ and $u(k)$ are n- and r-dimensional state and control vectors, respectively.

Also, in (7.134), we assume that F and Q are each $n \times n$ dimensional positive semidefinite symmetric matrices, and R is an $r \times r$ positive-definite symmetric matrix.

The initial condition is given as $x(k_0) = x_0$ and the final condition $x(k_f)$ is free with k_f being fixed.

First the Hamiltonian is defined as

$$\mathcal{H}(x(k), u(k), \lambda(k+1))$$
$$= \frac{1}{2} \sum_{k=k_0}^{k_f-1} \left[[Cx(k) - r(k)]^\top Q[Cx(k) - r(k)] + [u(k)]^\top Ru(k) \right]$$
$$+ [\lambda(k+1)]^\top [Ax(k) + Bu(k)]. \tag{7.135}$$

Using (7.62c), (7.62a), and (7.62b) for the state, costate, and control, respectively, we obtain the state and costate equations given as

$$x^*(k+1) = \frac{\partial \mathcal{H}(x^*(k), u^*(k), \lambda^*(k+1))}{\partial \lambda(k+1)} = Ax^*(k) + Bu^*(k), \tag{7.136a}$$

$$\lambda^*(k) = \frac{\partial \mathcal{H}(x^*(k), u^*(k), \lambda^*(k+1))}{\partial x(k)}$$
$$= A^\top \lambda^*(k+1) + C^\top QCx^*(k) - C^\top Qr(k), \tag{7.136b}$$

and the optimal control equation as

$$\frac{\partial \mathcal{H}(x^*(k), u^*(k), \lambda^*(k+1))}{\partial u(k)} = B^\top \lambda^*(k+1) + Ru^*(k) = 0, \tag{7.137}$$

which implies that

$$u^*(k) = -R^{-1}B^\top \lambda^*(k+1). \tag{7.138}$$

By the above optimal control solution and the state and costate equations specified in (7.136a) and (7.136b), respectively, we can obtain

$$\begin{bmatrix} x^*(k+1) \\ \lambda^*(k) \end{bmatrix} = \begin{bmatrix} A & -E \\ C^\top QC & A^\top \end{bmatrix} \begin{bmatrix} x^*(k) \\ \lambda^*(k+1) \end{bmatrix} + \begin{bmatrix} 0 \\ -C^\top Q \end{bmatrix} r(k), \tag{7.139}$$

with $E \equiv BR^{-1}B^\top$.

The boundary condition (7.139) is specified as follows:

$$\lambda(k_f) = C^\top FCx(k_f) - C^\top Fr(k_f). \tag{7.140}$$

Next, we shall specify the optimal control with respect to the state. Firstly, by observing the boundary condition (7.140), suppose that

$$\lambda^*(k) = P(k)x^*(k) - \eta(k), \tag{7.141}$$

for some matrix $P(k)$ and a vector $\eta(k)$.

Thus by (7.139) and (7.141), we get

$$x^*(k+1) = Ax^*(k) - EP(k+1)x^*(k+1) + E\eta(k+1), \tag{7.142}$$

which implies

$$x^*(k+1) = [\mathbb{I} + EP(k+1)]^{-1} \left[Ax^*(k) + E\eta(k+1) \right]. \tag{7.143}$$

By applying (7.143) and (7.141) to (7.139), we have

$$\begin{aligned} & \left[-P(k) + A^\top P(k+1)[\mathbb{I} + EP(k+1)]^{-1}A + C^\top QC \right] x(k) + \eta(k), \\ & + A^\top P(k+1)[\mathbb{I} + EP(k+1)]^{-1}E\eta(k+1) - A^\top \eta(k+1) - C^\top Qr(k) = 0. \end{aligned} \tag{7.144}$$

Since the above equation holds for all values of $x^*(k)$, it implies that both the coefficient of $x^*(k)$ and the rest parts in (7.144) must vanish. Thus

$$P(k) = A^\top P(k+1)[\mathbb{I} + EP(k+1)]^{-1}A + C^\top QC, \tag{7.145a}$$

$$\eta(k) = A^\top \left[\mathbb{I} - \left[P^{-1}(k+1) + E \right]^{-1} E \right] \eta(k+1) + C^\top Qr(k), \tag{7.145b}$$

or equivalently the following inequalities hold:

$$P(k) = A^\top \left[P^{-1}(k+1) + E \right]^{-1} A + C^\top QC, \tag{7.146a}$$

$$\eta(k) = \left[A^\top - A^\top P^{-1}(k+1)[\mathbb{I} + EP(k+1)]^{-1}E \right] \eta(k+1) + C^\top Qr(k). \tag{7.146b}$$

By (7.140) and (7.141), we obtain the boundary conditions for (7.145) and (7.146) given as follows:

$$P(k_f) = C^\top FC, \tag{7.147a}$$

$$\eta(k_f) = C^\top Fr(k_f). \tag{7.147b}$$

Notice that (7.145) or (7.146) is the nonlinear, matrix difference Riccati equation (DRE) to be solved backward using the final condition (7.147a), and the linear, vector difference equation (7.146) is solved backward using the final condition (7.147b).

As $P(k)$ can be implemented as mentioned above, by applying (7.141) to (7.138), we can get

$$u^*(k) = -R^{-1}B^\top [P(k+1)x^*(k+1) - \eta(k+1)], \qquad (7.148)$$

and then by applying the state equation (7.136a) to (7.148), we have

$$u^*(k) = -R^{-1}B^\top P(k+1)\left[Ax^*(k) + Bu^*(k)\right] + R^{-1}B^\top \eta(k+1). \qquad (7.149)$$

Now premultiplying by R and solving for the optimal control $u^*(k)$, we have

$$u^*(k) = -L(k)x^*(k) + L_f(k)\eta(k+1), \qquad (7.150)$$

where the feedback gain $L(k)$ and the feed forward gain $L_f(k)$ are given by

$$L(k) = \left[R + B^\top P(k+1)B\right]^{-1} B^\top P(k+1)A \qquad (7.151a)$$
$$L_f(k) = \left[R + B^\top P(k+1)B\right]^{-1} B^\top. \qquad (7.151b)$$

Finally by (7.136a) and (7.150), the optimal state trajectory is given as

$$x^*(k+1) = [A - BL(k)]x^*(k) + BL_f(k)\eta(k+1). \qquad (7.152)$$

Example 7.9 Consider the state system studied in Example 7.7.
Implement the optimal control subject to the performance cost given as

$$J(x, u) = \frac{1}{2}[x_1(k_f) - 2]^2 + [x_2(k_f)]^2 + \sum_{k=k_0}^{k_f-1}\left[\frac{1}{2}[x_1(k)]^2 + \frac{1}{2}[x_2(k)]^2 + [u(k)]^2\right].$$
$$(7.153)$$

Solution. By the specifications of the tracking problems, we have

$$A = \begin{bmatrix} 1 & 1 \\ 0 & 1 \end{bmatrix}, B = \begin{bmatrix} 1 \\ 1 \end{bmatrix}, C = \begin{bmatrix} 1 & 0 \\ 0 & 1 \end{bmatrix}, F = \begin{bmatrix} 1 & 0 \\ 0 & 2 \end{bmatrix}, Q = \begin{bmatrix} 1 & 0 \\ 0 & 1 \end{bmatrix}, R = 2. \quad (7.154)$$

The reference trajectory $r(\cdot)$ is given as $r(k) = 0$ for all $k \neq k_f$, and $r(k_f) = \begin{bmatrix} 2 \\ 0 \end{bmatrix}$.
Thus, the optimal control can be implemented by applying the results developed in this section.   □

7.5   Discrete-Time Pontryagin's Minimum Principle

In this part, consider a discrete-time state system specified as follows:

$$x(k+1) = f(x(k), u(k), k), \text{ with } k = k_0, \cdots, k_f - 1, \qquad (7.155)$$

where $x(k) \in \mathbb{R}^n$, $u(k) \in \mathbb{R}^n$, and $u(k) \in U$ with U are a closed and bounded set, and the state at final time satisfies the following equality constraint:

$$\phi(x(k_f), k_f) = 0, \qquad (7.156)$$

which is a i-dimensional-valued function, with $i \leq n$.

We will implement the optimal control $u^* \equiv (u^*(k_0), \cdots, u^*(k_f - 1))$ for the system given above such that the following cost function is minimized:

$$J(x, u) \triangleq h(x(k_f), k_f) + \sum_{k=k_0}^{k_f-1} g(x(k), u(k), k). \qquad (7.157)$$

Theorem 7.5 will develop the specification for the optimal control solution to the underlying discrete-time optimal control problems. Before that, firstly define some notations below.

Define the Hamiltonian $\mathscr{H}(x(k), u(k), \lambda(k+1), k)$ such that

$$\mathscr{H}(x(k), u(k), \lambda(k+1), k) \triangleq g(x(k), u(k), k) + [\lambda(k+1)]^\top f(x(k), u(k), k). \qquad (7.158)$$

Consider certain assumptions below which will be used to support the proof of Theorem 7.5.

Assumption 7.1 For any pair of states $x(k)$ and $\widehat{x}(k)$ for all k, and any control $u(k)$, the following hold:

$$\|f(x(k), u(k), k) - f(\widehat{x}(k), u(k), k)\| \leq \alpha \|x(k) - \widehat{x}(k)\|, \qquad (7.159a)$$
$$\|g(x(k), u(k), k) - g(\widehat{x}(k), u(k), k)\| \leq \alpha \|x(k) - \widehat{x}(k)\|, \qquad (7.159b)$$

for some positive-valued α, and for any pair of controls $u(k)$ and $\widehat{u}(k)$ for all k, and any state $x(k)$, the following hold:

$$\|f(x(k), u(k), k) - f(x(k), \widehat{u}(k), k)\| \leq \beta \|u(k) - \widehat{u}(k)\|, \qquad (7.160a)$$
$$\|g(x(k), u(k), k) - g(x(k), \widehat{u}(k), k)\| \leq \beta \|u(k) - \widehat{u}(k)\|, \qquad (7.160b)$$

for some positive-valued β. $\square$

Theorem 7.5 *Under Assumption 7.1, the optimal control solution* (x^*, u^*, λ^*) *satisfies the following:*

$$x^*(k+1) = f(x^*(k), u^*(k), k), \tag{7.161a}$$

$$\lambda^*(k) = \frac{\partial \mathcal{H}(x^*(k), u^*(k), \lambda^*(k+1), k)}{\partial x(k)}, \tag{7.161b}$$

$$\mathcal{H}(x^*(k), u^*(k), \lambda^*(k+1), k) = \min_{u(k) \in U} \{\mathcal{H}(x^*(k), u(k), \lambda^*(k+1), k)\}, \tag{7.161c}$$

for all $k = k_0, \cdots, k_f - 1$, *with the boundary conditions for the state* $x^*(k_0) = x_0$ *and* $\phi(x^*(k_f), k_f) = 0$, *and the following for the costate*

$$\lambda^*(k_f) = \frac{\partial h(x^*(k_f), k_f)}{\partial x(k_f)} + \frac{\partial [\phi(x^*(k_f), k_f)]^\top}{\partial x(k_f)} \mu, \tag{7.162}$$

with certain vector-valued $\mu \in \mathbb{R}^i$.

Proof Firstly define the Lagrange function, denoted by $\mathcal{L}(u)$, for the underlying constrained optimization problems such that

$$\mathcal{L}(u) \triangleq J(x, u) + \mu^\top \phi(x(k_f), k_f); \tag{7.163}$$

then by the specification of the state equation (7.155), the cost function (7.157) and the Hamiltonian (7.158), we can obtain

$$\mathcal{L}(u) = h(x(t_f), t_f) + \mu^\top \phi(x(k_f), k_f)$$

$$+ \sum_{k=k_0}^{k_f-1} \Big[\mathcal{H}(x(k), u(k), \lambda(k+1), k) - [\lambda(k+1)]^\top x(k+1) \Big]. \tag{7.164}$$

Suppose that $u^* \equiv (u^*(k_0), \cdots, u^*(k_f - 1))$ is the optimal control, and $x^* \equiv (x^*(k_0 + 1), \cdots, x^*(k_f))$ is the optimal state subject to u^*.

Consider a variation of the control $\delta u \equiv (\delta u(k_0), \cdots, \delta u(k_f - 1))$ with respect to u^*, and denote by $\delta x \equiv (\delta x(k_0 + 1), \cdots, \delta x(k_f))$ the corresponding variation of the state, say

$$x(k) = x^*(k) + \delta x(k),$$
$$u(k) = u^*(k) + \delta u(k),$$

for all k.

Thus the increment of the Lagrange function, denoted by $\Delta\mathscr{L}(u)$, is specified as

$$\Delta\mathscr{L}(u) \triangleq \mathscr{L}(u) - \mathscr{L}(u^*) \tag{7.165}$$
$$= [h(x(t_f), t_f) - h(x^*(t_f), t_f)] + [\mu^\top \phi(x(k_f), k_f) - \mu^\top \phi(x^*(k_f), k_f)]$$
$$+ \sum_{k=k_0}^{k_f-1} \left[\mathscr{H}(x(k), u(k), \lambda^*(k+1), k) - \mathscr{H}(x^*(k), u(k), \lambda^*(k+1), k) \right]$$
$$+ \sum_{k=k_0}^{k_f-1} \left[\mathscr{H}(x^*(k), u(k), \lambda^*(k+1), k) - \mathscr{H}(x^*(k), u^*(k), \lambda^*(k+1), k) \right]$$
$$+ \sum_{k=k_0}^{k_f-1} \left[[\lambda^*(k+1)]^\top x^*(k+1) - [\lambda^*(k+1)]^\top x(k+1) \right]$$
$$= \left[\frac{\partial h(x^*(k_f), k_f)}{\partial x(k_f)} \right]^\top \delta x(k_f) + \left[\frac{\partial [\phi(x^*(k_f), k_f)]^\top}{\partial x(k_f)} \mu \right]^\top \delta x(k_f)$$
$$+ \sum_{k=k_0}^{k_f-1} \left[\frac{\partial \mathscr{H}(x^*(k), u(k), \lambda^*(k+1), k)}{\partial x(k)} \right]^\top \delta x(k)$$
$$+ \sum_{k=k_0}^{k_f-1} \left[\mathscr{H}(x^*(k), u(k), \lambda^*(k+1), k) - \mathscr{H}(x^*(k), u^*(k), \lambda^*(k+1), k) \right]$$
$$- \sum_{k=k_0}^{k_f-1} [\lambda^*(k+1)]^\top \delta x(k+1)$$
$$+ \sum_{k=k_0}^{k_f-1} \mathcal{O}(\delta x(k)), \tag{7.166}$$

where $\mathcal{O}(\delta x(k))$ denotes the higher orders of $\delta x(k)$.

Furthermore, we reorganize the above and obtain

$$\Delta\mathscr{L}(u) = \left[\frac{\partial h(x^*(k_f), k_f)}{\partial x(k_f)} + \frac{\partial [\phi(x^*(k_f), k_f)]^\top}{\partial x(k_f)} \mu - \lambda^*(k_f) \right]^\top \delta x(k_f)$$
$$+ \sum_{k=k_0}^{k_f-1} \left[\frac{\partial \mathscr{H}(x^*(k), u(k), \lambda^*(k+1), k)}{\partial x(k)} - \lambda^*(k) \right]^\top \delta x(k)$$
$$+ \sum_{k=k_0}^{k_f-1} \left[\mathscr{H}(x^*(k), u(k), \lambda^*(k+1), k) - \mathscr{H}(x^*(k), u^*(k), \lambda^*(k+1), k) \right]$$
$$+ \sum_{k=k_0}^{k_f-1} \mathcal{O}(\delta x(k)). \tag{7.167}$$

Thus by setting a costate λ^* satisfying (7.161b) for all $k = k_0 + 1, \cdots, k_f - 1$ and the boundary condition (7.162) at k_f, we can obtain that

$$
\Delta \mathscr{L}(u) = \sum_{k=k_0}^{k_f - 1} \left[\frac{\partial \mathscr{H}(x^*(k), u(k), \lambda^*(k+1), k)}{\partial x(k)} \right.
$$

$$
\left. - \frac{\partial \mathscr{H}(x^*(k), u^*(k), \lambda^*(k+1), k)}{\partial x(k)} \right]^{\top} \delta x(k)
$$

$$
+ \sum_{k=k_0}^{k_f - 1} \left[\mathscr{H}(x^*(k), u(k), \lambda^*(k+1), k) - \mathscr{H}(x^*(k), u^*(k), \lambda^*(k+1), k) \right]
$$

$$
+ \sum_{k=k_0}^{k_f - 1} \mathscr{O}(\delta x(k))
$$

$$
= \sum_{k=k_0}^{k_f - 1} \left[\mathscr{H}(x^*(k), u(k), \lambda^*(k+1), k) - \mathscr{H}(x^*(k), u^*(k), \lambda^*(k+1), k) \right]
$$

$$
+ \sum_{k=k_0}^{k_f - 1} \mathscr{O}(\delta x(k)), \tag{7.168}
$$

where the last equality holds due to the assumption that $\dfrac{\partial \mathscr{H}(x(k), u(k), \lambda(k+1), k)}{\partial x(k)}$ is continuous with respect to the control $u(k)$.

Next consider a specific variation of control $\delta u(k)$ such that

$$
\delta u(k) = \begin{cases} \theta, & \text{in case } k = j \\ 0, & \text{otherwise} \end{cases}, \tag{7.169}
$$

for some j with $k_0 \leq j \leq k_f - 1$, where θ is infinitesimally positive valued; then it results as follows:

$$
\Delta \mathscr{L}(u) = \mathscr{H}(x^*(k), u(k), \lambda^*(k+1), k) - \mathscr{H}(x^*(k), u^*(k), \lambda^*(k+1), k)
$$

$$
+ \sum_{k=k_0}^{k_f - 1} \mathscr{O}(\delta x(k)). \tag{7.170}
$$

Also by the state equation given in (7.155), We have

$$
\delta x(k+1) \triangleq x(k+1) - x^*(k+1)
$$

$$
= [f(x(k), u(k), k) - f(x^*(k), u(k), k)]
$$

$$
+ [f(x^*(k), u(k), k) - f(x^*(k), u^*(k), k)]; \tag{7.171}
$$

then by the assumptions on the function f, we can obtain

$$\|\delta x(k+1)\| \le \alpha_k \xi, \tag{7.172}$$

in case $\|\delta u(k)\| \le \xi$, for all $k = k_0, \cdots, k_f - 1$, with some small positive-valued ξ.

Thus, we can claim that considering an infinitesimal variation $\delta u(k)$, the sign $\Delta \mathscr{L}(u)$ specified in (7.170) is determined by

$$\mathscr{H}(x^*(j), u(j), \lambda^*(j+1), j) - \mathscr{H}(x^*(j), u^*(j), \lambda^*(j+1), j).$$

As a consequence, in order to set $\Delta \mathscr{L}(u) \ge 0$, the following should hold:

$$\mathscr{H}(x^*(j), u^*(j), \lambda^*(j+1), j) \le \mathscr{H}(x^*(j), u(j), \lambda^*(j+1), j). \tag{7.173}$$

We can obtain the same analysis for all $j = k_0, k_0 + 1, \cdots, k_f - 1$, and then get

$$\mathscr{H}(x^*(k), u^*(k), \lambda^*(k+1), k) \le \mathscr{H}(x^*(k), u(k), \lambda^*(k+1), k), \tag{7.174}$$

for all $k = k_0, k_0 + 1, \cdots, k_f - 1$, which implies that

$$\mathscr{H}(x^*(k), u^*(k), \lambda^*(k+1), k) = \min_{u(k) \in U} \mathscr{H}(x^*(k), u(k), \lambda^*(k+1), k), \tag{7.175}$$

for all $k = k_0, k_0 + 1, \cdots, k_f - 1$, say $\mathscr{H}(x^*(k), u(k), \lambda^*(k+1), k)$ reaches a minima at $u^*(k)$. $\qquad\square$

We will apply the results proposed in Theorem 7.5 to implement the optimal control solution for a specific numerical example.

Example 7.10 Consider a one-dimensional state system such that

$$x(k+1) = x(k) + \beta u(k), \tag{7.176}$$

with fixed k_0 and k_f, and a given β. We also consider that $u(k) \le 1, x(k_0) = x_0$, and $x(k_f) = x_f$.

Implement the optimal control u^* such that the following cost function:

$$J(u) = \sum_{k=k_0}^{k_f-1} u^2(k) \tag{7.177}$$

is minimized.

Solution. Firstly define the Hamiltonian $\mathscr{H}$

$$\mathscr{H}(x(k), u(k), \lambda(k+1), k) \triangleq u^2(k) + \lambda(k+1)[x(k) + \beta u(k)], \tag{7.178}$$

and

$$\widehat{h}(x(k_f), k_f) = \eta[x(k_f) - x_f], \tag{7.179}$$

where η is the Lagrange multiplier.

By (7.161b), the costate equation is given as follows:

$$\lambda^*(k) = \frac{\partial \mathscr{H}(x^*(k), u^*(k), \lambda^*(k+1), k)}{\partial x(k)} = \lambda^*(k+1), \tag{7.180}$$

for all $k = k_0 + 1, k_0 + 2, \cdots, k_f - 1$, and, by (7.162), the boundary condition as

$$\lambda^*(k_f) = \frac{\partial \left[\widehat{h}(x(k_f), k_f)\right]^{\top}}{\partial x(k_f)} = \eta. \tag{7.181}$$

Thus we obtain that the costate is constant valued, such that

$$\lambda^*(k) = \eta, \tag{7.182}$$

for all $k = k_0 + 1, k_0 + 2, \cdots, k_f$, and then we have

$$\mathscr{H}(x(k), u(k), \lambda^*(k+1), k) \triangleq u^2(k) + \lambda^*(k+1)[x(k) + \beta u(k)]$$
$$= u^2(k) + \eta[x(k) + \beta u(k)]. \tag{7.183}$$

Since, in this example, we consider that the admissible control set is

$$U(k) = \{u : |u| \leq 1\}, \tag{7.184}$$

for all $k = k_0, k_0 + 1, \cdots, k_f - 1$, by which and (7.183), and by applying (7.161c), we have

$$u^*(k) = \begin{cases} -\text{sgn}\left(\frac{\beta\eta}{2}\right), & \text{in case } \left|\frac{\beta\eta}{2}\right| > 1 \\ -\frac{\beta\eta}{2}, & \text{otherwise} \end{cases}, \tag{7.185}$$

that is to say, the optimal control sequence $u^* \equiv u^*(k_0, \cdots, k_f - 1)$ is specified as

$$u^* = (\zeta, \zeta, \cdots, \zeta), \tag{7.186}$$

with $\zeta = 1, -1$, or $-\dfrac{\beta\eta}{2}$.

Hence by the state equation and its boundary condition, we have

$$x^*(t_f) = x_f = x^*(t_f - 1) + \beta\zeta = x_0 + [k_f - k_0]\beta\zeta; \tag{7.187}$$

then we can obtain, in case $\left| \dfrac{x_f - x_0}{[k_f - k_0]\beta} \right| \leq 1,$

$$u^*(k) = \zeta = \frac{x_f - x_0}{[k_f - k_0]\beta}, \tag{7.188}$$

for all $k = k_0, k_0 + 1, \cdots, k_f - 1$.

However in case that $\left| \dfrac{x_f - x_0}{[k_f - k_0]\beta} \right| > 1$, there does not exist any admissible control solution such that the state system can be driven from x_0 at k_0 to x_f at k_f. □

7.6 Discrete-Time Dynamic Programming

In this part, consider the optimal control of discrete-time systems by applying the dynamic programming method.

7.6.1 Optimal Control Problems with Discrete State Values

In this section, we will firstly study a class of the shortest path problems.

The shortest path problems can be defined for graphs whether undirected, directed, or mixed. Here, we only consider directed graphs where a path is composed of a collection of consecutive vertices which are connected by an appropriate directed edge.

Denote by $v(k)$ a vertex at stage k. Two vertices $v(k)$ and $v(k + 1)$ are adjacent to each other when they are both connected to a common edge denoted by $(v(k), v(k + 1))$. A path with length k_f from $v(0)$ to $v(k_f)$ in a directed graph $\mathcal{G}$ is a sequence of vertices

$$P = (v(0), v(1), \cdots, v(k_f)) \in \mathcal{V} \times \mathcal{V} \times \cdots \times \mathcal{V}. \tag{7.189}$$

More specifically, for analytical simplicity, here consider that each path starts from a common vertex v_0 at the initial stage 0 and terminates at a common vertex v_f at the final stage k_f.

Denote by a real-valued weighting cost $\psi(v(k), v(k + 1))$ with respect to an edge $(v(k), v(k + 1))$.

See Fig. 7.1 for an illustration of a specific directed graph with the weighting cost for each edge.

Thus the performance cost with respect to a path $P \equiv (v(0), v(1), \cdots, v(k_f))$, denoted by $J(P)$, is defined as

Fig. 7.1 A directed graph

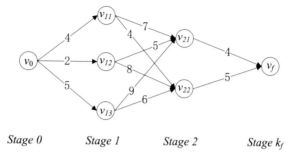

Stage 0   Stage 1   Stage 2   Stage k_f

$$J(P) = \sum_{k=0}^{k_f-1} \psi(v(k), v(k+1)). \tag{7.190}$$

The shortest path from v_0 to v_f is a path $P^* \equiv (v^*(0), v^*(1), \cdots, v^*(k_f))$ such that $(v^*(k), v^*(k+1))$, for all k, is an edge of the graph $\mathcal{G}$, $v^*(0) = v_0$ and $v^*(k_f) = v_f$ and

$$J(P^*) = \min_{P} J(P). \tag{7.191}$$

Also denote by $J_k^*(v)$ the minimum cost from a vertex v at stage k to the final stage k_f, say

$$J_k^*(v) = \min_{\substack{v(k)=v \\ (v(i),v(i+1)) \text{ is an edge,} \forall i}} \sum_{i=k}^{k_f-1} \psi(v(i), v(i+1)). \tag{7.192}$$

The principle of optimality for the underlying problem is given as

$$J_k^*(v) \triangleq \min_{\hat{v}} \{\psi(v, \hat{v}) + J_{k+1}^*(\hat{v})\}, \tag{7.193}$$

for all $k = 0, 1, \cdots, k_f - 1$.

Example 7.11 Implement the optimal solution for a shortest path problem as specified in Fig. 7.1 by applying the principle of optimality (7.193).

Solution. As observed in Fig. 7.1, we consider the final stage $k_f = 3$ and implement the optimal values firstly, and then work backward for $k = 2, 1, 0$, respectively.

At the last stage before the final stage $k_f - 1 = 2$, the minimum cost with each of the vertices v_{21} and v_{22} is given as

$$J_2^*(v_{21}) = \min \left\{\psi(v_{21}, v_f)\right\} = 4, \tag{7.194a}$$

$$J_2^*(v_{22}) = \min \left\{\psi(v_{22}, v_f)\right\} = 5; \tag{7.194b}$$

Fig. 7.2 The shortest path
for Example 7.11

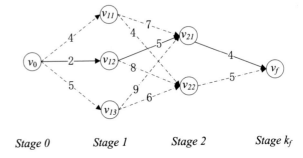

Stage 0   *Stage 1*   *Stage 2*   *Stage k_f*

then at stage 1, the minimum cost with each of the vertices at stage 1, v_{1j}, with
$j = 1, 2, 3$, is given as follows:

$$J_1^*(v_{11}) = \min \left\{ \psi(v_{11}, v_{21}) + J_2^*(v_{21}); \psi(v_{11}, v_{22}) + J_2^*(v_{22}) \right\} = 9, \quad (7.195)$$

with $\arg\min\{J_1^*(v_{11})\} = v_{22}$.

$$J_1^*(v_{12}) = \min \left\{ \psi(v_{12}, v_{21}) + J_2^*(v_{21}); \psi(v_{12}, v_{22}) + J_2^*(v_{22}) \right\} = 9, \quad (7.196)$$

with $\arg\min\{J_1^*(v_{12})\} = v_{21}$.

$$J_1^*(v_{13}) = \min \left\{ \psi(v_{13}, v_{21}) + J_2^*(v_{21}); \psi(v_{13}, v_{22}) + J_2^*(v_{22}) \right\} = 11, \quad (7.197)$$

with $\arg\min\{J_1^*(v_{13})\} = v_{22}$.
Finally at the initial stage $k_0 = 0$, the minimum cost with the state at stage 0 is
given as

$$J_0^*(v_0) = \min \left\{ \psi(v_0, v_{11}) + J_1^*(v_{11}); \psi(v_0, v_{12}) + J_1^*(v_{12}); \psi(v_0, v_{13}) + J_1^*(v_{13}) \right\}$$
$$= \min \left\{ 13; 11; 16 \right\} = 11, \quad (7.198)$$

with $\arg\min\{J_0^*(v_0)\} = v_{12}$.
Thus in summary by the above analysis, the shortest path is given as

$$J(P^*) = 11, \quad \text{with } P^* = (v_0, v_{12}, v_{21}, v_f); \quad (7.199)$$

see the solid-line path as the implemented shortest path displayed in Fig. 7.2. □

In Example 7.12 below, we will study how to solve another discrete-time optimal control problem where the control is constrained in a set composed of a finite number of components by applying the dynamic programming method.

Example 7.12 Consider a scalar-valued state system as follows:

$$x(k + 1) = x(k) + u(k), \tag{7.200}$$

and define a performance cost to be minimized as

$$J = \frac{1}{2}x^2(k_f) + \frac{1}{2}\sum_{k=k_0}^{k_f-1} u^2(k), \tag{7.201}$$

where, for simplicity of calculations, consider $k_f = 2$.

Here suppose that the control and state satisfy the following constraints, respectively:

$$u(k) \in \mathscr{U} \equiv \{-1, -0.5, 0, 0.5, 1\}, \tag{7.202a}$$
$$x(k) \in \mathscr{X} \equiv \{0, 0.5, 1, 1.5, 2\}, \tag{7.202b}$$

for all $k = 0, 1, 2$.

Implement the optimal control u^* subject to which the performance cost (7.201) is minimized.

Solution. Same as the last example, to apply the principle of optimality, we first consider the final time $k_f = 2$ and implement the optimal values, and then work backward for $k = 1, 0$, respectively.

At the final time $k_f = 2$, by (7.201), the minimum cost with the final state $x(2)$ is given as

$$J_2^*(x(2)) = \frac{1}{2}x^2(2) = \begin{cases} 2, & \text{with } x(2) = 2 \\ 1.125, & \text{with } x(2) = 1.5 \\ 0.5, & \text{with } x(2) = 1 \\ 0.125 & \text{with } x(2) = 0.5 \\ 0, & \text{with } x(2) = 0 \end{cases} . \tag{7.203}$$

The state at the final time $k_f = 2$, $x(2)$, is specified by applying the state transition given in (7.200); see an illustration displayed in Fig. 7.3, for all admissible values of $x(k)$ and $u(k)$ given by (7.202a) and (7.202b).

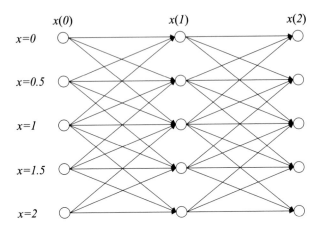

Fig. 7.3 The state transition subject to control u

For instance, suppose that $x(1) = 0.5$; then by applying

$$u(1) \in \mathcal{U} \equiv \{-1, -0.5, 0, 0.5, 1\},$$

we obtain

$$x(2) = x(1) + u(1) = \begin{cases} -0.5, & \text{with } u = -1 \\ 0, & \text{with } u = -0.5 \\ 0.5, & \text{with } u = 0 \\ 1, & \text{with } u = 0.5 \\ 1.5, & \text{with } u = 1 \end{cases} . \qquad (7.204)$$

However with $x(1) = 0.5$, the control value of $-1 \in \mathcal{U}$ is inadmissible since the resulting state is -0.5 which does not satisfy the constraint of (7.202b), that is to say, the admissible set of the controls is dependent on the value of the state.

By the principle of optimality and (7.201), we have

$$J_k^*(x(k)) = \min_{u(k)\in\mathcal{U}} \left\{ \frac{1}{2} u^2(k) + J_{k+1}^*(x(k+1)) \right\}, \qquad (7.205)$$

with $k = 0, 1$, and the boundary condition $J_{k_f}^*(x(k_f))$ given in (7.203).

Then at time $k_f - 1 = 1$, by (7.205), the minimum cost with the state at time $k = 1, x(1)$, is given as

$$J_1^*(x(1)) = \min_{u(1) \in \mathcal{U}} \left\{ \frac{1}{2} u^2(1) + J_2^*(x(2)) \right\}$$

$$= \min_{u(1) \in \mathcal{U}} \left\{ \frac{1}{2} u^2(1) + \frac{1}{2} x^2(2) \right\}; \qquad (7.206)$$

thus, by which, together with (7.204), we obtain

$$J_1^*(x(1) = 0.5) = \min_{u(1) \in \mathcal{U}} \left\{ \frac{1}{2} u^2(1) + J_2^*(x(2)) \right\}$$

$$= \min \left\{ \frac{1}{2} \times [-0.5]^2 + J_2^*(0); \frac{1}{2} \times 0^2 + J_2^*(0.5); \frac{1}{2} \times 0.5^2 + J_2^*(1); \frac{1}{2} \times 1^2 + J_2^*(1.5) \right\}$$

$$= \min \{0.125; 0.125; 0.625; 1.625\}$$

$$= 0.125. \qquad (7.207)$$

As a consequence, with $x(1) = 0.5$, we have

$$J_1^*(x(1) = 0.5) = 0.125, \text{ with } u^*(1; x(1) = 0.5) = -0.5 \text{ or } 0. \qquad (7.208)$$

Following the same analysis, we can obtain the minimum cost at time 1 for any other state at this time. Skip the procedure and list the results in Table 7.1.

Next, at initial time k_0, by (7.201) and (7.204), the minimum cost with the state at time $k = 0$, $x(0)$, is given as

$$J_0^*(x(0)) = \min_{u(0) \in \mathcal{U}} \left\{ \frac{1}{2} u^2(0) + J_1^*(x(0) + u(0)) \right\}, \qquad (7.209)$$

thus, by which, together with (7.203), (7.204) and Table 7.1, we obtain

$$J_0^*(x(0) = 1) = \min_{u(0) \in \mathcal{U}} \left\{ \frac{1}{2} u^2(0) + J_1^*(x(0) + u(0)) \right\}$$

$$= \min \left\{ \frac{1}{2} \times [-1]^2 + J_1^*(0); \frac{1}{2} \times [-0.5]^2 + J_1^*(0.5); \frac{1}{2} \times 0^2 + J_1^*(1); \right.$$

$$\left. \frac{1}{2} \times 0.5^2 + J_1^*(1.5); \frac{1}{2} \times 1^2 + J_1^*(2) \right\}$$

$$= \min \{0.5; 0.25; 0.25; 0.75; 1.5\}$$

$$= 0.25. \qquad (7.210)$$

As a consequence, we can get

$$J_0^*(x(0) = 1) = 0.25, \text{ with } u^*(0, x(0) = 1) = -0.5 \text{ or } 0. \qquad (7.211)$$

Table 7.1 The minimum cost function at time $k = 1$ and the associated optimal control at this time

State $x(1)$	Admissible Control $u(1)$	$J_1(x(1), u(1))$	$u^*(1, x(1))$	$J_1^*(x(1))$
	0	0	0	0
0	0.5	0.25		
	1	1		
	−0.5	0.125	−0.5 or 0	0.125
0.5	0	0.125		
	0.5	0.625		
	1	1.625		
	−1	0.5	−0.5	0.25
	−0.5	0.25		
1	0	0.5		
	0.5	1.25		
	1	2.5		
	−1	0.625	−1 or −0.5	0.625
1.5	−0.5	0.625		
	0	1.125		
	0.5	2.125		
	−1	1	−1	1
2	−0.5	1.25		
	0	2		

As earlier, following the same analysis, we can obtain the minimum cost at the initial time 0 for any other state at this time. Skip the procedure and list the results in Table 7.2.

In summary, by the above analysis, we can obtain the optimal control and the associated state trajectories for any initial state $x(0)$; see an illustration displayed in Fig. 7.4.

Thus following the arrows of the solid lines in Fig. 7.4, we can give the optimal trajectories, for any initial state $x(0)$, which may not be unique.

For instance, suppose that $x(0) = 0.5$; there exist two optimal state trajectories:

$$x^* = (0.5, 0.5, 0), \text{ with } u^* = (0, -0.5), \tag{7.212a}$$

$$x^* = (0.5, 0, 0), \text{ with } u^* = (-0.5, 0), \tag{7.212b}$$

both of which share an identical minimum cost

$$J_0^*(x(0) = 0.5) = 0.125, \tag{7.213}$$

as listed in Table 7.2.   □

Table 7.2 The minimum cost function at time $k = 1$ and the associated optimal control at this time

State $x(0)$	Admissible Control $u(0)$	$J_0(x(0), u(0))$	$u^*(0, x(0))$	$J_0^*(x(0))$
	0	0	0	0
0	0.5	0.25		
	1	0.75		
	−0.5	0.125	−0.5 or 0	0.125
0.5	0	0.125		
	0.5	0.375		
	1	1.125		
	−1	0.5	−0.5 or 0	0.25
	−0.5	0.25		
1	0	0.25		
	0.5	0.75		
	1	1.5		
	−1	0.625	−0.5	0.375
1.5	−0.5	0.375		
	0	0.625		
	0.5	1.125		
	−1	0.75	−1 or −0.5	0.75
2	−0.5	0.75		
	0	1		

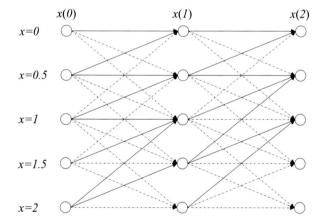

Fig. 7.4 The optimal control and associated state trajectories

Notice that as demonstrated in Example 7.12 where the sets of states and controls are composed of finite components as defined in (7.202a) and (7.202b), respectively, the admissible control set may be dependent upon the given state values, since the transited state subject to a certain control value may be beyond the defined set of states, e.g., the control $u = 0.5$ or $u = 1$ for state $x = 2$ is inadmissible since the resulting state is 2.5 or 3 which does not belong to $\mathscr{X}$ defined in (7.202b).

7.6.2 Optimal Control Problems with Continuous State Values

In the last section, as demonstrated in Example 7.12, we studied discrete-time optimal control problems where the state and control are all constrained in a set composed of finite number of components, respectively.

In the following, we will study a specific discrete-time optimal control problem with the state and control constrained in a compact set.

Consider that the state system is given as

$$x(k+1) = f(x(k), u(k), k), \tag{7.214}$$

and the performance cost function is defined as

$$J(x, u) \triangleq h(x(k_f)) + \sum_{k=k_0}^{k_f-1} g(x(k), u(k)), \tag{7.215}$$

where the state $x(k)$ and the control $u(k)$ are the n-dimensional and r-dimensional state and control vectors, respectively.

Denote by $\mathscr{X}$ the set of states, and denote by $\mathscr{U}(x, k)$ the set of admissible controls of a state x at time k, such that

$$\mathscr{U}(x, k) \triangleq \{u \in U; \text{ s.t. } f(x, u, k) \in \mathscr{X}\}, \tag{7.216}$$

where U represents a set of controls.

Also define a collection of performance cost functions denoted by $J_k(x(k), u)$, such that

$$J_k(x(k), u) \triangleq h(x(k_f)) + \sum_{i=k}^{k_f-1} g(x(i), u(i)), \tag{7.217}$$

for all $k = k_0, \cdots, k_f - 1$.

Notice that as observed in (7.217), we can show that $J_k(x(k), u)$ represents the accumulated costs over the interval $\{k, k + 1, \cdots, k_f - 1\}$ with the initial state value at time k as $x(k)$ subject to the control $u \equiv (u(k), u(k + 1), \cdots, u(k_f - 1))$.

We will specify the optimal control u^* to minimize the performance cost function defined in (7.215) by applying the principle of optimality.

Suppose that we have evaluated the optimal control, state, and cost for all values starting from time $k + 1$ to time k_f; then, at any time k, by the principle of optimality, we have

$$J_k^*(x(k)) = \min_{u(k) \in \mathscr{U}(x(k))} \left\{ g(x(k), u(k)) + J_{k+1}^*(x(k + 1)) \right\}, \qquad (7.218)$$

with the boundary condition as

$$J_{k_f}^*(x) = h(x). \qquad (7.219)$$

Thus (7.218) means that if the optimal control, state, and cost from time $k + 1$ to the final time k_f, then the optimal values for a single time from k to $k + 1$ can be specified by (7.218).

In the following example, we will demonstrate how to implement the optimal control by applying the principle of optimality.

Example 7.13 Consider a state system such that

$$x(k + 1) = x(k) + u(k), \qquad (7.220)$$

with $k = 0, 1$, where $x(k) \in \mathscr{X} \equiv [0, 2]$ and $u(k) \in U \equiv [-1, 1]$, and the performance cost function to be minimized is given as

$$J \triangleq \frac{1}{2} x^2(k_f) + \frac{1}{2} \sum_{k=k_0}^{k_f-1} u^2(k). \qquad (7.221)$$

Implement the optimal control u^* and the associated state x^* which minimize the performance cost function given in (7.221).

Solution. For the underlying problem, for any state $x \in \mathscr{X}$, define the admissible control set, denoted by $\mathscr{U}(x)$, such that

$$\mathscr{U}(x) \triangleq \{u \in U; \text{ s.t. } u + x \in \mathscr{X}\}; \qquad (7.222)$$

then by $\mathscr{X} \equiv [0, 2]$ and $U \equiv [-1, 1]$,

$$\mathscr{U}(x) = [\max\{-x, -1\}, \min\{2 - x, 1\}]. \qquad (7.223)$$

To apply the principle of optimality, we consider the final time $k_f = 2$ and implement the optimal values firstly, and then work backward for $k = 1, 0$, respectively.

At the final time $k_f = 2$, by (7.221), the minimum cost with the final state $x(2)$ is given as

$$J_2^*(x(2)) = \frac{1}{2}x^2(2).\tag{7.224}$$

Thus at time $k_f - 1 = 1$, by (7.221), the minimum cost with the state at time 1, $x(1)$, is given as

$$
\begin{aligned}
J_1^*(x(1)) &= \min_{u(1)\in\mathcal{U}(x(1))} \left\{\frac{1}{2}u^2(1) + J_2^*(x(2))\right\}\\
&= \min_{u(1)\in\mathcal{U}(x(1))} \left\{\frac{1}{2}u^2(1) + \frac{1}{2}x^2(2)\right\}\\
&= \min_{u(1)\in\mathcal{U}(x(1))} \left\{\frac{1}{2}u^2(1) + \frac{1}{2}[x(1) + u(1)]^2\right\}, \text{ by (7.200)}\\
&= \frac{1}{4}x^2(1) + \min_{u(1)\in\mathcal{U}(x(1))} \left[u(1) + \frac{1}{2}x(1)\right]^2,\tag{7.225}
\end{aligned}
$$

thus, by which and by (7.223), we have

$$J_1^*(x(1)) = \frac{1}{4}x^2(1), \text{ with } u^*(1) = -\frac{1}{2}x(1).\tag{7.226}$$

Finally at the initial time $k_0 = 0$, by (7.221), the minimum cost with the state at time 1, $x(1)$, is given as

$$
\begin{aligned}
J_0^*(x(0)) &= \min_{u(0)\in\mathcal{U}(x(0))} \left\{\frac{1}{2}u^2(0) + J_1^*(x(1))\right\}\\
&= \min_{u(0)\in\mathcal{U}(x(0))} \left\{\frac{1}{2}u^2(0) + \frac{1}{4}x^2(1)\right\}\\
&= \min_{u(0)\in\mathcal{U}(x(0))} \left\{\frac{1}{2}u^2(0) + \frac{1}{4}[x(0) + u(0)]^2\right\}, \text{ by (7.220)}\\
&= \frac{1}{6}x^2(0) + \frac{3}{4}\min_{u(0)\in\mathcal{U}(x(0))} \left[u(0) + \frac{1}{3}x(0)\right]^2,\tag{7.227}
\end{aligned}
$$

thus, by which and by (7.223), we have

$$J_0^*(x(0)) = \frac{1}{6}x^2(0), \text{ with } u^*(0) = -\frac{1}{3}x(0).\tag{7.228}$$

In summary, we have specified the optimal control u^* as follows:

$$u^*(0) = -\frac{1}{3}x(0), \qquad (7.229)$$

$$u^*(1) = -\frac{1}{2}x(1), \qquad (7.230)$$

and the associated state trajectory x^*.

For example, suppose that $x(0) = 1$; then by applying the optimal control we have $u^*(x(0)) = -\frac{1}{3}$; then

$$x^*(1) = x(0) + u^*(0) = \frac{2}{3}, \qquad (7.231)$$

and $u^*(x^*(1)) = -\frac{1}{2}x^*(1) = -\frac{1}{3}$; then

$$x^*(2) = x^{(}1) + u^*(1) = \frac{2}{3} - \frac{1}{3} = \frac{1}{3}. \qquad (7.232)$$

As a result, we get $J_0^*(x(0)) = \frac{1}{3}$. □

7.6.3 Discrete-Time Linear-Quadratic Problems

Here, we are to derive the optimal feedback control of a discrete-time system using the dynamic programming method.

Consider a linear, time-invariant, discrete-time state system,

$$x(k+1) = Ax(k) + Bu(k), \qquad (7.233)$$

and the associated performance cost

$$J_k(x(k)) = \frac{1}{2}[x(k_f)]^\top F x(k_f) + \frac{1}{2}\sum_{i=k}^{k_f-1}\left[[x(i)]^\top Qx(i) + [u(i)]^\top Ru(i)\right], \qquad (7.234)$$

where $x(k)$ and $u(k)$ are n- and r-dimensional state and control vectors, and $A(k)$ and $B(k)$ are matrices of $n \times n$ and $n \times r$ dimensions, respectively.

Further, F and Q are $n \times n$, symmetric, positive semidefinite matrices, respectively, and R is an $r \times r$ symmetric, positive-definite matrix.

For the present discussion, assume that there are no constraints on the state or control.

The problem is to determine the optimal control $u^*(k)$, for all $k = i, i + 1, \cdots,$ $k_f - 1$, subject to which the performance cost J_k is minimized by applying the principle of optimality.

Suppose further that the initial state $x(k_0)$ is fixed and the final state $x(k_f)$ is free.

As demonstrated earlier in this section with a numerical example, by applying the dynamic programming method, we start with the final time k_f and then work backward. As a result, we implement the optimal control and state at each time.

- At the final time k_f, the cost function is given as

$$J_{k_f}(x(k_f)) = \frac{1}{2}[x(k_f)]^\top F x(k_f). \tag{7.235}$$

- Next at time $k_f - 1$, the cost function defined in (7.234) becomes

$$J_{k_f-1}(x(k_f - 1), u(k_f - 1)) = \frac{1}{2}[x(k_f - 1)]^\top Q x(k_f - 1)$$
$$+ \frac{1}{2}[u(k_f - 1)]^\top R u(k_f - 1) + \frac{1}{2}[x(k_f)]^\top F x(k_f). \tag{7.236}$$

According to the principle of optimality (7.218), we need to find the optimal control $u^*(k_f - 1)$ to minimize the cost function (7.236).

Before that, rewrite the relation (7.236) to make all the terms in (7.236) to belong to time $k_f - 1$. For this, by (7.233) and (7.236), we have

$$J_{k_f-1}(x(k_f - 1), u(k_f - 1))$$
$$= \frac{1}{2}[x(k_f - 1)]^\top Q x(k_f - 1) + \frac{1}{2}[u(k_f - 1)]^\top R u(k_f - 1)$$
$$+ \frac{1}{2}[Ax(k_f - 1) + Bu(k_f - 1)]^\top F[Ax(k_f - 1) + Bu(k_f - 1)]. \tag{7.237}$$

Since, in this part, we do not consider any constraints on states or controls, it is direct to implement the minimum value of (7.237) with respect to $u(k_f - 1)$ by the following:

$$\frac{\partial J_{k_f-1}(x(k_f - 1), u^*(k_f - 1))}{\partial u(k_f - 1)}$$
$$= Ru^*(k_f - 1) + B^\top F[Ax(k_f - 1) + Bu^*(k_f - 1)] = 0; \tag{7.238}$$

then by solving the above equation for $u^*(k_f - 1)$, we have

$$u^*(k_f - 1) = -[R + B^\top F B]^{-1} B^\top F Ax(k_f - 1) \equiv -L(k_f - 1)x(k_f - 1), \tag{7.239}$$

with $L(k_f - 1)$ specified as follows:

$$L(k_f - 1) \triangleq [R + B^\top F B]^{-1} B^\top F A, \tag{7.240}$$

which is also called the Kalman gain.

Now the optimal cost $J^*_{k_f-1}(x(k_f - 1))$ for this time $k_f - 1$ is found by substituting the optimal control $u^*(k_f - 1)$ from (7.239) into the cost function (7.237) to get

$$
\begin{aligned}
J^*_{k_f-1}(x(k_f - 1)) &= \frac{1}{2}[x(k_f - 1)]^\top \left[[A - BL(k_f - 1)]^\top F[A - BL(k_f - 1)] \right. \\
&\quad \left. + [L(k_f - 1)]^\top RL(k_f - 1) + Q \right] x(k_f - 1) \\
&= \frac{1}{2}[x(k_f - 1)]^\top P(k_f - 1)x(k_f - 1),
\end{aligned}
\tag{7.241}
$$

with $P(k_f - 1)$ defined as

$$
\begin{aligned}
P(k_f - 1) \triangleq [A - BL(k_f - 1)]^\top F[A - BL(k_f - 1)] \\
+ [L(k_f - 1)]^\top RL(k_f - 1) + Q.
\end{aligned}
\tag{7.242}
$$

- Now consider the analysis for the time $k_f - 2$ in this part.
  With $k = k_f - 2$ in the cost function (7.234), we have

$$
\begin{aligned}
& J_{k_f-2}(x(k_f - 2), u(k_f - 2), u(k_f - 1)) \\
&= \frac{1}{2}[x(k_f)]^\top Fx(k_f) + \frac{1}{2}[x(k_f - 2)]^\top Qx(k_f - 2) \\
&\quad + \frac{1}{2}[u(k_f - 2)]^\top Ru(k_f - 2) + \frac{1}{2}[x(k_f - 1)]^\top Qx(k_f - 1) \\
&\quad + \frac{1}{2}[u(k_f - 1)]^\top Ru(k_f - 1).
\end{aligned}
\tag{7.243}
$$

Thus, by (7.233), (7.239), (7.242), and (7.243), we can obtain

$$
\begin{aligned}
& J_{k_f-2}(x(k_f - 2), u(k_f - 2), u^*(k_f - 1)) \\
&= \frac{1}{2}[x(k_f - 2)]^\top Qx(k_f - 2) + \frac{1}{2}[u(k_f - 2)]^\top Ru(k_f - 2) \\
&\quad + \frac{1}{2}[x(k_f - 1)]^\top P(k_f - 1)x(k_f - 1),
\end{aligned}
\tag{7.244}
$$

where $P(k_f - 1)$ is given by (7.242).

At this time, we need to express all functions at time $k_f - 2$. Then, once again, for this time, to determine $u^*(k_f - 2)$ according to the principle of optimality (7.218), we minimize $J_{k_f-2}(x(k_f - 2), u(k_f - 2))$ in (7.244) with respect to $u(k_f - 2)$ and get relations similar to (7.239), (7.240), (7.241), and (7.242).

For example, the optimal cost function becomes

$$
J^*_{k_f-2}(x(k_f - 2)) = \frac{1}{2}[x(k_f - 2)]^\top P(k_f - 2)x(k_f - 2),
\tag{7.245}
$$

where $P(k_f - 2)$ is obtained similar to (7.242) except that it replaces $k_f - 1$ by $k_f - 2$.

We can continue the analysis given above for all other instants $k_f - 3$, $k_f - 4$, $\cdots$, k_0, and can generalize them for any time k. Thus the optimal control is given by

$$u^*(k) = -L(k)x^*(k), \tag{7.246}$$

where the Kalman gain $L(k)$ is given as

$$L(k) = [R + B^\top P(k+1)B]^{-1} B^\top P(k+1)A, \tag{7.247}$$

where the matrix $P(k)$, also called the Riccati matrix, is the backward solution of

$$P(k) = [A - BL(k)]^\top P(k+1)[A - BL(k)] + [L(k)]^\top RL(k) + Q, \tag{7.248}$$

with the final condition $P(k_f) = F$, and the optimal cost function as

$$J_k^*(x(k)) = \frac{1}{2}[x^*(k)]^\top P(k)x^*(k). \tag{7.249}$$

Notice that up to now we have established the sufficient condition of the uniqueness of optimal control solution for the discrete-time linear-quadratic problem defined in this section, say F and Q positive semidefinite, and R positive definite, respectively.

In Corollary 7.1 below, we will give the necessary and sufficient condition for the uniqueness of the optimal control solution as well.

Corollary 7.1 *There exists a unique optimal control solution for the discrete-time linear-quadratic problem with the state equation and the cost function specified in (7.233) and (7.234), respectively, if and only if F and R are positive semidefinite and positive definite, respectively, and*

$$R + B^\top P(k+1)B \tag{7.250}$$

is positive definite.

Proof The proof is directly from the above analysis.   □

Notice that these are the same set of relations we obtained in Chap. 5 by using Pontryagin's minimum principle.

7.7 Discrete-Time Noncooperative Dynamic Games

7.7.1 Formulation of Discrete-Time Noncooperative Dynamic Games

In this part, we introduce a class of discrete-time noncooperative dynamic games. Basically, in each of these games, there are a set of individual players denoted by $\mathcal{N} \equiv \{1, 2, \cdots, N\}$.

Moreover, we consider that each of these players tries to maximize his own payoff function or minimize his own cost function.

We will formulate this class of dynamic games in the following.

Denote by $u_n(\cdot)$ the control of player n, respectively, over the discrete-time interval $\{k_0, \cdots, k_f - 1\}$.

Further, define $\mathcal{U}_n(\{k_0, \cdots, k_f - 1\})$ as *the sets of control strategies* for player n, such that

$$\mathcal{U}_n(\{k_0, \cdots, k_f - 1\}) \triangleq \{u_n(\cdot) : \{k_0, \cdots, k_f - 1\} \to U_n, \text{ s.t. } u_n(\cdot) \text{ is measurable}\}, \tag{7.251}$$

where U_n is *the admissible control set* of player n.

Consider a state system denoted by $x(\cdot)$ which is driven by the controls $u_n(t)$, with $n \in \mathcal{N}$, such that

$$x(k + 1) = f(x(k), u_1(k), \cdots, u_N(k), k), \tag{7.252}$$

for all $k \in \{k_0, k_f - 1\}$, with an initial value of $x(k_0) = x_0$, where $x(k) \in \mathbb{R}^n$.

Define a cost function of individual player n, denoted by $J_n(x, k, u)$ subject to a collection of control strategies of players $u \equiv (u_n; n \in \mathcal{N})$, such that

$$J_n(u_n, u_{-n}) \triangleq \sum_{k=k_0}^{k_f - 1} g_n(x(k + 1), x(k), u_1(k), \cdots, u_N(k), k), \tag{7.253}$$

with $u_{-n} \equiv (u_1, \cdots, u_{n-1}, u_{n+1}, \cdots, u_N)$.

It is called an N-person, *discrete-time noncooperative dynamic game* if the objective of each individual player is to minimize the cost function (7.253), respectively.

Definition 7.1 (*Nash Equilibrium of Noncooperative Dynamic Game*) Call a strategy denoted by $u^*(\cdot) \equiv (u_n^*(\cdot); n \in \mathcal{N})$ as a Nash equilibrium (NE) of the discrete-time noncooperative dynamic game, if the following holds:

$$J_n(u_n^*, u_{-n}^*) \leq J_n(u_n, u_{-n}^*), \tag{7.254}$$

for all $u_n \equiv u_n(k_0, k_0 + 1, \cdots, k_f - 1) \in \mathcal{U}_n$, for all $n \in \mathcal{N}$. $\qquad\qquad \square$

7.7.2 NE of Discrete-Time Noncooperative Dynamic Games

In Theorem 7.6 below, we will give a necessary condition for an NE of a discrete-time noncooperative dynamic game; before that we define a local Hamiltonian for an individual player n, with $n \in \mathcal{N}$, in the following:

$$\mathcal{H}_n(x(k), u_1(k), \cdots, u_N(k), \lambda_n(k+1), k)$$
$$\triangleq g_n(f(x(k), u_1(k), \cdots, u_N(k), k), x(k), u_1(k), \cdots, u_N(k), k)$$
$$+ [\lambda_n(k+1)]^\top f(x(k), u_1(k), \cdots, u_N(k), k), \tag{7.255}$$

for each time $k \in \{k_0, k_0 + 1, \cdots, k_f - 1\}$, where λ represents the costate of the noncooperative dynamic game system.

Theorem 7.6 *Suppose that u^* is an NE of a discrete-time noncooperative dynamic game, and x^* is the corresponding state trajectory subject to this NE strategy, say*

$$x^*(k+1) = f(x^*(k), u_1^*(k), \cdots, u_N^*(k), k), \tag{7.256}$$

for all $k \in \{k_0, \cdots, k_f - 1\}$, with an initial value of $x(k_0) = x_0$.

Then under Assumption 7.1, there exists a costate trajectory for each player $n \in \mathcal{N}$, denoted by $\lambda_n^* \equiv (\lambda_n^*(k_0), \cdots, \lambda_n^*(k_f - 1))$, such that

$$u_n^*(k, x_0) = \underset{u_n(k)}{argmin}\ \mathcal{H}_n(x^*(k), u_n(k), u_{-n}^*(k), \lambda_n^*(k+1), k) \tag{7.257a}$$

$$\lambda_n^*(k) = \left[\lambda_n^*(k+1) + \frac{\partial g_n(x^*(k+1), x^*(k), u^*(k), k)}{\partial x(k+1)} \right]^\top \frac{\partial f(x^*(k), u^*(k), k)}{\partial x(k)}$$
$$+ \frac{\partial g_n(x^*(k+1), x^*(k), u^*(k), k)}{\partial x(k)}. \tag{7.257b}$$

Proof Suppose that $u^* \equiv (u_1^*, \cdots, u_N^*)$ is an NE for the underlying dynamic game; then by Definition 7.1, the following holds:

$$J_n(u_n^*, u_{-n}^*) \leq J_n(u_n, u_{-n}^*), \tag{7.258}$$

for each individual player n, that is to say, for each given control of other players $u_{-n} \equiv (u_1, \cdots, u_{n-1}, u_{n+1}, \cdots, u_N)$, u_n^* is an optimal control for $J_n(u_n, u_{-n}^*)$, which is the cost function of player n, defined in (7.253) subject to the state equation

$$x(k+1) = f(x(k), u_1^*(k), \cdots, u_{n-1}^*(k), u_n(k), u_{n+1}^*(k), \cdots, u_N^*(k), k),$$

for all $k \in \{k_0, k_f - 1\}$.

Then by Theorem 7.5, there exists a costate trajectory for each player $n \in \mathcal{N}$, denoted by $\lambda_n^* \equiv (\lambda_n^*(k_0), \cdots, \lambda_n^*(k_f - 1))$, such that (7.257a) holds by (7.161c), and

$$\lambda^*(k) = \frac{\partial \mathcal{H}_n(x^*(k), u^*(k), \lambda_n^*(k+1), k)}{\partial x(k)}, \tag{7.259}$$

for all $k = k_0, \cdots, k_f - 1$.

Thus by the specification of $\mathcal{H}_n$ given in (7.255), we can get

$$\lambda_n^*(k) = \frac{\partial g_n(x^*(k+1), x^*(k), u^*(k), k)}{\partial x(k)}$$

$$+ \left[\frac{\partial g_n(x^*(k+1), x^*(k), u^*(k), k)}{\partial x(k+1)}\right]^\top \frac{\partial f(x^*(k), u^*(k), k)}{\partial x(k)}$$

$$+ [\lambda_n^*(k+1)]^\top \frac{\partial f(x^*(k), u^*(k), k)}{\partial x(k)}, \tag{7.260}$$

which implies the conclusion of (7.257b).   □

7.7.3   Discrete-Time Linear-Quadratic Noncooperative Dynamic Games

In the following, we further study a class of *discrete-time noncooperative linear-quadratic dynamic games* in Definition 7.2.

Definition 7.2 (*Discrete-Time Noncooperative Linear-Quadratic Dynamic Games*) A game is called a discrete-time noncooperative linear-quadratic dynamic game if $f(x(k), u_1(k), \cdots, u_N(k), k)$ in the state equation (7.252) and $g_n(x(k+1), x(k), u_1(k), \cdots, u_N(k), k)$ in the local cost function of each individual player n (7.253) satisfy the following, respectively:

$$f(x(k), u_1(k), \cdots, u_N(k), k) = A(k)x(k) + \sum_{n \in \mathcal{N}} B_n(k)u_n(k), \tag{7.261a}$$

$$g_n(x(k+1), x(k), u_1(k), \cdots, u_N(k), k)$$

$$= \frac{1}{2}\left[[x(k+1)]^\top Q_n(k+1)x(k+1) + \sum_{\hat{n} \in \mathcal{N}} [u_{\hat{n}}(k)]^\top R_{n\hat{n}}(k)u_{\hat{n}}(k)\right], \tag{7.261b}$$

where $A(k), B_n(k), Q_n(k+1)$, and $R_{n\hat{n}}(k)$ are matrices with appropriate dimensions, respectively.

We further suppose that $Q_n(k+1)$ is symmetric and positive definite, and $R_{nn}(k)$ is strictly positive definite, for all $k \in \{k_0, \cdots, k_f - 1\}$ and all $n \in \mathcal{N}$.   □

In Theorem 7.7 below, we show the existence and uniqueness of NE for the linear-quadratic noncooperative dynamic game specified in Definition 7.2 above. Before that, firstly define notations of $\Lambda(k)$ and $\mathfrak{M}_n(k)$, respectively, as follows:

$$\Lambda(k) \triangleq \mathbb{I} + \sum_{n \in \mathcal{N}} B_n(k)[R_{nn}(k)]^{-1}[B_n(k)]^\top \mathfrak{M}_n(k+1), \tag{7.262a}$$

$$\mathfrak{M}_n(k) = Q_n(k) + [A(k)]^\top \mathfrak{M}_n(k+1)\Lambda^{-1}(k)A(k), \tag{7.262b}$$

for all $n \in \mathcal{N}$, with the boundary condition $\mathfrak{M}_n(k_f) = Q_n(k_f)$.

Theorem 7.7 *(Uniqueness and Existence of NE of Discrete-Time Linear-Quadratic Noncooperative Dynamic Game) Consider an N-person linear-quadratic dynamic game such that the matrices given in (7.262a) are invertible; then the game system has a unique NE specified as follows:*

$$u_n^*(k) = -[R_{nn}(k)]^{-1}[B_n(k)]^\top \left[\mathfrak{M}_n(k+1)\Lambda^{-1}(k)A(k)x^*(k) + \xi_n(k)\right], \tag{7.263}$$

for all $k \in \{k_0, \cdots, k_f - 1\}$ and $n \in \mathcal{N}$, where $x^*(\cdot)$ satisfies the following:

$$x^*(k+1) = \Lambda^{-1}(k)[A(k)x^*(k) + \eta(k)], \tag{7.264}$$

with the initial condition $x^*(k_0) = x_0$, and $\xi_n(k)$ and $\eta(k)$ are specified, respectively, as follows:

$$\xi_n(k) = \mathfrak{M}_n(k+1)\Lambda^{-1}(k)\eta(k) + \mathfrak{m}_n(k+1), \tag{7.265a}$$

$$\eta(k) = -\sum_{n \in \mathcal{N}} B_n(k)[R_{nn}(k)]^{-1}[B_n(k)]^\top \mathfrak{m}_n(k+1), \tag{7.265b}$$

where $\mathfrak{m}_n(k)$ are recursively generated as

$$\mathfrak{m}_n(k) = [A(k)]^\top \left[\mathfrak{m}_n(k+1) + \mathfrak{M}_n(k+1)\Lambda^{-1}(k)\eta(k)\right], \tag{7.266}$$

for all $k \in \{k_0, \cdots, k_f - 1\}$ and $n \in \mathcal{N}$, and with the boundary condition as $\mathfrak{m}_n(k_f) = 0$.

Proof Firstly $J_n(u_1, \cdots, u_n)$ is strictly convex with respect to u_n, since $Q(k+1)$ is positive semidefinite and $R_{nn}(k)$ is strictly positive definite.

Thus, by Theorem 7.6, an (open-loop) NE strategy is a control satisfying (7.256), (7.257a), and (7.257b).

Hence, we can claim that the NE is unique if it can show that the control strategy given in (7.263) is the only candidate.

Define the local Hamiltonian for player n as

$$\mathcal{H}_n(k) \equiv \mathcal{H}_n(x(k), u_1(k), \cdots, u_N(k), \lambda(k+1), k)$$

of player n for the underlying game as

$$\mathscr{H}_n(k) \triangleq \frac{1}{2}[x(k+1)]^\top Q_n(k+1)x(k+1)$$

$$+ \frac{1}{2}\sum_{\widehat{n}\in\mathscr{N}} [u_{\widehat{n}}(k)]^\top R_{n\widehat{n}}(k)u_{\widehat{n}}(k) + [\lambda_n(k+1)]^\top x(k+1) \qquad (7.267)$$

$$= \frac{1}{2}\left[A(k)x(k) + \sum_{\widehat{n}\in\mathscr{N}} B_{\widehat{n}}(k)u_{\widehat{n}}(k)\right]^\top Q_n(k+1)\left[A(k)x(k) + \sum_{\widehat{n}\in\mathscr{N}} B_{\widehat{n}}(k)u_{\widehat{n}}(k)\right]$$

$$+ \frac{1}{2}\sum_{\widehat{n}\in\mathscr{N}} \left[[u_{\widehat{n}}(k)]^\top R_{n\widehat{n}}(k)u_{\widehat{n}}(k)\right]$$

$$+ [\lambda_n(k+1)]^\top \left[A(k)x(k) + \sum_{\widehat{n}\in\mathscr{N}} B_{\widehat{n}}(k)u_{\widehat{n}}(k)\right], \qquad (7.268)$$

for all $k = k_0, \cdots, k_f - 1$.

Due to the fact that $Q(k+1)$ is positive semidefinite and $R_{nn}(k)$ is strictly positive definite, it is straightforward to obtain that the optimal control to minimize the local Hamiltonian $\mathscr{H}_n(k)$ is specified as

$$u_n^*(k) = -[R_{nn}(k)]^{-1}[B_n(k)]^\top \left[\lambda_n^*(k+1) + Q_n(k+1)x^*(k+1)\right], \qquad (7.269)$$

where $x^*(k+1)$ is given as

$$x^*(k+1) = A(k)x^*(k) + \sum_{\widehat{n}\in\mathscr{N}} B_{\widehat{n}}(k)u_{\widehat{n}}^*(k), \qquad (7.270)$$

with its boundary condition $x^*(k_0) = x_0$.

Also, by (7.257b), the costate equation becomes

$$\lambda_n^*(k) = [A(k)]^\top \left[\lambda_n^*(k+1) + Q_n(k+1)x^*(k+1)\right], \qquad (7.271)$$

with its boundary condition $\lambda_n^*(k_f) = 0$.

As observed in (7.269), the optimal control is in the form of $\lambda_n^*(k+1)$ and $x^*(k+1)$ which should be specified.

Next, we will show by induction that the unique optimal control defined in (7.269)–(7.271) can be specified by (7.263), (7.264), and

$$\lambda_n^*(k) = [A(k)]^\top \left[\mathfrak{M}_n(k+1)x^*(k+1) + \mathfrak{m}_n(k+1)\right],$$

for all $n \in \mathscr{N}$ and $k = k_0, \cdots, k_f - 1$.

Firstly deal with the time of $k = k_f - 1$; then by (7.269), we have

$$u_n^*(k_f - 1) = -[R_{nn}(k_f - 1)]^{-1}[B_n(k_f - 1)]^\top \mathfrak{M}_n(k_f)x^*(k+1), \qquad (7.272)$$

since $\lambda_n(k_f) = 0$ and $Q_n(k_f) = \mathfrak{M}_n(k_f)$.

By (7.272), together with the specification of $\Lambda(k)$ given in (7.262a) and (7.270), we can get

$$x^*(k_f) - A(k_f - 1)x^*(k_f - 1) = [\mathbb{I} - \Lambda(k_f - 1)]x^*(k_f),$$

which implies that

$$x^*(k_f) = \Lambda^{-1}(k_f - 1)A(k_f - 1)x^*(k_f - 1), \tag{7.273}$$

by the assumed invertibility of $\Lambda(k)$, which is (7.264) with $k = k_f - 1$.

Thus we can obtain (7.263) with $k = k_f - 1$ by (7.273) and (7.272).

To show by induction, we assume that the unique optimal control defined in (7.269)–(7.271) can be specified by (7.263) and (7.264) and

$$\lambda_n^*(k) = [A(k)]^\top \left[\mathfrak{M}_n(k + 1)x^*(k + 1) + \mathfrak{m}_n(k + 1) \right],$$

for all $n \in \mathcal{N}$ and $k = \widehat{k} + 1$, for some $\widehat{k}$, with $k_0 \leq \widehat{k} \leq k_f - 2$, and then show that this conclusion holds for $k = \widehat{k}$ as well.

By the assumption, we have the following:

$$\lambda_n^*(\widehat{k} + 1) = [A(\widehat{k} + 1)]^\top \left[\mathfrak{M}_n(\widehat{k} + 2)x^*(\widehat{k} + 2) + \mathfrak{m}_n(\widehat{k} + 2) \right]; \tag{7.274}$$

then by applying this to (7.269) in case $k = \widehat{k}$, we can get

$$u_n^*(\widehat{k}) = - \left[R_{nn}(\widehat{k}) \right]^{-1} \left[B_n(\widehat{k}) \right]^\top \left[\mathfrak{M}_n(\widehat{k} + 1)x^*(\widehat{k} + 1) + \mathfrak{m}_n(\widehat{k} + 1) \right]. \tag{7.275}$$

Moreover, by applying the same technique used in the verification of (7.273) with $k = k_f$, we can obtain

$$x^*(\widehat{k} + 1) = \left[\Lambda(\widehat{k}) \right]^{-1} \left[A(\widehat{k})x^*(\widehat{k}) + \eta(\widehat{k}) \right], \tag{7.276}$$

which is (7.264).

By applying (7.276) to (7.275), we can obtain (7.263) with $k = \widehat{k}$.

Furthermore, by applying (7.274) and (7.276) to (7.271), we can also obtain the following:

$$\lambda_n^*(\widehat{k}) = \left[A(\widehat{k}) \right]^\top \left[\mathfrak{M}_n(\widehat{k} + 1)x^*(\widehat{k} + 1) + \mathfrak{m}_n(\widehat{k} + 1) \right],$$

with $\mathfrak{m}_n(\widehat{k} + 1)$ given by (7.266).

Thus we have verified the situation with $k = \widehat{k}$, and then can get the conclusion of the theorem. $\square$

Here we revisit the game studied in Example 6.3, Chap. 6, in discrete-time case below.

Example 7.14 Consider a dynamic game with the state system given as

$$x_1(k+1) = a_1 x_1(k) + b_1 u_1(k), \tag{7.277a}$$
$$x_2(k+1) = a_2 x_2(k) + b_2 u_2(k), \tag{7.277b}$$

with a given performance cost function

$$J_n(u) = \sum_{k=k_0}^{k_f-1} \left[\frac{1}{2} q_n [x_n(k+1)]^2 + \frac{1}{2} r_n [u_n(k)]^2 \right], \tag{7.278}$$

with $r_n > 0$, $n = 1, 2$.

Implement the NE strategy $u^* \equiv (u_1^*, u_2^*)$ for the dynamic game system.
Solution. Firstly, by (7.261a), and (7.261b), we obtain

$$A(k) = \begin{bmatrix} a_1 & 0 \\ 0 & a_2 \end{bmatrix}, \quad B_1(k) = \begin{bmatrix} b_1 \\ 0 \end{bmatrix}, \quad B_2(k) = \begin{bmatrix} 0 \\ b_2 \end{bmatrix}, \quad C(k) = \begin{bmatrix} 0 \\ 0 \end{bmatrix},$$

$$Q_1(k) = \begin{bmatrix} q_1 & 0 \\ 0 & 0 \end{bmatrix}, \quad Q_2(k) = \begin{bmatrix} 0 & 0 \\ 0 & q_2 \end{bmatrix},$$

$$R_{11}(k) = r_1, \quad R_{22}(k) = r_2, \quad R_{12}(k) = R_{21}(k) = 0,$$

for all $k \in \{k_0, \cdots, k_f - 1\}$ and with $n = 1, 2$.
By (7.262a) and (7.262b),

$$\Lambda(k) \triangleq \mathbb{I}_{2\times2} + \begin{bmatrix} \frac{[b_1]^2}{r_1} & 0 \\ 0 & 0 \end{bmatrix} \mathfrak{M}_1(k+1) + \begin{bmatrix} 0 & 0 \\ 0 & \frac{[b_2]^2}{r_2} \end{bmatrix} \mathfrak{M}_2(k+1), \tag{7.280a}$$

$$\mathfrak{M}_n(k) = Q_n(k) + \begin{bmatrix} a_1 & 0 \\ 0 & a_2 \end{bmatrix} \mathfrak{M}_n(k+1)\Lambda^{-1}(k) \begin{bmatrix} a_1 & 0 \\ 0 & a_2 \end{bmatrix}, \tag{7.280b}$$

with $n = 1, 2$, and the boundary condition $\mathfrak{M}_n(k_f) = Q_n(k_f)$.
Thus by Theorem 7.7, the NE strategy u^* is given as

$$u_n^*(k) = -\frac{1}{r_n}[B_n(k)]^\top \left[\mathfrak{M}_n(k+1)\Lambda^{-1}(k) \begin{bmatrix} a_1 & 0 \\ 0 & a_2 \end{bmatrix} x^*(k) + \xi_n(k) \right], \tag{7.281}$$

for all $k \in \{k_0, \cdots, k_f - 1\}$ and with $n = 1, 2$, and

$$x^*(k+1) = \Lambda^{-1}(k) \left[\begin{bmatrix} a_1 & 0 \\ 0 & a_2 \end{bmatrix} x^*(k) + \eta(k) \right], \tag{7.282}$$

with the initial condition $x^*(k_0) = x_0$, and $\xi(k)$ and $\eta(k)$ are given as

$$\xi_n(k) = \mathfrak{M}_n(k+1)\Lambda^{-1}(k)\eta(k) + \mathfrak{m}_n(k+1), \tag{7.283a}$$

$$\eta(k) = -\begin{bmatrix} \frac{[b_1]^2}{r_1} & 0 \\ 0 & 0 \end{bmatrix}\mathfrak{m}_1(k+1) - \begin{bmatrix} 0 & 0 \\ 0 & \frac{[b_2]^2}{r_2} \end{bmatrix}\mathfrak{m}_2(k+1), \tag{7.283b}$$

with $\mathfrak{m}_n(\cdot)$ given as

$$\mathfrak{m}_n(k) = \begin{bmatrix} a_1 & 0 \\ 0 & a_2 \end{bmatrix}\big[\mathfrak{m}_n(k+1) + \mathfrak{M}_n(k+1)\Lambda^{-1}(k)\eta(k)\big], \tag{7.284}$$

with the boundary condition $\mathfrak{m}_n(k_f) = 0$. □

7.8 Discrete-Time Two-Person Zero-Sum Dynamic Games

7.8.1 Formulation of Discrete-Time Two-Person Zero-Sum Dynamic Games

In this part, we will introduce a specific class of discrete-time noncooperative dynamic games, say two-person zero-sum games, such that the cost functions of player 1 and player 2 satisfy the following relationship:

$$g(\cdot, k) = g_1(\cdot, k) = -g_2(\cdot, k), \tag{7.285}$$

for all instants $k = k_0, \cdots, k_f - 1$; then by the specification of the cost function, the Nash equilibrium of the underlying zero-sum games degenerates, from the inequality given in Definition 7.1, into the following so-called saddle point inequality

$$J(u_1^*, u_2) \le J(u_1^*, u_2^*) \le J(u_1, u_2^*), \tag{7.286}$$

where $J(u_1, u_2) \triangleq J_1(u_1, u_2) = -J_2(u_1, u_2)$, for all $u_1 \in \mathcal{U}_1$ and $u_2 \in \mathcal{U}_2$.

Denote by (u_1^*, u_2^*) an (open-loop) saddle point solution, and

$$x^*(\cdot) \equiv \big(x^*(k_0), \cdots, x^*(k_f)\big)$$

represents the state trajectory subject to (u_1^*, u_2^*), say

$$x^*(k+1) = f(x^*(k), u_1^*(k), u_2^*(k), k), \tag{7.287}$$

with the boundary condition $x^*(k_0) = x_0$.

7.8.2   Saddle Point of Discrete-Time Two-Person Zero-Sum Dynamic Games

In Theorem 7.6 below, we will give a necessary condition for an NE of a two-person zero-sum discrete-time dynamic game; before that define a Hamiltonian, denoted by $\mathcal{H}(x(k), u_1(k), u_2(k), \lambda_n(k+1), k)$, in the following:

$$
\begin{aligned}
&\mathcal{H}(x(k), u_1(k), u_2(k), \lambda(k+1), k) \\
&\triangleq g(f(x(k), u_1(k), u_2(k), k), x(k), u_1(k), u_2(k), k) \\
&\quad + [\lambda(k+1)]^\top f(x(k), u_1(k), u_2(k), k), \quad\quad\quad (7.288)
\end{aligned}
$$

for each time $k \in \{k_0, k_0 + 1, \cdots, k_f - 1\}$, where λ represents the costate of the zero-sum game system.

Theorem 7.8 *Consider Assumption 7.1 for a two-person zero-sum discrete-time dynamic game; then there exists a costate trajectory $\lambda^*(\cdot) \equiv \left(\lambda^*(k_0), \cdots, \lambda^*(k_f)\right)$ satisfying the following:*

$$
\begin{aligned}
&\mathcal{H}(x^*(k), u_1^*(k), u_2(k), \lambda^*(k+1), k) \\
&\leq \mathcal{H}(x^*(k), u_1^*(k), u_2^*(k), \lambda^*(k+1), k) \\
&\leq \mathcal{H}(x^*(k), u_1(k), u_2^*(k), \lambda^*(k+1), k), \forall u_1(k) \in U_1, u_2(k) \in U_2, \quad (7.289a)
\end{aligned}
$$

$$
\begin{aligned}
\lambda^*(k) = &\left[\lambda(k+1) + \frac{\partial g(x^*(k+1), x^*(k), u^*(k), k)}{\partial x(k+1)}\right]^\top \frac{\partial f(x^*(k), u^*(k), k)}{\partial x(k)} \\
&+ \frac{\partial g(x^*(k+1), x^*(k), u^*(k), k)}{\partial x(k)}, \quad\quad\quad (7.289b)
\end{aligned}
$$

for all $k = k_0, \cdots, k_f - 1$, with the boundary condition $\lambda^*(k_f) = 0$.

Proof By Theorem 7.5, there exists a costate trajectory for each player $n \in \mathcal{N}$, denoted by $\lambda_n^* \equiv (\lambda_n^*(k_0), \cdots, \lambda_n^*(k_f - 1))$, such that the following holds by (7.161c):

$$
\mathcal{H}(x^*(k), u_1^*(k), u_2^*(k), \lambda^*(k+1), k) = \min_{u_1(k)}\{\mathcal{H}(x^*(k), u_1(k), u_2^*(k), \lambda^*(k+1), k)\},
$$
$$
\mathcal{H}(x^*(k), u_1^*(k), u_2^*(k), \lambda^*(k+1), k) = \max_{u_2(k)}\{\mathcal{H}(x^*(k), u_1^*(k), u_2(k), \lambda^*(k+1), k)\},
$$

for all $k = k_0, \cdots, k_f - 1$, which implies the conclusion of (7.289a).

Moreover by (7.161b), we get

$$
\lambda^*(k) = \frac{\partial \mathcal{H}(x^*(k), u^*(k), \lambda_n^*(k+1), k)}{\partial x(k)}, \quad\quad\quad (7.290)
$$

for all $k = k_0, \cdots, k_f - 1$.

Thus by the specification of $\mathscr{H}$ given in (7.288), we can obtain

$$
\lambda^*(k) = \frac{\partial g(x^*(k+1), x^*(k), u^*(k), k)}{\partial x(k)}
$$
$$
+ \left[\frac{\partial g(x^*(k+1), x^*(k), u^*(k), k)}{\partial x(k+1)} \right]^\top \frac{\partial f(x^*(k), u^*(k), k)}{\partial x(k)}
$$
$$
+ [\lambda^*(k+1)]^\top \frac{\partial f(x^*(k), u^*(k), k)}{\partial x(k)}, \tag{7.291}
$$

which implies the conclusion of (7.289b).   □

7.8.3 Discrete-Time Linear-Quadratic Two-Person Zero-Sum Dynamic Games

Similar to the analysis on the discrete-time noncooperative dynamic game, we also introduce a class of linear-quadratic two-person zero-sum dynamic games such that

$$
x(k+1) = A(k)x(k) + B_1(k)u_1(k) + B_2(k)u_2(k), \tag{7.292}
$$

where $A(k)$, $B_1(k)$, and $B_2(k)$ are matrices with appropriate dimensions, respectively, and the cost function is given as

$$
J(u_1, u_2) \triangleq \frac{1}{2} \sum_{k=k_0}^{k_f-1} \left[[x(k+1)]^\top Q(k+1)x(k+1) \right.
$$
$$
\left. + [u_1(k)]^\top u_1(k) - [u_2(k)]^\top u_2(k) \right], \tag{7.293}
$$

where $Q(k+1)$ is a matrix with an appropriate dimension.

Theorem 7.9 shows the existence and uniqueness of NE for the discrete-time linear-quadratic two-person zero-sum game specified above under certain conditions which will be discussed in Lemma 7.1 below.

Lemma 7.1 (Sufficient and Necessary Condition of the Strict Concavity on u_2 in Two-Person Linear-Quadratic Discrete-Time Zero-Sum Dynamic Game) *The performance cost function $J(u_1, u_2)$ of a two-person linear-quadratic discrete-time zero-sum dynamic game is strictly concave on u_2, the control of player 2, for all given u_1, if and only if the following inequality holds:*

$$
\Gamma(k) \equiv \mathbb{I} - [B_2(k)]^\top S(k+1)B_2(k) > 0, \tag{7.294}
$$

for all $k = k_0, \cdots, k_f - 1$, *say* $\Gamma(k)$ *is positive definite for all k, where* $S(k)$ *is given as*

$$
\begin{aligned}
S(k) = Q(k) &+ [A(k)]^\top S(k+1) A(k) \\
&+ [A(k)]^\top S(k+1) B_2(k) \Gamma^{-1}(k) [B_2(k)]^\top S(k+1) A(k),
\end{aligned}
\tag{7.295}
$$

with the boundary condition $S(k_f) = Q(k_f)$.

Proof Firstly since, as specified in (7.293), $J(u_1, u_2)$ is a quadratic function of u_2, the property of strict concavity on u_2 is equivalent to the existence and uniqueness of the optimal solution to the following problem:

$$
\min_{u_2 \in \mathbb{R}^{\dim_2}} \left\{ -J(u_1, u_2) \right\},
\tag{7.296}
$$

for any given $u_1 \in \mathbb{R}^{\dim_1}$, with the state equation given as (7.292), where $\dim_n$ denotes the dimension of the control u_n, with $n = 1, 2$.

Also since the second partial derivatives of the cost function $J(\cdot)$ on u_2 are independent of u_2, we can state that it is equivalent to study of the optimal control solution of player 2 for the following problem:

$$
\min_{u_2 \in \mathbb{R}^{\dim_2}} \left\{ -[u_2(k)]^\top u_2(k) - [x(k+1)]^\top Q(k+1) x(k+1) \right\},
\tag{7.297}
$$

with the state equation given as

$$
x(k+1) = A(k) x(k) + B_2(k) u_2(k),
\tag{7.298}
$$

for all $k = k_0, \cdots, k_f - 1$.

Thus by following the analysis of Corollary 7.1, we can obtain the conclusion. □

Also define notations of $\Lambda(k)$ and $\mathfrak{M}_n(k)$, respectively, below which will be used in Theorem 7.9,

$$
\Lambda(k) \triangleq \mathbb{I} + \left[B_1(k)[B_1(k)]^\top - B_2(k)[B_2(k)]^\top \right] \mathfrak{M}(k+1),
\tag{7.299a}
$$

$$
\mathfrak{M}(k) \triangleq Q(k) + [A(k)]^\top \mathfrak{M}(k+1) \Lambda^{-1}(k) A(k),
\tag{7.299b}
$$

with $\mathfrak{M}(k_f) = Q(k_f)$.

Define the Hamiltonian $\mathcal{H}(x(k), u_1(k), u_2(k), \lambda(k+1), k)$ for the specified linear-quadratic discrete-time zero-sum game in the following:

$$\mathscr{H}(x(k), u_1(k), u_2(k), \lambda(k+1), k)$$

$$\triangleq \frac{1}{2}\left[[x(k+1)]^\top Q(k+1)x(k+1) + [u_1(k)]^\top u_1(k) - [u_2(k)]^\top u_2(k)\right]$$

$$+ [\lambda(k+1)]^\top x(k+1)$$

$$= \frac{1}{2}[A(k)x(k) + B_1(k)u_1(k) + B_2(k)u_2(k)]^\top$$

$$Q(k+1)[A(k)x(k) + B_1(k)u_1(k) + B_2(k)u_2(k)]$$

$$+ \frac{1}{2}\left[[u_1(k)]^\top u_1(k) - [u_2(k)]^\top u_2(k)\right]$$

$$+ [\lambda(k+1)]^\top [A(k)x(k) + B_1(k)u_1(k) + B_2(k)u_2(k)], \tag{7.300}$$

for all $k = k_0, \cdots, k_f - 1$.

Theorem 7.9 (Uniqueness and Existence of Saddle Point Solution of Discrete-Time Linear-Quadratic Two-Person Zero-Sum Dynamic Game) *Consider a two-person linear-quadratic discrete-time zero-sum dynamic game such that $Q(k)$ is positively semidefinite for all $k = k_0, \cdots, k_f - 1$, and (7.294) is satisfied; then the matrices $\Lambda(k)$, for all $k = k_0, \cdots, k_f - 1$, given in (7.299a) are invertible, and the game system has a unique saddle point control strategy specified as follows:*

$$u_n^*(k) = [-1]^n [B_n(k)]^\top \left[\mathfrak{M}(k+1)\Lambda^{-1}(k)A(k)x^*(k) + \xi(k)\right], \tag{7.301}$$

with $n = 1, 2$, for all $k \in \{k_0, \cdots, k_f - 1\}$, where $x^*(\cdot)$ is given as

$$x^*(k+1) = \Lambda^{-1}(k)[A(k)x^*(k) + \eta(k)], \tag{7.302}$$

with the initial condition $x^*(k_0) = x_0$, and $\xi(k)$ and $\eta(k)$ are specified as follows:

$$\xi(k) = \mathfrak{M}(k+1)\Lambda^{-1}(k)\eta(k) + \mathfrak{m}_n(k+1), \tag{7.303a}$$

$$\eta(k) = -\left[B_1(k)[B_1(k)]^\top - B_2(k)[B_2(k)]^\top\right]\mathfrak{m}(k+1), \tag{7.303b}$$

where $\mathfrak{m}(k)$ are recursively generated as

$$\mathfrak{m}(k) = [A(k)]^\top \left[\Lambda^{-1}(k)\right]^\top \mathfrak{m}(k+1), \tag{7.304}$$

for all $k \in \{k_0, \cdots, k_f - 1\}$ and $n \in \mathcal{N}$, and with the boundary condition as $\mathfrak{m}_n(k_f) = 0$.

Proof As assumed, (7.294) holds; then by applying Lemma 7.1, we have that $J(u_1, u_2)$ is strictly concave on u_2 for all given u_1. Also due to the positive semidefinite property of $Q(k)$ for all k, it is straightforward to show that $J(u_1, u_2)$ is strictly convex on u_1 for all given u_2.

Thus by Corollary 4.5 in Dynamic Noncooperative Game Theory by Tamer Basar and Geert Olsder, 1995, the underlying system is a static strictly convex–concave-quadratic zero-sum game which possesses a unique saddle point control strategy.

Hence by Theorem 7.8, (7.287) becomes the following:

$$x^*(k+1) = A(k)x^*(k) + \sum_{\widehat{n} \in \mathcal{N}} B_{\widehat{n}}(k)u^*_{\widehat{n}}(k), \qquad (7.305a)$$

with its boundary condition $x^*(k_0) = x_0$.

While (7.289a) and (7.289b) in Theorem 7.8 can be reorganized as (7.306) and (7.307), respectively,

$$\mathcal{H}(x^*(k), u^*_1(k), u_2(k), \lambda^*(k+1), k)$$
$$\leq \mathcal{H}(x^*(k), u^*_1(k), u^*_2(k), \lambda^*(k+1), k)$$
$$\leq \mathcal{H}(x^*(k), u_1(k), u^*_2(k), \lambda^*(k+1), k), \qquad (7.306)$$

where the above inequalities hold for all $u_1(k) \in \mathbb{R}^{\dim_1}$ and $u_2(k) \in \mathbb{R}^{\dim_2}$, for all instants $k = k_0, \cdots, k_f - 1$.

$$\lambda^*(k) = [A(k)]^\top \left[\lambda^*(k+1) + Q(k+1)x^*(k+1)\right], \qquad (7.307)$$

for all instants $k = k_0, \cdots, k_f - 1$, with the boundary condition $\lambda^*(k_f) = 0$.

Then following the same technique used in the proof of Theorem 7.7, we can obtain

$$u^*_n(k) = [-1]^n [B_n(k)]^\top \left[\mathfrak{M}(k+1)x^*(k+1) + \mathfrak{m}(k+1)\right], \qquad (7.308)$$

with $n = 1, 2$, for all instants $k = k_0, \cdots, k_f - 1$, and

$$\lambda^*(k) = [A(k)]^\top \left[\mathfrak{M}(k+1)x^*(k+1) + \mathfrak{m}(k+1)\right], \qquad (7.309)$$

for all instants $k = k_0, \cdots, k_f - 1$, where the state trajectory satisfies the following:

$$\Lambda(k)x^*(k+1) = A(k)x^*(k) + \eta(k). \qquad (7.310)$$

As verified, there is a unique saddle point control strategy $u^* \equiv (u^*_1, u^*_2)$; then there exists a unique state trajectory x^* subject to u^*. Consequently, by (7.310), the $N \times N$ matrix $\Lambda(k)$ is full rank for each k, and hence it is invertible for each k. Thus by (7.310), we obtain the conclusion of (7.302).

By using (7.302) to (7.308), we can obtain (7.301) with $\xi(k)$, $\eta(k)$, and $\mathfrak{m}(k)$ given in (7.303a), (7.303b), and (7.304), respectively. □

7.9 Summary

In this chapter, the optimal solutions for discrete-time control problems via the variational method are developed. Moreover, for the purpose of comparison, how to implement the optimal solution for discrete-time control problems by applying the dynamic programming method is also introduced.

7.10 Exercises

Exercise 7.1 Specify the extreme solution for the problem formulated in Example 7.1 with $k_f = 8$ and $x(k_f) = 6$.

Exercise 7.2 Specify the optimal control solution for the problem defined in Example 7.9.

Exercise 7.3 Show that the coefficient matrix $P(k)$ in the matrix difference Riccati equation (7.93)

$$P(k) = A^{\mathsf{T}}(k)P(k+1)[\mathbb{I} + E(k)P(k+1)]^{-1}A(k) + Q(k)$$

is positive definite.

Exercise 7.4 Consider a one-dimensional state system given as

$$x(k+1) = x(k) + u(k),$$

and the performance cost to be minimized as

$$J(x, u) = \frac{1}{2} \sum_{k=k_0}^{k_f - 1} u^2(k), \tag{7.311}$$

with $k_0 = 0$, $k_f = 6$, and the boundary conditions $x(k_0) = 1$ and $x(k_f) = 0$.

Implement the optimal control u^* and the state x^* such that the cost function of (7.311) is minimized.

Exercise 7.5 A two-order discrete-time state system is given by

$$x_1(k+1) = 2x_1(k) + 0.2x_2(k), \text{ with } x_1(0) = 5$$
$$x_2(k+1) = 2x_2(k) + 0.2u(k), \text{ with } x_2(0) = 0.$$

The performance cost to be minimized is given as follows:

$$J(x, u) = \frac{1}{2} \sum_{k=k_0}^{k_f - 1} \left[[x_1(k)]^2 + u^2(k) \right],$$

where $k_0 = 0$ and $k_f = 8$.

- Implement the open-loop optimal control, and
- Implement the closed-loop optimal control

such that the state system is driven to the final state $x_1(k_f) = x_2(k_f) = 0$ and the cost function is minimized.

Exercise 7.6 Implement the open-loop optimal control u^* for a two-order discrete-time state system given such that

$$x(k + 1) = \begin{bmatrix} 0 & 1 \\ -1 & 1 \end{bmatrix} x(k) + \begin{bmatrix} 0 \\ 1 \end{bmatrix} u(k),$$

with $x(0) = \begin{bmatrix} 1 \\ 2 \end{bmatrix}$, and the performance cost given as

$$J(x, u) = \sum_{k=k_0}^{k_f - 1} \left[x_1^2(k) + u^2(k) \right],$$

where $k_f = 5$, $x_1(5)$ being unspecified and $x_2(5) = 0$.

Exercise 7.7 Derive the relation for the optimal cost function given by (7.108) as

$$J^* = \frac{1}{2} [x^*(k_0)]^T P(k_0) x^*(k_0).$$

Exercise 7.8 Given a state stem such that

$$x(k + 1) = A(k)x(k) + B(k)u(k),$$

with the fixed boundary conditions as $x(k_0)$, $x(k_f)$, and k_f being fixed.

Also consider a performance cost as follows:

$$J(x, u) = \frac{1}{2} [x(k_f)]^T F(k_f) x(k_f) + \frac{1}{2} \sum_{k=k_0}^{k_f - 1} \left[[x(k)]^T Q(k)x(k) + [u(k)]^T R(k)u(k) \right].$$

Specify the closed-loop, optimal control solution.

Exercise 7.9 Implement the closed-loop optimal controls for the problem specified in Exercise 7.6 with the final time as $k_f = 3$ and $k_f = \infty$, respectively.

Exercise 7.10 For the state system given in Exercise 7.6, implement the optimal control u^* such that the following performance cost

$$J(x, u) = \sum_{k=k_0}^{k_f-1} \left[[x(k) - z(k)]^2 + x_1^2(k) + u^2(k) \right],$$

$r(k) = \begin{bmatrix} 0.5k \\ 0 \end{bmatrix}$, is minimized, where $k_f = 5$, $x_1(5)$ is unspecified and $x_2(5) = 0$.

Exercise 7.11 Find out the shortest path solution for the Example 1.15 introduced in Chap. 1.

Exercise 7.12 Formulate Example 1.16 introduced in Chap. 1 as an optimal control problem and solve it by applying the dynamic programming method.

Exercise 7.13 Consider a state system such that

$$x(k + 1) = x(k) + u(k),$$

with the following constraints on the control and state:

$$u(k) \in \{-1, -0.5, 0, 0.5, 1\}, \text{ with } k = 0, 1, \tag{7.312a}$$
$$x(k) \in \{0, 0.5, 1, 1.5, 2\}, \text{ with } k = 0, 1, 2. \tag{7.312b}$$

Implement the optimal control u^* and the associated state x^* for the system specified above such that the performance cost defined in Exercise 7.4 is minimized.

Exercise 7.14 Consider a one-dimensional state system

$$x(k + 1) = x(k) + u(k), \tag{7.313}$$

and the performance cost to be minimized as

$$J(x, u) = \frac{1}{2}x^2(2) + \frac{1}{2}u^2(0) + \frac{1}{2}u^2(1), \tag{7.314}$$

with the constraints on the control and state given as

$$u(k) \in \{-1, -0.5, 0, 0.5, 1\}, \text{ with } k = 0, 1, \tag{7.315a}$$
$$x(k) \in \{0, 0.5, 1, 1.5\}, \text{ with } k = 0, 1, 2. \tag{7.315b}$$

Implement the optimal control u^* and the state x^* such that the cost function is minimized.

Exercise 7.15 Consider a one-order state system as follows:

$$x(k+1) = x(k) + u(k),\qquad(7.316)$$

and the performance cost to be minimized as

$$J(x, u) = \frac{1}{2}x^2(2) + \frac{1}{2}u^2(0) + \frac{1}{2}u^2(1),\qquad(7.317)$$

with the constraints on the control and state as given in Exercise 7.14.

Implement the optimal control u^* and the state x^* such that the performance cost is minimized.

Exercise 7.16 Consider a state system such that

$$x(k+1) = x(k) + u(k),$$

with $k = 0, 1$, where $x(k) \in [0, 2]$ and $u(k) \in [-1, 1]$, and the performance cost given as

$$J \triangleq x^2(k_f) + \sum_{k=k_0}^{k_f-1} \left[x^2(k) + u^2(k)\right].$$

Implement the optimal control u^* and the associated state x^* for the problem specified above.

Exercise 7.17 Consider a state system such that

$$x(k+1) = x(k) + u(k),$$

with $k = 0, 1$, where $x(k) \in [-3, 3]$ and $u(k) \in [-1, 1]$.

Implement the optimal control u^* and the associated state x^* for the system specified above such that the performance cost defined in Exercise 7.4 is minimized.

Chapter 8
Conclusions

Chapter 1, briefly introduces the background, the motivation, and the organization of this book.

Chapter 2, firstly introduces some fundamental terms related to functions and functionals, respectively, and then specifies the sufficient and necessary condition for the extrema of functionals via the variational method. Based upon the results of the variational method, it gives the necessary and sufficient conditions for the extrema of functionals with respect to multiple functions which are independent of each other. it further introduces the extrema problems of functions and functionals, respectively, considering constraints which are solved by the elimination/direct method and the Lagrange method.

Chapter 3, studies the optimal control problems by applying the developed results of the extremal of functional via the variational method. More specially, it gives the necessary and sufficient conditions for the optimal solution to the optimal control problems with unbounded controls. Based upon this result, it develops the optimal solution to the optimal control problems with different boundary conditions on the final time and final state, respectively. Then it analyzes linear-quadratic regulation and tracking problems.

So far, it was studied the optimal control problems with the assumption that the admissible controls and states are not constrained by any boundaries. However, such constraints certainly will occur in realistic systems. Thus in Chap. 4, Pontryagin's minimum principle for optimal control problems with constrained control and constrained system state, respectively, is developed; then by applying the proposed Pontryagin's minimum principle, it is studied how to implement the optimal control solution for specific interesting optimal control problems with constraints, say minimum time, minimum fuel, and minimum energy problems, respectively. For the purpose of comparison, it is further studied the optimal controls for problems concerning the tradeoff between the elapsed time and the consumed fuel/energy.

Z. Ma and S. Zou, *Optimal Control Theory*,
https://doi.org/10.1007/978-981-33-6292-5_8

Besides, Pontryagin's minimum principle described in the last chapter, in Chap. 5, introduces another key branch of optimal control methods, say the dynamic programming. Also for the purpose of comparison, the relationship between these two optimal control methods is studied.

Based on the results developed in previous chapters in this book, Chap. 6, introduces some games, such as noncooperative differential games and two-person zerosum differential games, where the system is driven by individual players each of which would like to minimizes its own performance cost function. The NE strategies are solved and analyzed by applying the variational method.

Chapter 7 studies different classes of optimal control problems in discrete-time case.

Printed in the United States
by Baker & Taylor Publisher Services